Selected Titles in This Series

47 **A. D. Elmendorf, I. Kriz, M. A. Mandell, and J. P. May (with an appendix by M. Cole),** Rings, modules, and algebras in stable homotopy theory, 1997
46 **Stephen Lipscomb,** Symmetric inverse semigroups, 1996
45 **George M. Bergman and Adam O. Hausknecht,** Cogroups and co-rings in categories of associative rings, 1996
44 **J. Amorós, M. Burger, K. Corlette, D. Kotschick, and D. Toledo,** Fundamental groups of compact Kähler manifolds, 1996
43 **James E. Humphreys,** Conjugacy classes in semisimple algebraic groups, 1995
42 **Ralph Freese, Jaroslav Ježek, and J. B. Nation,** Free lattices, 1995
41 **Hal L. Smith,** Monotone dynamical systems: an introduction to the theory of competitive and cooperative systems, 1995
40.2 **Daniel Gorenstein, Richard Lyons, and Ronald Solomon,** The classification of the finite simple groups, number 2, 1995
40.1 **Daniel Gorenstein, Richard Lyons, and Ronald Solomon,** The classification of the finite simple groups, number 1, 1994
39 **Sigurdur Helgason,** Geometric analysis on symmetric spaces, 1993
38 **Guy David and Stephen Semmes,** Analysis of and on uniformly rectifiable sets, 1993
37 **Leonard Lewin, Editor,** Structural properties of polylogarithms, 1991
36 **John B. Conway,** The theory of subnormal operators, 1991
35 **Shreeram S. Abhyankar,** Algebraic geometry for scientists and engineers, 1990
34 **Victor Isakov,** Inverse source problems, 1990
33 **Vladimir G. Berkovich,** Spectral theory and analytic geometry over non-Archimedean fields, 1990
32 **Howard Jacobowitz,** An introduction to CR structures, 1990
31 **Paul J. Sally, Jr. and David A. Vogan, Jr., Editors,** Representation theory and harmonic analysis on semisimple Lie groups, 1989
30 **Thomas W. Cusick and Mary E. Flahive,** The Markoff and Lagrange spectra, 1989
29 **Alan L. T. Paterson,** Amenability, 1988
28 **Richard Beals, Percy Deift, and Carlos Tomei,** Direct and inverse scattering on the line, 1988
27 **Nathan J. Fine,** Basic hypergeometric series and applications, 1988
26 **Hari Bercovici,** Operator theory and arithmetic in H^∞, 1988
25 **Jack K. Hale,** Asymptotic behavior of dissipative systems, 1988
24 **Lance W. Small, Editor,** Noetherian rings and their applications, 1987
23 **E. H. Rothe,** Introduction to various aspects of degree theory in Banach spaces, 1986
22 **Michael E. Taylor,** Noncommutative harmonic analysis, 1986
21 **Albert Baernstein, David Drasin, Peter Duren, and Albert Marden, Editors,** The Bieberbach conjecture: Proceedings of the symposium on the occasion of the proof, 1986
20 **Kenneth R. Goodearl,** Partially ordered abelian groups with interpolation, 1986
19 **Gregory V. Chudnovsky,** Contributions to the theory of transcendental numbers, 1984
18 **Frank B. Knight,** Essentials of Brownian motion and diffusion, 1981
17 **Le Baron O. Ferguson,** Approximation by polynomials with integral coefficients, 1980
16 **O. Timothy O'Meara,** Symplectic groups, 1978
15 **J. Diestel and J. J. Uhl, Jr.,** Vector measures, 1977
14 **V. Guillemin and S. Sternberg,** Geometric asymptotics, 1977
13 **C. Pearcy, Editor,** Topics in operator theory, 1974
12 **J. R. Isbell,** Uniform spaces, 1964

(Continued in the back of this publication)

Rings, Modules, and Algebras in Stable Homotopy Theory

Mathematical Surveys and Monographs

Volume 47

Rings, Modules, and Algebras in Stable Homotopy Theory

A. D. Elmendorf
I. Kriz
M. A. Mandell
J. P. May

with an Appendix by
M. Cole

American Mathematical Society

Kriz and May were supported in part by NSF Grants, and Cole was supported by NSF and GAANN Fellowships.

1991 *Mathematics Subject Classification.* Primary 19D99, 55N20, 55P42;
Secondary 19L99, 55N22, 55T25.

Library of Congress Cataloging-in-Publication Data

Rings, modules, and algebras in stable homotopy theory / A. D. Elmendorf ... [et al.] ; with an appendix by M. Cole.
p. cm. — (Mathematical surveys and monographs ; v. 47)
Includes bibliographical references and index.
ISBN 0-8218-0638-6 (hard : alk. paper)
1. Homotopy theory. 2. Groups of points. I. Elmendorf, Anthony D., 1952– . II. Series: Mathematical surveys and monographs ; no. 47.
QA612.7.R56 1996
514′.24—dc20   96-35999
CIP

∞ The paper used in this book is acid-free and falls within the guidelines
established to ensure permanence and durability.

10 9 8 7 6 5 4 3 2 1   02 01 00 99 98 97

Abstract

Let S be the sphere spectrum. We construct an associative, commutative, and unital smash product in a complete and cocomplete category $\mathscr{M}_S$ of "S-modules" whose derived category $\mathscr{D}_S$ is equivalent to the classical stable homotopy category. This allows a simple and algebraically manageable definition of "S-algebras" and "commutative S-algebras" in terms of associative, or associative and commutative, products $R \wedge_S R \longrightarrow R$. These notions are essentially equivalent to the earlier notions of A_∞ and E_∞ ring spectra, and the older notions feed naturally into the new framework to provide plentiful examples. There is an equally simple definition of R-modules in terms of maps $R \wedge_S M \longrightarrow M$. When R is commutative, the category $\mathscr{M}_R$ of R-modules also has an associative, commutative, and unital smash product, and its derived category $\mathscr{D}_R$ has properties just like the stable homotopy category.

Working in the derived category $\mathscr{D}_R$, we construct spectral sequences that specialize to give generalized universal coefficient and Künneth spectral sequences. Classical torsion products and Ext groups are obtained by specializing our constructions to Eilenberg-Mac Lane spectra and passing to homotopy groups, and the derived category of a discrete ring R is equivalent to the derived category of its associated Eilenberg-Mac Lane S-algebra.

We also develop a homotopical theory of R-ring spectra in $\mathscr{D}_R$, analogous to the classical theory of ring spectra in the stable homotopy category, and we use it to give new constructions as MU-ring spectra of a host of fundamentally important spectra whose earlier constructions were both more difficult and less precise.

Working in the module category $\mathscr{M}_R$, we show that the category of finite cell modules over an S-algebra R gives rise to an associated algebraic K-theory spectrum KR. Specialized to the Eilenberg-Mac Lane spectra of discrete rings, this recovers Quillen's algebraic K-theory of rings. Specialized to suspension spectra $\Sigma^\infty(\Omega X)_+$ of loop spaces, it recovers Waldhausen's algebraic K-theory of spaces.

Replacing our ground ring S by a commutative S-algebra R, we define R-algebras and commutative R-algebras in terms of maps $A \wedge_R A \longrightarrow A$, and we show that the categories of R-modules, R-algebras, and commutative R-algebras are all topological model categories. We use the model structures to study Bousfield localizations of R-modules and R-algebras. In particular, we prove that KO and KU are commutative ko and ku-algebras and therefore commutative S-algebras.

We define the topological Hochschild homology R-module $THH^R(A; M)$ of A with coefficients in an (A, A)-bimodule M and give spectral sequences for the calculation of its homotopy and homology groups. Again, classical Hochschild homology and cohomology groups are obtained by specializing the constructions to Eilenberg-Mac Lane spectra and passing to homotopy groups.

Contents

Introduction

The last thirty years have seen the importation of more and more algebraic techniques into stable homotopy theory. Throughout this period, most work in stable homotopy theory has taken place in Boardman's stable homotopy category [6], or in Adams' variant of it [2], or, more recently, in Lewis and May's variant [38]. That category is analogous to the derived category obtained from the category of chain complexes over a commutative ring k by inverting the quasi-isomorphisms. The sphere spectrum S plays the role of k, the smash product $\wedge$ plays the role of the tensor product, and weak equivalences play the role of quasi-isomorphisms. A fundamental difference between the two situations is that the smash product on the underlying category of spectra is not associative and commutative, whereas the tensor product between chain complexes of k-modules is associative and commutative. For this reason, topologists generally work with rings and modules in the stable homotopy category, with their products and actions defined only up to homotopy. In contrast, of course, algebraists generally work with differential graded k-algebras that have associative point-set level multiplications.

We here introduce a new approach to stable homotopy theory that allows one to do point-set level algebra. We construct a new category $\mathscr{M}_S$ of S-modules that has an associative, commutative, and unital smash product $\wedge_S$. Its derived category $\mathscr{D}_S$ is obtained by inverting the weak equivalences; $\mathscr{D}_S$ is equivalent to the classical stable homotopy category, and the equivalence preserves smash products. This allows us to rethink all of stable homotopy theory: all previous work in the subject might as well have been done in $\mathscr{D}_S$. Working on the point-set level, in $\mathscr{M}_S$, we define an S-algebra to be an S-module R with an associative and unital product $R \wedge_S R \longrightarrow R$; if the product is also commutative, we call R a commutative S-algebra. Although the definitions are now very simple, these are not new notions: they are refinements of the A_∞ and E_∞ ring spectra that were introduced over twenty years ago by May, Quinn, and Ray [48]. In general, the latter need not satisfy the precise unital property that is enjoyed by our new S-algebras, but it is a simple matter to construct a weakly equivalent S-algebra from an A_∞ ring spectrum and a weakly equivalent commutative S-algebra from an E_∞ ring spectrum.

It is tempting to refer to (commutative) S-algebras as (commutative) ring spectra. However, this would introduce confusion since the term "ring spectrum" has had a definite meaning for thirty years as a stable homotopy category level notion. Ring spectra in the classical homotopical sense are not rendered obsolete by our theory since there are many examples that admit no S-algebra structure. In any case, the term S-algebra more accurately describes our new concept. With our theory, and the new possibilities that it opens up, it becomes vitally important to keep track of when one is working on the point-set level and when one is working up to homotopy. In the absence (or ignorance) of a good point-set level category of spectra, topologists have tended to be sloppy about this. The dichotomy will run through our work. *The terms* "ring spectrum" *and* "module spectrum" *will always refer to the classical homotopical notions. The terms* "S-algebra" *and* "S-module" *will always refer to the strict point-set level notions.*

We define a (left) module M over an S-algebra R to be an S-module M with an action $R \wedge_S M \longrightarrow M$ such that the standard diagrams commute. We obtain a category $\mathscr{M}_R$ of (left) R-modules and a derived category $\mathscr{D}_R$. There is a smash product $M \wedge_R N$ of a right R-module M and a left R-module N, which is an S-module. For left R-modules M and N, there is a function S-module $F_R(M, N)$ that enjoys properties just like modules of homomorphisms in algebra. Each $F_R(M, M)$ is an S-algebra. If R is commutative, then $M \wedge_R N$ and $F_R(M, N)$ are R-modules, and in this case $\mathscr{M}_R$ and $\mathscr{D}_R$ enjoy all of the properties of $\mathscr{M}_S$ and $\mathscr{D}_S$. Thus each commutative S-algebra R determines a derived category of R-modules that has all of the structure that the stable homotopy category has. These new categories are of substantial intrinsic interest, and they give powerful new tools for the investigation of the classical stable homotopy category.

When we restrict to Eilenberg-Mac Lane spectra, our topological theory subsumes a good deal of classical homological algebra. For a discrete ring R and R-modules M and N, we have

$$\mathrm{Tor}^R_n(M, N) \cong \pi_n(HM \wedge_{HR} HN) \quad \text{and} \quad \mathrm{Ext}^n_R(M, N) \cong \pi_{-n} F_{HR}(HM, HN).$$

Here $\wedge_R$ and F_R must be interpreted in the derived category; that is, HM must be a CW HR-module. Moreover, the algebraic derived category $\mathscr{D}_R$ is equivalent to the topological derived category $\mathscr{D}_{HR}$.

In general, for an S-algebra R, approximation of R-modules M by weakly equivalent cell R-modules is roughly analogous to forming projective resolutions in algebra. There is a much more precise analogy that involves developing the derived categories of modules over rings or, more generally, DGA's in terms of cell modules. It is presented in [35], which gives an algebraic theory of A_∞ and E_∞ k-algebras that closely parallels the present topological theory.

When we restrict to the sphere spectrum S, the derived smash products $M \wedge_S N$ and function spectra $F_S(M, N)$ have as their homotopy groups the homology and cohomology groups $N_*(M)$ and $N^*(M)$. This suggests the alternative notations

$$\mathrm{Tor}^R_n(M, N) = \pi_n(M \wedge_R N) = N^R_n(M)$$

and

$$\mathrm{Ext}^n_R(M, N) = \pi_{-n} F_R(M, N) = N^n_R(M)$$

for R-modules M and N. When R is connective, there are ordinary homology and cohomology theories on R-modules, represented by Eilenberg-Mac Lane spectra that are R-modules, and there are Atiyah-Hirzebruch spectral sequences for the computation of generalized homology and cohomology theories on R-modules.

The realization of algebraic Tor and Ext groups via Eilenberg-Mac Lane spectra generalizes to spectral sequences

$$E^2_{p,q} = \mathrm{Tor}^{R_*}_{p,q}(M_*, N_*) \Longrightarrow \mathrm{Tor}^R_{p+q}(M, N)$$

and

$$E_2^{p,q} = \mathrm{Ext}^{p,q}_{R^*}(M^*, N^*) \Longrightarrow \mathrm{Ext}^{p+q}_R(M, N).$$

These specialize to give Künneth and universal coefficient spectral sequences in classical generalized homology and cohomology theories. There are also Eilenberg-Moore type spectral sequences for the calculation of $E_*(M \wedge_R N)$ under appropriate hypotheses on R and E.

Thinking of $\mathscr{D}_R$ as a new stable homotopy category, where R is a commutative S-algebra, we can realize the action of an element $x \in R_n$ on an R-module M as a map of R-modules $x : \Sigma^n M \longrightarrow M$. We define M/xM to be the cofiber of x, and we define the localization $M[x^{-1}]$ to be the telescope of a countable iterate of desuspensions of x, starting with $M \longrightarrow \Sigma^{-n}M$. By iteration, we can construct quotients by sequences of elements and localizations at sequences of elements. We define R-ring spectra, associative R-ring spectra, and commutative R-ring spectra in the homotopical sense, with products $A \wedge_R A \longrightarrow A$ defined via maps in the derived category $\mathscr{D}_R$, and it turns out to be quite simple to study when quotients and localizations of R-ring spectra are again R-ring spectra.

When we take $R = MU$, we find easy direct constructions as MU-modules of all of the various spectra $(MU/X)[Y^{-1}]$ that are usually obtained by means of the Baas-Sullivan theory of manifolds with singularities or the Landweber exact functor theorem. When their homotopy groups are integral domains concentrated in degrees congruent to zero mod 4, these MU-modules all admit canonical structures of associative and commutative MU-ring spectra. Remarkably, it is far simpler to prove the sharper statements that apply in the derived category of MU-modules than the much weaker stable homotopy category level analogs that were obtainable before our theory.

Thinking of $\mathscr{M}_R$ as a new category of point-set level modules, where R is again a commutative S-algebra, we can define R-algebras A via point-set level products $A \wedge_R A \longrightarrow A$ such that the appropriate diagrams commute. For example, $F_R(M, M)$ is an R-algebra for any R-module M. These have all of the good formal properties of S-algebras. We repeat the dichotomy for emphasis: *The terms* "R-ring spectrum" *and* "R-module spectrum" *will always refer to the homotopical notions defined in the derived category* $\mathscr{D}_R$. *The terms* "R-algebra" *and* "R-module" *will always refer to the strict, point-set, level notions.*

We shall construct Bousfield localizations of R-modules at a given R-module E. In principle, this is a derived category notion, but we shall obtain precise point-set level constructions. Using different point-set level constructions, we shall prove that the Bousfield localizations of R-algebras can be constructed to be R-algebras and the Bousfield localizations of commutative R-algebras can be constructed to be commutative R-algebras. In particular, the localization R_E of R at E is a commutative R-algebra, and we shall see that the category of R_E-modules plays an intrinsically central role in the study of Bousfield localizations.

As a very special case, this theory will imply that the spectra KO and KU that represent real and complex periodic K-theory can be constructed as commutative algebras over the S-algebras ko and ku that represent real and complex connective K-theory. Therefore KO and KU are commutative S-algebras, as had long been conjectured in the earlier context of E_∞ ring spectra. Again, it is far simpler to prove the sharper ko and ku-algebra statements than to construct S-algebra structures directly.

For an R-algebra A, we define the enveloping R-algebra $A^e = A \wedge_R A^{op}$, and we define the topological Hochschild homology of A with coefficients in an (A, A)-bimodule M to be the derived smash product

$$THH^R(A; M) = M \wedge_{A^e} A.$$

This is the correct generalization from algebra to topology since, if R is a discrete commutative ring and A is an R-algebra that is flat as an R-module, then the algebraic and topological Hochschild homology are isomorphic:

$$HH_n^R(A; M) \equiv \mathrm{Tor}_n^{A^e}(M, A) \cong \mathrm{Tor}_n^{HA^e}(HM, HA) \equiv \pi_n(THH^R(A; M)).$$

In general, for a commutative S-algebra R, an R-algebra A, and an (A, A)-bimodule M, there is a spectral sequence

$$E^2_{p,q} = HH^{R_*}_{p,q}(M_*, A_*) \Longrightarrow \pi_{p+q}(THH^R(A; M))$$

under suitable flatness hypotheses. More generally, there are similar spectral sequences converging to $E_*(THH^R(A; M))$ for a commutative ring spectrum E.

There is also a point-set level version $thh^R(A; M)$ of topological Hochschild homology. It is obtained by mimicking topologically the standard complex for the calculation of algebraic Hochschild homology. When $M = A$, this construction has particularly nice formal properties, as was observed in [54] and as we shall explain: it is isomorphic to the tensor $A \otimes S^1$. A key technical point is that the derived category and point-set level definitions become equivalent after replacing R and A by suitable weakly equivalent approximations.

Our S-algebras and their modules are enough like ordinary rings and modules that we can construct the algebraic K-theory spectrum KR associated to an S-algebra R by applying Waldhausen's $S_\bullet$-construction to the category of finite cell R-modules. Applied to the Eilenberg-Mac Lane spectrum HR of a discrete ring R, this gives a new construction of Quillen's algebraic K-theory. Applied to the suspension spectrum $\Sigma^\infty(\Omega X)_+$, this gives a new construction of Waldhausen's algebraic K-theory of the space X. The resulting common framework for topological Hochschild homology and Quillen and Waldhausen algebraic K-theory opens up several new directions and appears to bring a number of standing conjectures within reach. We merely lay the foundations here.

The technical heart of our theory is the problem of keeping our formal point-set level constructions under homotopical control. While we shall show by essentially formal categorical arguments that our various categories of R-modules, R-algebras, and commutative R-algebras are cocomplete and complete, tensored and cotensored, topological model categories, this formal structure does not in itself address the problem: forgetful functors from more to less structured spectra rarely preserve cofibrant objects, and may well not do so even up to homotopy type. The problem requires deeper analysis, and a crucial aspect of our work is that our discussion of model categories gives sufficient control on the underlying homotopy types of cofibrant R-algebras and cofibrant commutative R-algebras to allow the calculational use of bar constructions and topological Hochschild homology complexes. This is also crucial to our proof that Bousfield localizations of R-algebras can be constructed as R-algebras.

Another tool in keeping homotopical control is the category of "tame" spectra. It is an intermediate category between the ground category of spectra, which is well-designed for formal point-set level work but not for homotopical analysis, and the category of CW spectra, which is well-designed for homotopical analysis but not for formal work. The homotopy category of tame spectra is symmetric monoidal under the smash product, and we can approximate any structured spectrum by a weakly equivalent tame structured spectrum by means of a "cylinder construction" defined using homotopy colimits. Actually, this tool will only be needed in Chapter I, since the smash product of S-modules turns out to better behaved under weak equivalences than the smash product of spectra.

The basic construction underlying all of our work is the "twisted half-smash product" $A \ltimes E$ of a suitable space A and a spectrum E. This construction is defined with respect to a given map α from A to an appropriate space of linear isometries. After the penultimate draft of this book was completed, Michael Cole came up with a new construction of twisted half-smash products, one that is much easier to understand than the original construction of Lewis and May and that allows much simpler proofs of some of our main technical results. In particular, we proved and Cole reproved that a homotopy equivalence $A' \longrightarrow A$, with homotopy inverse unrelated to α, induces a homotopy equivalence $A' \ltimes E \longrightarrow A \ltimes E$ when E is tame. This invariance statement is the technical lynchpin of our theory. We have discarded our original proof in favor of Cole's, and he will present his new treatment of twisted half-smash products in the Appendix of this book.

The construction of thh, of bar constructions needed in our work, and of functorial homotopy colimits of spectra all require geometric realizations of simplicial spectra. This raises another technical problem. To understand geometric realization homotopically, the given simplicial spectra must satisfy certain cofibration conditions, and it is hard to verify that a map of spectra is a cofibration (satisfies the homotopy extension property). The solution to this problem is basic to the homotopical understanding of cofibrant R-algebras and commutative R-algebras.

The reader interested in using our theory need not be concerned with these matters, and most of the technical proofs are deferred until the last few chapters and the appendix. The first three chapters explain the foundations needed for the applications of the next three, which are independent of one another. Chapter VII explains the foundations needed for Chapters VIII and IX, which are independent of each other. Each chapter has its own brief introduction. References within a chapter are of the form "Lemma 3.4"; references to results in other chapters are of the form "I.3.4" or, in the case of the Appendix, A.3.4.

Our work is not independent of earlier work: the groundwork was laid in [38], and all of our ring, module, and algebra spectra are spectra in the sense of Lewis and May with additional structure. In view of Cole's new treatment of twisted half-smash products, only the first 75 pages or so of [38] are relevant, and we have given much more readable expositions of this background material in [23] and [52]. In [38], the focus was on equivariant stable homotopy theory, the study of spectra with actions by compact Lie groups G. We have chosen to write this book nonequivariantly in the interests of readability. However, we have kept a close eye on the equivariant generalization, and we have been careful to use only arguments that directly generalize to the equivariant setting. We state a metatheorem.

THEOREM 0.1. *All of the definitions and all of the general theory in this paper apply to G-spectra for any compact Lie group G.*

This has been used by Greenlees and May [28] to prove a completion theorem for the calculation of $M_*(BG)$ and $M^*(BG)$ for any MU-module spectrum M. Some of that work, together with some of ours, was described in the announcement [22]. The recent series of expository papers [23, 26, 27] gives a descriptive account of some of the present theory and its equivariant applications, including both equivariant and non-equivariant applications of the theory to localizations and completions of R-modules at ideals in $\pi_*(R)$. The survey volume [52], which is a companion volume to this one, gives a more leisurely and thorough expository account, together with full details of the definitional framework in the equivariant setting.

We warn the knowledgeable reader that this material has undergone several major revisions, and the final definitions and terminology are not those of earlier announcements and drafts. In particular, our S-modules enjoy a unital property that was not imposed on the S-modules, here called $\mathbb{L}$-spectra, of the earlier versions written by Elmendorf, Kriz, and May alone. The fact that one can impose this unital property and still retain homotopical control is one of many new insights contributed by Mandell. This substantially sharpens and simplifies the theory. Paradoxically, however, one cannot impose such a unital property in the parallel algebraic theory of [35]. Therefore, to facilitate a comparison of the algebraic and topological theories, we run through a little of the previous variant of our theory in the last chapter.

The chapter on algebraic K-theory has not been previously announced and is entirely work of Mandell; it is part of his Chicago PhD thesis in preparation. Similarly, the Appendix is entirely work of Cole and is part of his Chicago PhD thesis in preparation.

Two other Chicago students deserve thanks. Maria Basterra has carefully read several drafts and caught numerous soft spots of exposition; her Chicago PhD thesis in preparation will give a thorough treatment of the André-Quillen cohomology of commutative S-algebras. Jerome Wolbert has made many helpful comments, and his Chicago PhD thesis in preparation will analyze the new derived categories associated to the various K-theory spectra.

It is a pleasure to thank Mike Hopkins, Gaunce Lewis, and Jim McClure for many helpful conversations and e-mails. We owe a critical lemma, namely I.5.4, to Hopkins [32]. Although trivial to prove, it broke a psychological barrier and played a pivotal role in our thinking. We should acknowledge the pioneering work on A_∞ ring and module spectra of Alan Robinson, which gave precursors of many of the results of Chapter IV. We learned the material of IX§3, on thh, from the application of our theory given in the paper [54] of McClure, Schwänzl, and Vogt. We learned many of the results on cocomplete and complete, tensored and cotensored, topological model categories in II§7 and VII§§2,4 from Hopkins and McClure. Their foresight and insight have been inspirational.

CHAPTER I

Prologue: the category of $\mathbb{L}$-spectra

In this prologue, we construct a category whose existence was previously thought to be impossible by at least two of the authors: a complete and cocomplete category of spectra, namely the $\mathbb{L}$-spectra, with an associative and commutative smash product. This contrasts with the category constructed by Lewis and the fourth author in [48, 38], whose smash product is neither associative nor commutative (before passage to homotopy categories), and with the category constructed by the first author in [20], which is neither complete nor cocomplete. We will also give a function $\mathbb{L}$-spectrum construction that is right adjoint to the new smash product. The category of $\mathbb{L}$-spectra has all of the properties that we desire except that its smash product, denoted by $\wedge_{\mathscr{L}}$, is not unital. It has a natural unit map $\lambda : S \wedge_{\mathscr{L}} M \longrightarrow M$, which is often an isomorphism and always a weak equivalence.

The curtain will rise on our real focus of interest in the next chapter, where we will define an S-module to be an $\mathbb{L}$-spectrum M such that $\lambda : S \wedge_{\mathscr{L}} M \longrightarrow M$ is an isomorphism. Restricting $\wedge_{\mathscr{L}}$ to S-modules and renaming it $\wedge_S$, this will give us a symmetric monoidal category in which to develop stable topological algebra.

1. Background on spectra and the stable homotopy category

We begin by recalling the basic definitions in Lewis and May's approach to the stable category. We first recall the definition of a coordinate-free spectrum; see [38, I§2], [20, §2], or [52, Ch. XII] for further details. A coordinate-free spectrum is a spectrum that takes as its indexing set, instead of the integers, the set of finite dimensional subspaces of a "universe", namely a real inner product space $U \cong \mathbb{R}^\infty$. Thus, a spectrum E assigns a based space EV to each finite dimensional subspace V of U, with (adjoint) structure maps

$$\tilde{\sigma}_{V,W} : EV \xrightarrow{\cong} \Omega^{W-V} EW$$

when $V \subset W$. Here $W - V$ is the orthogonal complement of V in W and $\Omega^W X$ is the space of based maps $F(S^W, X)$, where S^W is the one-point compactification of W. These maps are required to be *homeomorphisms* and to satisfy an evident associativity relation. A map of spectra $f : E \longrightarrow E'$ is a collection of maps of based spaces $f_V : EV \longrightarrow E'V$ for which each of the following diagrams commutes:

$$\begin{array}{ccc} EV & \xrightarrow{f_V} & E'V \\ \Big\downarrow{\scriptstyle \tilde{\sigma}_{V,W}} & & \Big\downarrow{\scriptstyle \tilde{\sigma}'_{V,W}} \\ \Omega^{W-V}EW & \xrightarrow{\Omega^{W-V}f_W} & \Omega^{W-V}E'W. \end{array}$$

We obtain the category $\mathscr{S}U$ of spectra indexed on U. If we drop the requirement that the maps $\tilde{\sigma}_{V,W}$ be homeomorphisms, we obtain the notion of a prespectrum and the category $\mathscr{P}U$ of prespectra. The forgetful functor $\mathscr{S}U \longrightarrow \mathscr{P}U$ has a left adjoint L, details of which are given in [38, App]. Functors on prespectra that do not preserve spectra are extended to spectra by applying the functor L. For example, for a based space X and a prespectrum E, we have the prespectrum $E \wedge X$ specified by $(E \wedge X)(V) = EV \wedge X$. When E is a spectrum, the structure maps for this prespectrum level smash product are not homeomorphisms, and we understand the smash product $E \wedge X$ to be the spectrum $L(E \wedge X)$. For example, $\Sigma E = E \wedge S^1$. Function spectra are easier. We set $F(X,E)(V) = F(X,EV)$ and find that this functor on prespectra preserves spectra. For example, $\Omega E = F(S^1, E)$. The following result is discussed in [38, p.13].

PROPOSITION 1.1. *The category $\mathscr{S}U$ is complete and cocomplete.*

PROOF. Limits and colimits are computed on prespectra spacewise. Limits preserve spectra, and colimits of spectra are obtained by use of the functor L. □

A homotopy in the category of spectra is a map $E \wedge I_+ \longrightarrow E'$, and we let $h\mathscr{S}U$ denote the homotopy category of spectra; its objects are spectra and its morphisms are homotopy classes of maps between them. We have cofiber and fiber sequences that behave exactly as in the category of spaces. The cofiber Cf of a map $f : E \longrightarrow E'$ of spectra is the pushout $E' \cup_f CE$, where $CE = E \wedge I$. A cofibration of spectra is a map $i : E \longrightarrow E'$ that satisfies the homotopy extension property (HEP: a homotopy $h : E \wedge I_+ \longrightarrow F$ of a restriction of a map $f : E' \longrightarrow F$ extends to a homotopy $\tilde{h} : E' \wedge I_+ \longrightarrow F$ of f). The canonical maps $E \longrightarrow CE$ and $E' \longrightarrow Cf$ are examples. The fiber Ff of a map $f : E' \longrightarrow E$ is the pullback $E' \times_f PE$, where $PE = F(I, E)$. A fibration of spectra is a map $p : E \longrightarrow E'$ that satisfies the covering homotopy property (CHP: a homotopy $h : F \wedge I_+ \longrightarrow E'$ of a projection $p \circ f$, $f : F \longrightarrow E$, is covered by a homotopy $\tilde{h} : F \wedge I_+ \longrightarrow E$ of f). The canonical maps $PE \longrightarrow E$ and $Ff \longrightarrow E'$ are examples.

A map f of spectra is a weak equivalence if each of its component maps f_V is a weak equivalence of spaces. The stable homotopy category $\bar{h}\mathscr{S}U$ is constructed from the homotopy category $h\mathscr{S}U$ by adjoining formal inverses to the weak equivalences, a process that is made rigorous by CW approximation.

The Vth space functor from spectra to spaces has a left adjoint that we shall denote by Σ^∞_V, or Σ^∞_n when $V = \mathbb{R}^n$ [38, I§4]. Its definition will be recalled in X.4.5. When $V = \{0\}$, this is the suspension spectrum functor Σ^∞. For $n \geq 0$, the sphere spectrum S^n is the suspension spectrum $\Sigma^\infty S^n$ of the sphere space S^n. For $n > 0$, the sphere spectrum S^{-n} is $\Sigma^\infty_n S^0$. There are canonical isomorphisms $\Sigma^m S^n \cong S^{m+n}$ for $m \geq 0$ and integers n and there are canonical isomorphisms $\Sigma^\infty_m S^n \cong S^{n-m}$ for $m \geq 0$ and $n \geq 0$. Sphere spectra are used to define the homotopy groups of spectra, $\pi_n(E) = h\mathscr{S}U(S^n, E)$, and a map of spectra is a weak equivalence if and only if it induces an isomorphism of spectrum-level homotopy groups.

Although we shall not introduce different notations for space level and spectrum level spheres, we shall generally write S for the zero sphere spectrum, reserving the notation S^0 for the two-point space.

The theory of cell and CW spectra is developed by taking sphere spectra as the domains of attaching maps [38, I§5]. The stable homotopy category $\bar{h}\mathscr{S}U$ is equivalent to the homotopy category of CW spectra. It is important to remember that homotopy-preserving functors on spectra that do not preserve weak equivalences are transported to the stable category by first replacing their variables by weakly equivalent CW spectra.

2. External smash products and twisted half-smash products

The construction of our new smash product will start from the external smash product of spectra. This is an associative and commutative pairing

$$\mathscr{S}U \times \mathscr{S}U' \longrightarrow \mathscr{S}(U \oplus U')$$

for any pair of universes U and U'. It is constructed by starting with the prespectrum level definition

$$(E \wedge E')(V \oplus V') = EV \wedge E'V'.$$

The structure maps fail to be homeomorphisms when E and E' are spectra, and we apply the spectrification functor L to obtain the desired spectrum level smash product. This external smash product is the one used in [20].

There is an associated function spectrum functor

$$F : (\mathscr{S}U')^{op} \times \mathscr{S}(U \oplus U') \longrightarrow \mathscr{S}U$$

and an adjunction

$$\mathscr{S}(U \oplus U')(E \wedge E', E'') \cong \mathscr{S}U(E, F(E', E''))$$

for $E \in \mathscr{S}U$, $E' \in \mathscr{S}U'$, and $E'' \in \mathscr{S}(U \oplus U')$; see [38, p. 69].

Now let $\mathscr{I}$ denote the category whose objects are universes U and whose morphisms are linear isometries. Universes are topologized as the unions of their finite dimensional subspaces, and the set $\mathscr{I}(U, U')$ of linear isometries $U \longrightarrow U'$ is given the function space topology; it is a contractible space [38, II.1.5]. The category $\mathscr{S}$ constructed in [20] augments to the category $\mathscr{I}$. Since $\mathscr{I}$ fails to have limits and colimits (it even fails to have coproducts), $\mathscr{S}$ suffers from the same defects.

In order to obtain smash products internal to a single universe U, we shall exploit the "twisted half-smash product". The input data for this functor consist of two universes U and U' (which may be the same), an unbased space A with a given structure map $\alpha : A \longrightarrow \mathscr{I}(U, U')$, and a spectrum E indexed on U. The output is the spectrum $A \ltimes E$, which is indexed on U'. It must be remembered that the construction depends on α and not just on A. When A has CW homotopy type, different choices of α lead to equivalent functors on the level of stable categories, by A.7.6. The intuition is that the twisted half-smash product is a generalization to spectra of the "untwisted" functor $A_+ \wedge X$ on based spaces X. This intuition is made precise by the following "untwisting formula" that relates twisted half-smash products and suspension spectra. A proof will be given in A.5.5, where the result will be generalized to shift desuspension functors Σ^∞_V.

PROPOSITION 2.1. *For any map $A \longrightarrow \mathscr{I}(U, U')$ and any $n \geq 0$, there is an isomorphism of spectra*

$$A \ltimes \Sigma^\infty X \cong \Sigma^\infty(A_+ \wedge X)$$

that is natural in spaces A over $\mathscr{I}(U, U')$ and based spaces X.

Observe that the functor Σ^∞ implicitly refers to the universe U on the left and to the universe U' on the right. The twisted half-smash product enjoys the following formal properties, among others; their analogs for the space level functor $A_+ \wedge X$ are trivial. Proofs will be given in A.5.3 and A§6.

PROPOSITION 2.2. *The following statements hold.*

(i) *There is a canonical isomorphism $\{\mathrm{id}_U\} \ltimes E \cong E$.*

(ii) *Let $A \longrightarrow \mathscr{I}(U, U')$ and $B \longrightarrow \mathscr{I}(U', U'')$ be given; let $B \times A$ have the structure map given by the composite*

$$B \times A \longrightarrow \mathscr{I}(U', U'') \times \mathscr{I}(U, U') \xrightarrow{\ \circ\ } \mathscr{I}(U, U'').$$

Then there is a canonical isomorphism

$$(B \times A) \ltimes E \cong B \ltimes (A \ltimes E).$$

(iii) *Let $A \longrightarrow \mathscr{I}(U_1, U_1')$ and $B \longrightarrow \mathscr{I}(U_2, U_2')$ be given; let $A \times B$ have the structure map given by the composite*

$$A \times B \longrightarrow \mathscr{I}(U_1, U_1') \times \mathscr{I}(U_2, U_2') \xrightarrow{\ \oplus\ } \mathscr{I}(U_1 \oplus U_2, U_1' \oplus U_2').$$

Let E_1 and E_2 be spectra indexed on U_1 and U_2 respectively. Then there is a canonical isomorphism

$$(A \times B) \ltimes (E_1 \wedge E_2) \cong (A \ltimes E_1) \wedge (B \ltimes E_2).$$

(iv) *For $A \longrightarrow \mathscr{I}(U, U')$, $E \in \mathscr{S}U$, and a based space X, there is a canonical isomorphism*

$$A \ltimes (E \wedge X) \cong (A \ltimes E) \wedge X.$$

The functor $A \ltimes (-)$ is a left adjoint. Its right adjoint will be used in our construction of function S-modules.

PROPOSITION 2.3. *For any space A over $\mathscr{I}(U, U')$, the functor $A \ltimes (-)$ has a right adjoint, which is denoted by $F[A, -)$ and called a twisted function spectrum.*

The functor $A \ltimes E$ is homotopy-preserving in E, and it therefore preserves homotopy equivalences in the variable E. However, it only preserves homotopies over $\mathscr{I}(U, U')$ in A. Nevertheless, it very often preserves homotopy equivalences in the variable A. This fact will be essential in keeping control over the homotopical behavior of our point-set level constructions. To state it in proper generality, we need the following notion of a well-behaved spectrum.

DEFINITION 2.4. A prespectrum D is Σ-cofibrant if each of its structure maps $\sigma : \Sigma^W DV \longrightarrow D(V \oplus W)$ is a cofibration of based spaces. A spectrum E is Σ-cofibrant if it is isomorphic to one of the form LD, where D is a Σ-cofibrant prespectrum. A spectrum E is tame if it is homotopy equivalent to a Σ-cofibrant spectrum.

We shall discuss such spectra in X§4, where we shall see that all shift desuspensions of based spaces are Σ-cofibrant and that all CW spectra are tame. It follows that any spectrum is weakly equivalent to a tame one. We shall show in X§5 that structured ring or module spectra can be approximated functorially by weakly equivalent Σ-cofibrant spectra with the same structure. The following result will be restated and proven as A.7.4. It is central to our theory.

THEOREM 2.5. *Let $E \in \mathscr{S}U$ be tame and let A be a space over $\mathscr{I}(U, U')$. If $\phi : A' \longrightarrow A$ is a homotopy equivalence, then $\phi \ltimes \mathrm{id} : A' \ltimes E \longrightarrow A \ltimes E$ is a homotopy equivalence.*

Here, for a general spectrum E, we do not know whether or not $\phi \ltimes \mathrm{id}$ is even a weak equivalence. By A.7.3, the result has the following consequence.

COROLLARY 2.6. *Let $E \in \mathscr{S}U$ be a spectrum that has the homotopy type of a CW spectrum and let A be a space over $\mathscr{I}(U, U')$ that has the homotopy type of a CW complex. Then $A \ltimes E$ has the homotopy type of a CW spectrum.*

3. The linear isometries operad and internal smash products

For the rest of the paper, we restrict attention to a particular universe U; the reader is welcome to consider it as notation for $\mathbb{R}^\infty$. We agree to write $\mathscr{S}$ instead of $\mathscr{S}U$ for the category of spectra indexed on U. Except where explicitly stated otherwise, all given spectra, whatever extra structure they may have, will be in $\mathscr{S}$. We are especially interested in twisted half smash products defined in terms of the following spaces of linear isometries.

NOTATIONS 3.1. Let U^j be the direct sum of j copies of U and let $\mathscr{L}(j) = \mathscr{I}(U^j, U)$. The space $\mathscr{L}(0)$ is the point i, where $i : \{0\} \longrightarrow U$, and $\mathscr{L}(1)$ contains the identity map $1 = \mathrm{id}_U : U \longrightarrow U$. The left action of Σ_j on U^j by permutations induces a free right action of Σ_j on the contractible space $\mathscr{L}(j)$. Define maps

$$\gamma : \mathscr{L}(k) \times \mathscr{L}(j_1) \times \cdots \times \mathscr{L}(j_k) \longrightarrow \mathscr{L}(j_1 + \cdots + j_k)$$

by

$$\gamma(g; f_1, \ldots, f_k) = g \circ (f_1 \oplus \cdots \oplus f_k).$$

The spaces $\mathscr{L}(j)$ form an operad [45, p.1] with structural maps γ, called the linear isometries operad. Points $f \in \mathscr{L}(j)$ give inclusions $\{f\} \longrightarrow \mathscr{L}(j)$. The corresponding twisted half-smash product is denoted f_*; it sends spectra indexed on U^j to spectra indexed on U. Applied to a j-fold external smash product $E_1 \wedge \cdots \wedge E_j$, it gives an internal smash product $f_*(E_1 \wedge \cdots \wedge E_j)$. All of these smash products become equivalent in the stable homotopy category $\bar{h}\mathscr{S}$, but none of them are associative or commutative on the point set level. In fact, the following sharper version of this assertion holds.

THEOREM 3.2. *Let $\mathscr{S}_t \subset \mathscr{S}$ be the full subcategory of tame spectra and let $h\mathscr{S}_t$ be the category of tame spectra and homotopy classes of maps. The internal smash products $f_*(E \wedge E')$ determined by varying $f \in \mathscr{L}(2)$ are canonically isomorphic in $h\mathscr{S}_t$, and $h\mathscr{S}_t$ is symmetric monoidal under the internal smash product. For based spaces X and tame spectra E, there is a natural isomorphism $E \wedge X \simeq f_*(E \wedge \Sigma^\infty X)$ in $h\mathscr{S}_t$.*

PROOF. The external and internal smash products of Σ-cofibrant spectra are Σ-cofibrant by results in X§4. By Theorem 2.5, for any $f \in \mathscr{L}(j)$ and any spectra $E_i \in \mathscr{S}_t$, the map

$$f_*(E_1 \wedge \cdots \wedge E_j) \longrightarrow \mathscr{L}(j) \ltimes (E_1 \wedge \cdots \wedge E_j)$$

induced by the inclusion $\{f\} \longrightarrow \mathscr{L}(j)$ is a homotopy equivalence. Taking $j = 2$, this shows that the internal smash products obtained from varying f are homotopy equivalent. Replacing f by $f \circ \sigma$, where $\sigma \in \Sigma_2$ is the transposition, we obtain a natural homotopy equivalence

$$f_*(E_2 \wedge E_1) \longrightarrow \mathscr{L}(2) \ltimes E_1 \wedge E_2,$$

and this shows that the internal smash product is commutative up to homotopy. Similarly, for associativity, the inclusions of the points $\{f(1 \oplus f)\}$ and $\{f(f \oplus 1)\}$ in $\mathscr{L}(3)$ induce natural homotopy equivalences

$$f_*(E_1 \wedge f_*(E_2 \wedge E_3)) \longrightarrow \mathscr{L}(3) \ltimes (E_1 \wedge E_2 \wedge E_3) \longleftarrow f_*(f_*(E_1 \wedge E_2) \wedge E_3).$$

It is natural to think of based spaces as spectra indexed on the universe $\{0\}$. Then i_* and the suspension spectrum functor are both left adjoint to the zeroth space functor, hence $i_*X \cong \Sigma^\infty X$. The map $\mathscr{L}(2) \longrightarrow \mathscr{L}(1)$ that sends f to $f \circ (1 \oplus i)$ and the inclusion $\{1\} \longrightarrow \mathscr{L}(1)$ induce natural homotopy equivalences

$$f_*(E \wedge \Sigma^\infty X) \longrightarrow \mathscr{L}(1) \ltimes (E \wedge X) \longleftarrow E \wedge X.$$

Thus, up to natural isomorphisms, the internal smash product determined by f becomes commutative, associative, and unital with unit $S = \Sigma^\infty S^0$ on passage to $h\mathscr{S}_t$. The commutativity of coherence diagrams that is required for the assertion that $h\mathscr{S}_t$ is symmetric monoidal (see [43, p. 180]) can be checked by an elaboration of these arguments. □

The following consequence strengthens the assertion [38, I.6.1] that the stable homotopy category really is a stable category, in the sense that the suspension and loop functors Σ and Ω pass to inverse self-equivalences of $\bar{h}\mathscr{S}$.

COROLLARY 3.3. *For tame spectra E and $f \in \mathscr{L}(2)$, there is a natural homotopy equivalence between ΩE and $f_*(E \wedge S^{-1})$, and the unit $\eta : E \longrightarrow \Omega\Sigma E$ and counit $\varepsilon : \Sigma\Omega E \longrightarrow E$ of the (Σ, Ω)-adjunction are homotopy equivalences.*

PROOF. For based spaces X, $\Sigma^\infty X$ is naturally isomorphic to $\Sigma(\Sigma_1^\infty X)$ since the structural homeomorphism $E_0 \longrightarrow \Omega E_1$ gives a natural isomorphism between their right adjoints. Therefore, for $E \in \mathscr{S}_t$, there is a natural homotopy equivalence

$$E = E \wedge S^0 \simeq f_*(E \wedge \Sigma^\infty S^0) \cong f_*(E \wedge \Sigma(\Sigma_1^\infty S^0)) \cong \Sigma(f_*(E \wedge S^{-1})),$$

where the last isomorphism is given by Proposition 2.2(iv). It follows that, on $h\mathscr{S}_t$, the functor Σ is an adjoint equivalence with inverse given by the functor $f_*(E \wedge S^{-1})$. The rest is a formal consequence of the uniqueness of adjoints. □

Note that only actual homotopy equivalences, not weak ones, are relevant to these results. For this and other reasons, $h\mathscr{S}_t$ will be a technically convenient halfway house between $h\mathscr{S}$ and the stable homotopy category $\bar{h}\mathscr{S}$, which is obtained from either of these homotopy categories by inverting the weak equivalences.

We can deduce that cofiber sequences give rise to long exact sequences of homotopy groups.

COROLLARY 3.4. *Any cofiber sequence $E \xrightarrow{f} E' \longrightarrow Cf$ of tame spectra gives rise to a long exact sequence of homotopy groups*

$$\cdots \longrightarrow \pi_q(E) \longrightarrow \pi_q(E') \longrightarrow \pi_q(Cf) \longrightarrow \pi_{q-1}(E) \longrightarrow \cdots .$$

Therefore the natural map $Ff \longrightarrow \Omega Cf$ is a weak equivalence.

PROOF. Consider the diagram

$$\begin{array}{ccccccccc}
S^q & \xrightarrow{\mathrm{id}} & S^q & \longrightarrow & CS^q & \longrightarrow & \Sigma S^q & \xrightarrow{\mathrm{id}} & \Sigma S^q \\
{\scriptstyle \Sigma^{-1}\gamma}\downarrow & & \downarrow{\scriptstyle \alpha} & & \downarrow{\scriptstyle \beta} & & \downarrow{\scriptstyle \gamma} & & \downarrow{\scriptstyle \Sigma\alpha} \\
E & \xrightarrow{f} & E' & \xrightarrow{i} & Cf & \longrightarrow & \Sigma E & \xrightarrow{\Sigma f} & \Sigma E'.
\end{array}$$

Here α is given such that $i \circ \alpha \simeq 0$. A homotopy induces a map β such that the second square commutes. The usual cofiber sequence argument gives γ such that the right two squares homotopy commute. Since $\eta_E : E \longrightarrow \Omega\Sigma E$ is a weak equivalence, there is a map $\Sigma^{-1}\gamma : S^q \longrightarrow E$, unique up to homotopy, such that

$$\eta_E \circ \Sigma^{-1}\gamma \simeq \Omega\gamma \circ \eta_{S^q}.$$

Therefore

$$\eta_{E'} \circ f \circ \Sigma^{-1}\gamma = \Omega\Sigma f \circ \eta_E \circ \Sigma^{-1}\gamma \simeq \Omega\Sigma f \circ \Omega\gamma \circ \eta_{S^q} \simeq \Omega\Sigma\alpha \circ \eta_{S^q} = \eta_{E'} \circ \alpha.$$

Since $\eta_{E'}$ is a weak equivalence, this implies that $f \circ \Sigma^{-1}\gamma \simeq \alpha$. The long exact sequence follows by extending the given cofiber sequence to the right, as usual. The last statement follows by the five lemma and a comparison of our cofiber sequence with the fiber sequence associated to f. Details of this may be found in [38, pp 128-130]. For later use, observe that we only used that the maps η are weak equivalences, not that they are homotopy equivalences, in this proof. □

It follows that cofiber sequences are essentially equivalent to fiber sequences. More precisely, the cofibrations and fibrations give "triangulations" of the stable homotopy category such that the negative of a cofibration triangle is a fibration triangle, and conversely [38, pp 128-130].

COROLLARY 3.5. *Pushouts of tame spectra along cofibrations preserve weak equivalences. That is, for a commutative diagram of tame spectra*

$$\begin{array}{ccccc} E & \xleftarrow{\;i\;} & D & \xrightarrow{\;f\;} & F \\ \downarrow{\scriptstyle\beta} & & \downarrow{\scriptstyle\alpha} & & \downarrow{\scriptstyle\gamma} \\ E' & \xleftarrow[\;i'\;]{} & D' & \xrightarrow[\;f'\;]{} & F' \end{array}$$

in which i and i' are cofibrations and α, β, and γ are weak equivalences, the induced map $\delta : E \cup_D F \longrightarrow E' \cup_{D'} F'$ of pushouts is a weak equivalence.

PROOF. As for spaces, Ci is homotopy equivalent to E/D, the induced map $F \longrightarrow E \cup_D F$ is a cofibration, and the induced map $E/D \longrightarrow E \cup_D F/F$ is an isomorphism. The conclusion follows from the previous corollary by a diagram chase and the five lemma. □

PROPOSITION 3.6. *If E is a CW spectrum and $\phi : F \longrightarrow F'$ is a weak equivalence between tame spectra, then $f_*(\mathrm{id} \wedge \phi) : f_*(E \wedge F) \longrightarrow f_*(E \wedge F')$ is a weak equivalence.*

PROOF. The functor $f_*((-) \wedge F)$ preserves cofiber sequences. Therefore, by Corollary 3.5 and induction up the sequential filtration of E (see III.2.1), the result will hold for general E if it holds for $E = S^n$. When $E = S$, the conclusion holds by the unit equivalence $f_*(S \wedge F) \simeq F$ of Theorem 3.2. For $n > 0$, we easily deduce isomorphisms

$$f_*(S^n \wedge F) \cong \Sigma^n f_*(S \wedge F) \quad \text{and} \quad \Sigma^n f_*(S^{-n} \wedge F) \cong f_*(S \wedge F)$$

from Proposition 2.2(iv). In view of Corollary 3.3, the result for $E = S^{-n}$ and $E = S^n$ therefore follows from the result for S. □

It follows that for general spectra E and tame spectra F, the smash product $E \wedge F$ in the stable homotopy category $\bar{h}\mathscr{S}$ is represented by $f_*(\Gamma E \wedge F)$, where ΓE is a CW spectrum weakly equivalent to E. That is, we do not also have to apply CW approximation to F. The mild restriction to tame spectra serves to avoid pathological point-set behavior.

4. The category of $\mathbb{L}$-spectra

We think of $\mathscr{L}(j) \ltimes (E_1 \wedge \cdots \wedge E_j)$ as a canonical j-fold internal smash product. It is still not associative, but we shall construct a commutative and associative smash product by restricting to $\mathbb{L}$-spectra and shrinking the fat out of the construction. To define $\mathbb{L}$-spectra, we focus attention on a small part of the operad $\mathscr{L}$. Recall the notion of a monad in a category from [43, ch.VI] or [45, 2.1].

NOTATIONS 4.1. Let $\mathbb{L}$ denote the monad in the category $\mathscr{S}$ that is specified by $\mathbb{L}E = \mathscr{L}(1) \ltimes E$; the product

$$\mu : \mathbb{L}\mathbb{L}E \cong (\mathscr{L}(1) \times \mathscr{L}(1)) \ltimes E \longrightarrow \mathscr{L}(1) \ltimes E = \mathbb{L}E$$

is induced by the product $\gamma : \mathscr{L}(1) \times \mathscr{L}(1) \longrightarrow \mathscr{L}(1)$ and the unit

$$\eta : E \cong \{1\} \ltimes E \longrightarrow \mathscr{L}(1) \ltimes E = \mathbb{L}E$$

is induced by the inclusion $\{1\} \longrightarrow \mathscr{L}(1)$ of the identity element.

DEFINITION 4.2. An $\mathbb{L}$-spectrum is an $\mathbb{L}$-algebra M, that is, a spectrum M together with an action $\xi : \mathbb{L}M \longrightarrow M$ by the monad $\mathbb{L}$. Explicitly, the following diagrams are required to commute:

$$\begin{array}{ccc} \mathbb{L}\mathbb{L}M & \xrightarrow{\mu} & \mathbb{L}M \\ {\scriptstyle \mathbb{L}\xi}\downarrow & & \downarrow{\scriptstyle \xi} \\ \mathbb{L}M & \xrightarrow[\xi]{} & M \end{array} \qquad \text{and} \qquad \begin{array}{ccc} M & \xrightarrow{\eta} & \mathbb{L}M \\ & {\scriptstyle =}\searrow & \downarrow{\scriptstyle \xi} \\ & & M. \end{array}$$

A map $f : M \longrightarrow N$ of $\mathbb{L}$-spectra is a map of spectra such that the following diagram commutes:

$$\begin{array}{ccc} \mathbb{L}M & \xrightarrow{\mathbb{L}f} & \mathbb{L}N \\ {\scriptstyle \xi_M}\downarrow & & \downarrow{\scriptstyle \xi_N} \\ M & \xrightarrow[f]{} & N. \end{array}$$

We let $\mathscr{S}[\mathbb{L}]$ denote the category of $\mathbb{L}$-spectra.

There is a dual form of the definition that will occasionally be needed. It is based on the following standard categorical observation.

LEMMA 4.3. *Let $\mathbb{T}$ be a monad in a category $\mathscr{C}$, and suppose that the functor $\mathbb{T}$ has a right adjoint $\mathbb{T}^{\#}$. Then $\mathbb{T}^{\#}$ is a comonad such that the categories of $\mathbb{T}$-algebras and of $\mathbb{T}^{\#}$-coalgebras are isomorphic.*

We shall consistently use the notation $\mathbb{T}^{\#}$ for the comonad associated to a monad $\mathbb{T}$ that has a right adjoint. In particular, by Proposition 2.3, we now have a comonad $\mathbb{L}^{\#}$ such that an $\mathbb{L}^{\#}$-coalgebra is the same thing as an $\mathbb{L}$-spectrum. This implies the following result.

PROPOSITION 4.4. *The category of $\mathbb{L}$-spectra is complete and cocomplete, with both limits and colimits created in the underlying category $\mathscr{S}$. If X is a based space and M is an $\mathbb{L}$-spectrum, then $M \wedge X$ and $F(X, M)$ are $\mathbb{L}$-spectra, and the spectrum level fiber and cofiber of a map of $\mathbb{L}$-spectra are $\mathbb{L}$-spectra.*

PROOF. Since $\mathscr{S}[\mathbb{L}]$ is the category of algebras over the monad $\mathbb{L}$, the forgetful functor $\mathscr{S}[\mathbb{L}] \longrightarrow \mathscr{S}$ creates limits [43, VI.2, ex. 2]. Since $\mathscr{S}$ is complete, this implies the statement about limits. The statement about colimits follows similarly by use of the comonad $\mathbb{L}^{\#}$. The last statement is immediate from the canonical isomorphism

$$\mathscr{L}(1) \ltimes (M \wedge X) \cong (\mathscr{L}(1) \ltimes M) \wedge X$$

of Proposition 2.2(iv) and its analog (A.6.1)

$$F[\mathscr{L}(1), F(X, M)) \cong F(X, F[\mathscr{L}(1), M)). \quad \square$$

LEMMA 4.5. *The sphere spectrum S is an $\mathbb{L}$-spectrum. More generally, for based spaces X, $\Sigma^\infty X \cong S \wedge X$ is naturally an $\mathbb{L}$-spectrum.*

PROOF. Recall from the proof of Theorem 3.2 that a based space X may be viewed as a spectrum indexed on $\{0\}$ and that $\Sigma^\infty X \cong i_* X$, $i : \{0\} \longrightarrow U$. We may rewrite this as $\Sigma^\infty X = \mathscr{L}(0) \ltimes X$. Then the structure map is given by

$$\gamma \ltimes \mathrm{id} : \mathscr{L}(1) \ltimes (\mathscr{L}(0) \ltimes X) \cong (\mathscr{L}(1) \times \mathscr{L}(0)) \ltimes X \longrightarrow \mathscr{L}(0) \ltimes X.$$

In the middle, $\mathscr{L}(1) \times \mathscr{L}(0)$ is regarded as a space over $\mathscr{L}(0)$ via γ, and the isomorphism is given by an instance of Proposition 2.2(ii). Of course, γ here is just the unique map from $\mathscr{L}(1)$ to the one-point space $\mathscr{L}(0)$, and our structure map is just the composite

$$\mathscr{L}(1) \ltimes \Sigma^\infty X \cong \Sigma^\infty(\mathscr{L}(1)_+ \wedge X) \longrightarrow \Sigma^\infty(S^0 \wedge X) \cong \Sigma^\infty X,$$

where the first isomorphism is given by Proposition 2.1. $\square$

A homotopy in the category of $\mathbb{L}$-spectra is a map $M \wedge I_+ \longrightarrow N$. A map of $\mathbb{L}$-spectra is a weak equivalence if it is a weak equivalence as a map of spectra. The stable homotopy category $\bar{h}\mathscr{S}[\mathbb{L}]$ of $\mathbb{L}$-spectra is constructed from the homotopy category $h\mathscr{S}[\mathbb{L}]$ by adjoining formal inverses to the weak equivalences; again, the process is made rigorous by CW approximation. Since the theory of cell and CW $\mathbb{L}$-spectra is exactly like the theory of cell and CW spectra developed in [38, I§5], we shall not give details. The reader who would like to see an exposition is invited to look ahead to III§2. The theory of cell R-modules to be presented there applies (with minor simplifications) to give what is needed. It is formal that the monad $\mathbb{L}$ may be viewed as specifying the free functor from spectra to $\mathbb{L}$-spectra. The sphere $\mathbb{L}$-spectra that we take as the domains of attaching maps when defining cell $\mathbb{L}$-spectra are the free $\mathbb{L}$-spectra $\mathbb{L}S^n = \mathscr{L}(1) \ltimes S^n$. A weak equivalence of cell $\mathbb{L}$-spectra is a homotopy equivalence, any $\mathbb{L}$-spectrum is weakly equivalent to a CW $\mathbb{L}$-spectrum, and $\bar{h}\mathscr{S}[\mathbb{L}]$ is equivalent to the homotopy category of CW $\mathbb{L}$-spectra. We warn the reader that, although S itself is an $\mathbb{L}$-spectrum, it does not have the homotopy type of a CW $\mathbb{L}$-spectrum (see Warning

6.8 below). The following comparison between CW spectra and CW $\mathbb{L}$-spectra establishes an equivalence between $\bar{h}\mathscr{S}$ and $\bar{h}\mathscr{S}[\mathbb{L}]$.

THEOREM 4.6. *The following conclusions hold.*

(i) *The free functor $\mathbb{L} : \mathscr{S} \longrightarrow \mathscr{S}[\mathbb{L}]$ carries CW spectra to CW $\mathbb{L}$-spectra.*
(ii) *The forgetful functor $\mathscr{S}[\mathbb{L}] \longrightarrow \mathscr{S}$ carries $\mathbb{L}$-spectra of the homotopy types of CW $\mathbb{L}$-spectra to spectra of the homotopy types of CW spectra.*
(iii) *Every CW $\mathbb{L}$-spectrum M is homotopy equivalent as an $\mathbb{L}$-spectrum to $\mathbb{L}E$ for some CW spectrum E.*
(iv) *The unit $\eta : E \longrightarrow \mathbb{L}E$ of the adjunction*

$$\mathscr{S}[\mathbb{L}](\mathbb{L}E, M) \cong \mathscr{S}(E, M)$$

is a homotopy equivalence if $E \in \mathscr{S}_t$, for example if E is a CW spectrum.
(v) *The counit $\xi : \mathbb{L}M \longrightarrow M$ of the adjunction is a homotopy equivalence of spectra if M is tame and is a homotopy equivalence of $\mathbb{L}$-spectra if M has the homotopy type of a CW $\mathbb{L}$-spectrum.*

The free and forgetful functors establish an adjoint equivalence between the stable homotopy categories $\bar{h}\mathscr{S}$ and $\bar{h}\mathscr{S}[\mathbb{L}]$.

PROOF. Part (i) is immediate by induction up the sequential filtration (see III.2.1). Part (iv) is immediate from Theorem 2.5 and, applied to sphere spectra, it implies (ii). Since $\xi \circ \eta = \mathrm{id} : M \longrightarrow M$ for any M, (iv) and the Whitehead theorem in the category of $\mathbb{L}$-spectra imply (v). Part (iii) follows from (i) and (v) since there is a CW spectrum E and a homotopy equivalence of spectra $E \longrightarrow M$. It is a formal consequence of (i) that we have an induced adjunction

$$\bar{h}\mathscr{S}[\mathbb{L}](\mathbb{L}E, M) \cong \bar{h}\mathscr{S}(E, M)$$

(see [38, I.5.13]), and its unit and counit are natural isomorphisms. □

Observe that, dually, we can interpret $\mathbb{L}^{\#}$ as specifying the "cofree" functor from spectra to $\mathbb{L}$-spectra. That is, we have an adjunction

$$\mathscr{S}[\mathbb{L}](M, \mathbb{L}^{\#}E) \cong \mathscr{S}(M, E). \tag{4.7}$$

By part (ii) of the theorem and [38, I.5.13], there results an induced adjunction

$$\bar{h}\mathscr{S}[\mathbb{L}](M, \mathbb{L}^{\#}E) \cong \bar{h}\mathscr{S}(M, E).$$

It is an easy categorical observation that, in any adjoint equivalence of categories, the given left and right adjoints are also right and left adjoint to each other.

COROLLARY 4.8. *The functors $\mathbb{L} : \bar{h}\mathscr{S} \longrightarrow \bar{h}\mathscr{S}[\mathbb{L}]$ and $\mathbb{L}^{\#} : \bar{h}\mathscr{S} \longrightarrow \bar{h}\mathscr{S}[\mathbb{L}]$ are naturally isomorphic.*

5. The smash product of $\mathbb{L}$-spectra

Via instances of the structural maps γ of the operad $\mathscr{L}$, we have a left action of the monoid $\mathscr{L}(1)$ and a right action of the monoid $\mathscr{L}(1)\times\mathscr{L}(1)$ on $\mathscr{L}(2)$. These actions commute with each other. If M and N are $\mathbb{L}$-spectra, then $\mathscr{L}(1)\times\mathscr{L}(1)$ acts from the left on the external smash product $M\wedge N$ via the map

$$\xi : (\mathscr{L}(1)\times\mathscr{L}(1))\ltimes(M\wedge N)\cong(\mathscr{L}(1)\ltimes M)\wedge(\mathscr{L}(1)\ltimes N)\xrightarrow{\xi\wedge\xi} M\wedge N.$$

To form the twisted half smash product on the left, we think of $\mathscr{L}(1)\times\mathscr{L}(1)$ as mapping to $\mathscr{I}(U^2,U^2)$ via direct sum of linear isometries. The smash product over $\mathscr{L}$ of M and N is simply the balanced product of the two $\mathscr{L}(1)\times\mathscr{L}(1)$-actions.

DEFINITION 5.1. Let M and N be $\mathbb{L}$-spectra. Define the operadic smash product $M\wedge_{\mathscr{L}} N$ to be the coequalizer displayed in the diagram

$$(\mathscr{L}(2)\times\mathscr{L}(1)\times\mathscr{L}(1))\ltimes(M\wedge N)\underset{\mathrm{id}\ltimes\xi}{\overset{\gamma\ltimes\mathrm{id}}{\rightrightarrows}}\mathscr{L}(2)\ltimes(M\wedge N)\longrightarrow M\wedge_{\mathscr{L}} N.$$

Here we have implicitly used the isomorphism

$$(\mathscr{L}(2)\times\mathscr{L}(1)\times\mathscr{L}(1))\ltimes(M\wedge N)\cong\mathscr{L}(2)\ltimes[(\mathscr{L}(1)\times\mathscr{L}(1))\ltimes(M\wedge N)]$$

given by Proposition 2.2(ii). The left action of $\mathscr{L}(1)$ on $\mathscr{L}(2)$ induces a left action of $\mathscr{L}(1)$ on $M\wedge_{\mathscr{L}} N$ that gives it a structure of $\mathbb{L}$-spectrum.

We may mimic tensor product notation and write

$$M\wedge_{\mathscr{L}} N=\mathscr{L}(2)\ltimes_{\mathscr{L}(1)\times\mathscr{L}(1)}(M\wedge N).$$

We will freely use such notations for coequalizers below. The commutativity of this smash product is immediate.

PROPOSITION 5.2. *There is a natural commutativity isomorphism of* $\mathbb{L}$*-spectra*

$$\tau : M\wedge_{\mathscr{L}} N\longrightarrow N\wedge_{\mathscr{L}} M.$$

PROOF. The permutation $\sigma\in\Sigma_2$ acts on $\mathscr{L}(2)$ by $f\sigma=f\circ t$, where $t: U^2\longrightarrow U^2$ is the transposition isomorphism. We may regard σ as a map of spaces over $\mathscr{L}(2)$ from $\mathrm{id}:\mathscr{L}(2)\longrightarrow\mathscr{L}(2)$ to $\sigma:\mathscr{L}(2)\longrightarrow\mathscr{L}(2)$. We have an evident isomorphism $\iota: t_*(M\wedge N)\cong N\wedge M$ on external smash products and, by Proposition 2.2(ii), there results a canonical isomorphism

$$\sigma\ltimes\iota:\mathscr{L}(2)\ltimes M\wedge N\cong\mathscr{L}(2)\ltimes t_*(M\wedge N)\cong\mathscr{L}(2)\ltimes N\wedge M.$$

There is an analogous isomorphism

$$(\sigma\times t)\ltimes\iota:(\mathscr{L}(2)\times\mathscr{L}(1)\times\mathscr{L}(1))\ltimes(M\wedge N)\longrightarrow(\mathscr{L}(2)\times\mathscr{L}(1)\times\mathscr{L}(1))\ltimes(N\wedge M).$$

These maps induce an isomorphism of coequalizer diagrams

$$\begin{array}{ccccc}
(\mathscr{L}(2)\times\mathscr{L}(1)\times\mathscr{L}(1))\ltimes(M\wedge N) & \underset{\mathrm{id}\ltimes\xi}{\overset{\gamma\ltimes\mathrm{id}}{\rightrightarrows}} & \mathscr{L}(2)\ltimes(M\wedge N) & \longrightarrow & M\wedge_{\mathscr{L}} N \\
{\scriptstyle(\sigma\times t)\ltimes\iota}\Big\downarrow & & \Big\downarrow{\scriptstyle\sigma\ltimes\iota} & & \Big\downarrow{\scriptstyle\tau} \\
(\mathscr{L}(2)\times\mathscr{L}(1)\times\mathscr{L}(1))\ltimes(N\wedge M) & \underset{\mathrm{id}\ltimes\xi}{\overset{\gamma\ltimes\mathrm{id}}{\rightrightarrows}} & \mathscr{L}(2)\ltimes(N\wedge M) & \longrightarrow & N\wedge_{\mathscr{L}} M.
\end{array}$$

□

To show that this smash product is associative, we need some preliminary material on coequalizers. We first recall a standard categorical definition [43, VI.6].

DEFINITION 5.3. Working in an arbitrary category, suppose given a diagram

$$A \underset{f}{\overset{e}{\rightrightarrows}} B \xrightarrow{g} C$$

in which $ge = gf$. The diagram is called a split coequalizer if there are maps

$$h : C \longrightarrow B \quad \text{and} \quad k : B \longrightarrow A$$

such that $gh = \mathrm{id}_C$, $fk = \mathrm{id}_B$, and $ek = hg$. It follows that g is the coequalizer of e and f.

Observe that, while covariant functors need not preserve coequalizers in general, they clearly do preserve split coequalizers. The next observation is crucial; we learned it from Hopkins [32]. Note that, via structural maps γ, $\mathscr{L}(1)$ acts from the left on any $\mathscr{L}(i)$, hence $\mathscr{L}(1)\times\mathscr{L}(1)$ acts from the left on $\mathscr{L}(i)\times\mathscr{L}(j)$.

LEMMA 5.4 (HOPKINS). *For $i \geq 1$ and $j \geq 1$, the diagram*

$$\begin{array}{c}
\mathscr{L}(2)\times\mathscr{L}(1)\times\mathscr{L}(1)\times\mathscr{L}(i)\times\mathscr{L}(j) \\
{\scriptstyle\mathrm{id}\times\gamma^2}\Big\downarrow\Big\downarrow{\scriptstyle\gamma\times\mathrm{id}} \\
\mathscr{L}(2)\times\mathscr{L}(i)\times\mathscr{L}(j) \\
\Big\downarrow{\scriptstyle\gamma} \\
\mathscr{L}(i+j)
\end{array}$$

is a split coequalizer of spaces. Therefore,

$$\mathscr{L}(i+j) \cong \mathscr{L}(2)\times_{\mathscr{L}(1)\times\mathscr{L}(1)}\mathscr{L}(i)\times\mathscr{L}(j).$$

PROOF. Choose isomorphisms $s : U^i \longrightarrow U$ and $t : U^j \longrightarrow U$ and define

$$h(f) = (f\circ(s\oplus t)^{-1}, s, t)$$

and

$$k(f; g, g') = (f; g\circ s^{-1}, g'\circ t^{-1}; s, t).$$

It is trivial to check the identities of Definition 5.3. □

THEOREM 5.5. *There is a natural associativity isomorphism of $\mathbb{L}$-spectra*

$$(M \wedge_{\mathscr{L}} N) \wedge_{\mathscr{L}} P \cong M \wedge_{\mathscr{L}} (N \wedge_{\mathscr{L}} P).$$

PROOF. Note that, for any $\mathbb{L}$-spectrum N, $N \cong \mathscr{L}(1) \ltimes_{\mathscr{L}(1)} N$ since $\mathscr{L}(1) \ltimes N = \mathbb{L}N$ and, as with any monad [43, p. 148], we have a split coequalizer

$$\mathbb{L}\mathbb{L}N \rightrightarrows \mathbb{L}N \longrightarrow N.$$

We have the isomorphisms

$$\begin{aligned}(M \wedge_{\mathscr{L}} N) \wedge_{\mathscr{L}} P &\cong \mathscr{L}(2) \ltimes_{\mathscr{L}(1)^2} (\mathscr{L}(2) \ltimes_{\mathscr{L}(1)^2} (M \wedge N)) \wedge (\mathscr{L}(1) \ltimes_{\mathscr{L}(1)} P) \\ &\cong (\mathscr{L}(2) \times_{\mathscr{L}(1)^2} \mathscr{L}(2) \times \mathscr{L}(1)) \ltimes_{\mathscr{L}(1)^3} (M \wedge N \wedge P) \\ &\cong \mathscr{L}(3) \ltimes_{\mathscr{L}(1)^3} M \wedge N \wedge P.\end{aligned}$$

The symmetric argument shows that this is also isomorphic to $M \wedge_{\mathscr{L}} (N \wedge_{\mathscr{L}} P)$. □

In view of the generality of Lemma 5.4, the argument iterates to prove the following statement.

THEOREM 5.6. *For any j-tuple $M_1, \ldots, M_j$ of $\mathbb{L}$-spectra, there is a canonical isomorphism of $\mathbb{L}$-spectra*

$$M_1 \wedge_{\mathscr{L}} \cdots \wedge_{\mathscr{L}} M_j \cong \mathscr{L}(j) \ltimes_{\mathscr{L}(1)^j} (M_1 \wedge \cdots \wedge M_j),$$

where the iterated smash product on the left is associated in any fashion.

6. The equivalence of the old and new smash products

We here show that the smash product $\wedge_{\mathscr{L}}$ does in fact realize the classical smash product of spectra up to homotopy, in the sense that the equivalence between $\bar{h}\mathscr{S}$ and $\bar{h}\mathscr{S}[\mathbb{L}]$ preserves smash products.

Fix a linear isometric isomorphism $f : U^2 \longrightarrow U$ (not just an isometry) and use it to define the internal smash product of spectra in this section. We begin the comparison of smash products of $\mathbb{L}$-spectra with smash products of spectra with the following observation.

PROPOSITION 6.1. *For spectra X and Y, there are isomorphisms of $\mathbb{L}$-spectra*

$$\mathbb{L}X \wedge_{\mathscr{L}} \mathbb{L}Y \cong \mathscr{L}(2) \ltimes X \wedge Y \cong \mathbb{L}f_*(X \wedge Y).$$

For CW $\mathbb{L}$-spectra M and N, $M \wedge_{\mathscr{L}} N$ is a CW $\mathbb{L}$-spectrum with one $(p+q)$-cell for each p-cell of M and q-cell of N.

PROOF. The first isomorphism is immediate from the definition of $\wedge_{\mathscr{L}}$. Regarding f as a point in $\mathscr{L}(2)$, we see that $\gamma : \mathscr{L}(1) \times \{f\} \longrightarrow \mathscr{L}(2)$ is a homeomorphism since f is an isomorphism. It follows from Proposition 2.2(ii) that

$$\mathbb{L}f_*(X \wedge Y) = \mathscr{L}(1) \ltimes f_*(X \wedge Y) \cong \mathscr{L}(2) \ltimes (X \wedge Y).$$

When X and Y are sphere spectra, so is $f_*(X \wedge Y)$ [38, II.1.4]. The second statement now follows exactly as for the smash product of CW complexes or CW spectra. □

The crux of our comparison of smash products is the following proposition, which implies that $\mathbb{L}S$ is the unit for the smash product in the stable homotopy category $\bar{h}\mathscr{S}[\mathbb{L}]$. We defer the proof to XI§3.

PROPOSITION 6.2. *For $\mathbb{L}$-spectra N, there is a natural weak equivalence of $\mathbb{L}$-spectra $\omega : \mathbb{L}S \wedge_{\mathscr{L}} N \longrightarrow N$, and $\Sigma : \pi_n(N) \longrightarrow \pi_{n+1}(\Sigma N)$ is an isomorphism for all integers n.*

If we knew a priori that Σ preserved weak equivalences, we could derive the second clause from the first and the natural isomorphism of $\mathbb{L}$-spectra

$$\mathbb{L}S \wedge_{\mathscr{L}} N \cong \Sigma(\mathbb{L}S^{-1} \wedge_{\mathscr{L}} N)$$

by a formal uniqueness of adjoints argument (compare Corollary 3.3). It is a pleasant and surprising technical feature of our theory, immediate from the proposition, that Σ preserves weak equivalences of $\mathbb{L}$-spectra. That is, the $\mathbb{L}$ structure somehow has the effect of eliminating point-set pathology. Since Σ on homotopy groups is induced by $\eta : N \longrightarrow \Omega\Sigma N$, the proposition also has the following immediate consequence.

COROLLARY 6.3. *For $\mathbb{L}$-spectra N, the unit $\eta : N \longrightarrow \Omega\Sigma N$ and counit $\varepsilon : \Sigma\Omega N \longrightarrow N$ of the (Σ, Ω)-adjunction are weak equivalences.*

COROLLARY 6.4. *Any cofiber sequence $N \xrightarrow{f} N' \longrightarrow Cf$ of $\mathbb{L}$-spectra gives rise to a long exact sequence of homotopy groups*

$$\cdots \longrightarrow \pi_q(N) \longrightarrow \pi_q(N') \longrightarrow \pi_q(Cf) \longrightarrow \pi_{q-1}(N) \longrightarrow \cdots .$$

Therefore the natural map $Ff \longrightarrow \Omega Cf$ is a weak equivalence of $\mathbb{L}$-spectra.

PROOF. This follows from Corollary 6.3 via the proof of Corollary 3.4. □

COROLLARY 6.5. *Pushouts along cofibrations of $\mathbb{L}$-spectra preserve weak equivalences.*

PROOF. Since a cofibration of $\mathbb{L}$-spectra is a cofibration of spectra, by the retraction of mapping cylinders criterion, this follows from Corollary 6.4 via the proof of Corollary 3.5. □

PROPOSITION 6.6. *If M is a CW $\mathbb{L}$-spectrum and $\phi : N \longrightarrow N'$ is a weak equivalence of $\mathbb{L}$-spectra, then $\mathrm{id} \wedge_{\mathscr{L}} \phi : M \wedge_{\mathscr{L}} N \longrightarrow M \wedge_{\mathscr{L}} N'$ is a weak equivalence of $\mathbb{L}$-spectra.*

PROOF. The functor $(-) \wedge_{\mathscr{L}} N$ preserves cofiber sequences, hence the result for general M follows from Corollary 6.4 and the result for $M = \mathbb{L}S^n$. Here the result for $n = 0$ follows from Proposition 6.2 and the result for n and $-n$, $n > 0$, follows from the result for $n = 0$ as in the proof of Proposition 3.6. □

Thus, for $\mathbb{L}$-spectra M and N, the smash product $M \wedge_{\mathscr{L}} N$ in the stable homotopy category $\bar{h}\mathscr{S}[\mathbb{L}]$ is represented by $\Gamma M \wedge_{\mathscr{L}} N$, where ΓM is a CW $\mathbb{L}$-spectrum weakly equivalent to M; here we do not need to assume that N is tame. This is analogous to the situation in algebra. When transporting tensor products to algebraic derived categories, we need only apply cell approximation to one of the tensor factors, without any condition on the other [35].

THEOREM 6.7. *For $\mathbb{L}$-spectra M and N, there is a natural map of spectra $\alpha : f_*(M \wedge N) \longrightarrow M \wedge_{\mathscr{L}} N$, and α is a weak equivalence when M is a CW $\mathbb{L}$-spectrum and N is a tame spectrum. For any $\mathbb{L}$-spectrum N, the composite of the derived functor $(-) \wedge_{\mathscr{L}} N : \bar{h}\mathscr{S}[\mathbb{L}] \longrightarrow \bar{h}\mathscr{S}[\mathbb{L}]$ and the forgetful functor $\bar{h}\mathscr{S}[\mathbb{L}] \longrightarrow \bar{h}\mathscr{S}$ computes the derived internal smash product with N.*

PROOF. Define α to be the composite

$$f_*(M \wedge N) \longrightarrow \mathscr{L}(2) \ltimes M \wedge N \longrightarrow M \wedge_{\mathscr{L}} N$$

given by the the inclusion of $\{f\}$ in $\mathscr{L}(2)$ and the definition of $\wedge_{\mathscr{L}}$. Let M be a CW $\mathbb{L}$-spectrum throughout the proof. We first show that α is an equivalence when N is also a CW $\mathbb{L}$-spectrum. In this case, M and N have the homotopy types of CW spectra by Theorem 4.6 and are therefore tame by X.4.3. Thus the first map is a homotopy equivalence by Theorem 2.5. By Theorem 4.6(iii), we may assume without loss of generality that $M = \mathscr{L}(1) \ltimes X$ and $N = \mathscr{L}(1) \ltimes Y$ for CW spectra X and Y. The second arrow then reduces to the homotopy equivalence

$$\mathscr{L}(2) \ltimes (\mathscr{L}(1) \ltimes X) \wedge (\mathscr{L}(1) \ltimes Y) \longrightarrow \mathscr{L}(2) \ltimes X \wedge Y$$

induced by the homotopy equivalence $\gamma : \mathscr{L}(2) \times \mathscr{L}(1) \times \mathscr{L}(1) \longrightarrow \mathscr{L}(2)$ via Theorem 2.5. For a general $\mathbb{L}$-spectrum N, choose a weak equivalence $\gamma : \Gamma N \longrightarrow N$, where ΓN is a CW $\mathbb{L}$-spectrum. If N is tame, then Propositions 3.6 and 6.6 imply that the vertical arrows are weak equivalences in the commutative diagram

$$\begin{array}{ccc} f_*(M \wedge \Gamma N) & \xrightarrow{\alpha} & M \wedge_{\mathscr{L}} \Gamma N \\ {\scriptstyle \mathrm{id} \wedge \gamma}\downarrow & & \downarrow{\scriptstyle \mathrm{id} \wedge \gamma} \\ f_*(M \wedge N) & \xrightarrow[\alpha]{} & M \wedge_{\mathscr{L}} N. \end{array}$$

Thus the bottom arrow α is a weak equivalence since the top one is. For the last statement, simply note that the right-hand composite

$$(\mathrm{id} \wedge \gamma) \circ \alpha : f_*(M \wedge \Gamma N) \longrightarrow M \wedge_{\mathscr{L}} N$$

in the diagram is a weak equivalence even when N is not tame. □

WARNING 6.8. As said before, the sphere spectrum S does not have the homotopy type of a CW $\mathbb{L}$-spectrum. To see this, assume that it did. Then the action $\xi : \mathbb{L}S \longrightarrow S$ would be a homotopy equivalence of $\mathbb{L}$-spectra, by the Whitehead theorem, and the Σ_2-equivariant map

$$\xi \wedge_{\mathscr{L}} \xi : \mathbb{L}S \wedge_{\mathscr{L}} \mathbb{L}S \longrightarrow S \wedge_{\mathscr{L}} S$$

would be a Σ_2-equivariant homotopy equivalence of $\mathbb{L}$-spectra and thus of spectra. By Propositions 6.1 and 2.1, $\mathbb{L}S \wedge_{\mathscr{L}} \mathbb{L}S$ is isomorphic to $\Sigma^\infty(\mathscr{L}(2)_+)$, with Σ_2-action induced by that on $\mathscr{L}(2)$. By Proposition 8.2 below, $S \wedge_{\mathscr{L}} S$ is isomorphic to $S = \Sigma^\infty S^0$ and has trivial action by Σ_2. Under these isomorphisms, $\xi \wedge_{\mathscr{L}} \xi$ coincides with $\Sigma^\infty \pi$, where $\pi : \mathscr{L}(2)_+ \longrightarrow S^0$ sends all of $\mathscr{L}(2)$ to the non-basepoint. Since $\mathscr{L}(2)/\Sigma_2 \simeq B(\Sigma_2)$, our assumption implies that we obtain a homotopy equivalence $\Sigma^\infty B(\Sigma_2)_+ \longrightarrow \Sigma^\infty S^0$ on passage to orbits from $\xi \wedge_{\mathscr{L}} \xi$, which is absurd. A different perspective on this warning will be given in II.1.10.

7. Function $\mathbb{L}$-spectra

We here construct a functor $F_{\mathscr{L}}$ on $\mathbb{L}$-spectra that is related to the smash product $\wedge_{\mathscr{L}}$ by an adjunction of the usual form and consider its homotopical behavior.

THEOREM 7.1. *Let M, N, and P be $\mathbb{L}$-spectra. There is a function $\mathbb{L}$-spectrum functor $F_{\mathscr{L}}(M, N)$, contravariant in M and covariant in N, such that*

$$\mathscr{S}[\mathbb{L}](M \wedge_{\mathscr{L}} N, P) \cong \mathscr{S}[\mathbb{L}](M, F_{\mathscr{L}}(N, P)).$$

Given the adjunction, we can deduce the homotopical behavior of $F_{\mathscr{L}}$ from that of $\wedge_{\mathscr{L}}$. We run through this before turning to the construction. The following result is a formal consequence of Proposition 6.1; see [38, I.5.13].

PROPOSITION 7.2. *If M is a CW $\mathbb{L}$-spectrum and $\phi : N \longrightarrow N'$ is a weak equivalence of $\mathbb{L}$-spectra, then*

$$F_{\mathscr{L}}(\mathrm{id}, \phi) : F_{\mathscr{L}}(M, N) \longrightarrow F_{\mathscr{L}}(M, N')$$

is a weak equivalence of $\mathbb{L}$-spectra. There is an induced adjunction

$$\bar{h}\mathscr{S}[\mathbb{L}](M \wedge_{\mathscr{L}} N, P) \cong \bar{h}\mathscr{S}[\mathbb{L}](M, F_{\mathscr{L}}(N, P)).$$

As in Section 6, we fix a linear isometric isomorphism $f : U^2 \longrightarrow U$ and use the isomorphism $f_* : \mathscr{S}U^2 \longrightarrow \mathscr{S}U$ to define internal smash products $f_*(M \wedge N)$. Recall the external function spectrum $F(M, -)$ and the adjunction displayed for it at the start of Section 2. We use the inverse isomorphism $f^* = f_*^{-1} : \mathscr{S}U \longrightarrow \mathscr{S}U^2$ to define internal function spectra $F(M, f^*N)$, as in [38, II.3.11].

THEOREM 7.3. *For $\mathbb{L}$-spectra M and N, there is a natural map of spectra*

$$\tilde{\alpha} : F_{\mathscr{L}}(M, N) \longrightarrow F(M, f^*N),$$

and $\tilde{\alpha}$ is a weak equivalence if M is a CW $\mathbb{L}$-spectrum. Therefore the equivalence of categories $\bar{h}\mathscr{S}[\mathbb{L}] \longrightarrow \bar{h}\mathscr{S}$ induced by the forgetful functor from $\mathbb{L}$-spectra to spectra carries the function $\mathbb{L}$-spectrum functor $F_{\mathscr{L}}$ to the internal function spectrum functor F.

PROOF. In the category $\bar{h}\mathscr{S}[\mathbb{L}]$, $F_{\mathscr{L}}(M,N)$ means $F_{\mathscr{L}}(\Gamma M,N)$ where ΓM is a CW $\mathbb{L}$-spectrum weakly equivalent to M, hence the second statement will follow from the first. The desired map $\tilde{\alpha}$ is the adjoint of the composite

$$f_*(F_{\mathscr{L}}(M,N)\wedge M)\xrightarrow{\alpha} F_{\mathscr{L}}(M,N)\wedge_{\mathscr{L}} M\xrightarrow{\varepsilon} N,$$

where α is given by Theorem 6.7. By that result, if M is a CW $\mathbb{L}$-spectrum and X is a CW spectrum, then $\alpha : f_*(\mathbb{L}X\wedge M)\longrightarrow \mathbb{L}X\wedge_{\mathscr{L}} M$ is a weak equivalence of spectra, and it induces a weak equivalence of $\mathbb{L}$-spectra

$$\mathbb{L}(f_*(\mathbb{L}X\wedge M))\longrightarrow \mathbb{L}X\wedge_{\mathscr{L}} M.$$

Diagram chases show that the map

$$\tilde{\alpha}_* : \bar{h}\mathscr{S}(X, F_{\mathscr{L}}(M,N))\longrightarrow \bar{h}\mathscr{S}(X, F(M, f^*N))$$

coincides with the composite of the following chain of natural isomorphisms:

$$\begin{aligned}\bar{h}\mathscr{S}(X, F_{\mathscr{L}}(M,N)) &\cong \bar{h}\mathscr{S}[\mathbb{L}](\mathbb{L}X, F_{\mathscr{L}}(M,N))\\ &\cong \bar{h}\mathscr{S}[\mathbb{L}](\mathbb{L}X\wedge_{\mathscr{L}} M, N)\cong \bar{h}\mathscr{S}[\mathbb{L}](\mathbb{L}(f_*(\mathbb{L}X\wedge M)),N)\\ &\cong \bar{h}\mathscr{S}(f_*(\mathbb{L}X\wedge M),N)\cong \bar{h}\mathscr{S}(\mathbb{L}X\wedge M, f^*N)\\ &\cong \bar{h}\mathscr{S}(\mathbb{L}X, F(M,f^*N))\cong \bar{h}\mathscr{S}(X, F(M,f^*N)).\quad\square\end{aligned}$$

LEMMA 7.4. *The adjoint* $N\longrightarrow F_{\mathscr{L}}(\mathbb{L}S,N)$ *of the unit weak equivalence* $\omega : \mathbb{L}S\wedge_{\mathscr{L}} N\longrightarrow N$ *is a weak equivalence.*

PROOF. This is immediate from the natural isomorphisms

$$\bar{h}\mathscr{S}[\mathbb{L}](M,N)\cong \bar{h}\mathscr{S}[\mathbb{L}](\mathbb{L}S\wedge_{\mathscr{L}} M,N)\cong \bar{h}\mathscr{S}[\mathbb{L}](M, F_{\mathscr{L}}(\mathbb{L}S,N)).\quad\square$$

We must still prove Theorem 7.1. The desired adjunction dictates the definition of $F_{\mathscr{L}}$, and the reader is invited to skip to the next section. It will be simplest to construct $F_{\mathscr{L}}$ in two steps. Remember that

$$M\wedge_{\mathscr{L}} N = \mathscr{L}(2)\ltimes_{\mathscr{L}(1)\times\mathscr{L}(1)} M\wedge N.$$

In the first step we consider general spectra indexed on U^2 and acted upon by $\mathscr{L}(1)\times\mathscr{L}(1)$, thought of as a space over $\mathscr{I}(U^2,U^2)$ via direct sum of isometries. We call these $\mathscr{L}(1)\times\mathscr{L}(1)$-spectra and denote the category of such spectra by $\mathscr{S}[\mathscr{L}(1)\times\mathscr{L}(1)]$. Of course, the examples we have in mind are of the form $M\wedge N$. We use the twisted function spectrum construction $F[A,-)$ of Proposition 2.3.

LEMMA 7.5. *Let* N *be an* $\mathbb{L}$*-spectrum. There is an* $\mathscr{L}(1)\times\mathscr{L}(1)$*-spectrum* $F_{\mathscr{L}(1)}[\mathscr{L}(2),N)\in\mathscr{S}(U^2)$ *such that*

$$\mathscr{S}[\mathbb{L}](\mathscr{L}(2)\ltimes_{\mathscr{L}(1)\times\mathscr{L}(1)} P, N)\cong \mathscr{S}[\mathscr{L}(1)\times\mathscr{L}(1)](P, F_{\mathscr{L}(1)}[\mathscr{L}(2),N))$$

for $\mathscr{L}(1)\times\mathscr{L}(1)$*-spectra* P.

PROOF. We construct $F_{\mathscr{L}(1)}[\mathscr{L}(2), N)$ as the equalizer of two maps

$$F[\mathscr{L}(2), N) \rightrightarrows F[\mathscr{L}(1) \times \mathscr{L}(2), N).$$

The first is induced by $\gamma : \mathscr{L}(1) \times \mathscr{L}(2) \longrightarrow \mathscr{L}(2)$. The second is the composite

$$F[\mathscr{L}(2), N) \longrightarrow F[\mathscr{L}(1) \times \mathscr{L}(2), \mathscr{L}(1) \ltimes N) \xrightarrow{F[1,\xi)} F[\mathscr{L}(1) \times \mathscr{L}(2), N);$$

here the unlabelled arrow is adjoint to

$$(\mathscr{L}(1) \times \mathscr{L}(2)) \ltimes F[\mathscr{L}(2), N) \cong \mathscr{L}(1) \ltimes \mathscr{L}(2) \ltimes F[\mathscr{L}(2), N) \xrightarrow{\mathrm{id} \ltimes \varepsilon} \mathscr{L}(1) \ltimes N,$$

where ε is the counit of the adjunction. The left action of $\mathscr{L}(1) \times \mathscr{L}(1)$ on $F_{\mathscr{L}(1)}[\mathscr{L}(2), N)$ is induced by its right action on $\mathscr{L}(2)$. $\square$

The second step lands us back in the category of $\mathbb{L}$-spectra.

LEMMA 7.6. *Let N be an $\mathbb{L}$-spectrum and P be an $\mathscr{L}(1) \times \mathscr{L}(1)$-spectrum. There is an $\mathbb{L}$-spectrum $\hat{F}(N, P)$ such that*

$$\mathscr{S}[\mathscr{L}(1) \times \mathscr{L}(1)](M \wedge N, P) \cong \mathscr{S}[\mathbb{L}](M, \hat{F}(N, P))$$

for $\mathbb{L}$-spectra M.

PROOF. Again, we construct $\hat{F}(N, P)$ as an equalizer, this time of two maps

$$F(N, P) \rightrightarrows F(\mathbb{L}N, P).$$

The first is induced by the structure map $\mathbb{L}N \longrightarrow N$. The second is the composite

$$F(N, P) \longrightarrow F(\mathbb{L}N, (\{1\} \times \mathscr{L}(1)) \ltimes P) \longrightarrow F(\mathbb{L}N, P),$$

where the second arrow is induced by the structure map of P as an $\mathscr{L}(1) \times \mathscr{L}(1)$-module and the first arrow is adjoint to

$$F(N, P) \wedge \mathbb{L}N \cong (\{1\} \times \mathscr{L}(1)) \ltimes F(N, P) \wedge N \xrightarrow{\mathrm{id} \ltimes \varepsilon} (\{1\} \times \mathscr{L}(1)) \ltimes P.$$

The structure of $\hat{F}(N, P)$ as an $\mathbb{L}$-spectrum is induced by the action on P of the first factor of $\mathscr{L}(1)$ in $\mathscr{L}(1) \times \mathscr{L}(1)$; more precisely, the action $\mathbb{L}F(N, P) \longrightarrow F(N, P)$ is adjoint to the composite

$$(\mathbb{L}F(N, P)) \wedge N \cong (\mathscr{L}(1) \times \{1\}) \ltimes (F(N, P) \wedge N) \xrightarrow{\mathrm{id} \ltimes \varepsilon} (\mathscr{L}(1) \times \{1\}) \ltimes P \xrightarrow{\xi} P. \quad \square$$

We combine these two functorial constructions to define $F_{\mathscr{L}}$.

DEFINITION 7.7. For $\mathbb{L}$-spectra M and N, define

$$F_{\mathscr{L}}(M, N) = \hat{F}(M, F_{\mathscr{L}(1)}[\mathscr{L}(2), N)).$$

The adjunction of Theorem 7.1 is just the composite of the two adjunctions already obtained.

8. Unital properties of the smash product of $\mathbb{L}$-spectra

As we have already seen, $\mathbb{L}S$ is a unit for the smash product $\wedge_{\mathscr{L}}$ on $\bar{h}\mathscr{S}[\mathbb{L}]$. However, for precision in the consideration of algebraic structures, we wish to work in a category of spectra that is actually symmetric monoidal under its smash product, with a point-set level unit isomorphism. The appropriate candidate for a unit object is not $\mathbb{L}S$ but S itself, and at this point another special, and surprising, property of the linear isometries operad comes into play.

Consider the diagram

$$\mathscr{L}(2)\times\mathscr{L}(1)\times\mathscr{L}(1)\times\mathscr{L}(0)\times\mathscr{L}(0)\underset{\mathrm{id}\times\gamma^2}{\overset{\gamma\times\mathrm{id}}{\rightrightarrows}}\mathscr{L}(2)\times\mathscr{L}(0)\times\mathscr{L}(0)\xrightarrow{\gamma}\mathscr{L}(0).$$

This is not a split coequalizer, but it turns out to be a coequalizer. The coequalizer of the parallel pair of arrows is the orbit space $\mathscr{L}(2)/\mathscr{L}(1)\times\mathscr{L}(1)$.

LEMMA 8.1. *The orbit space $\mathscr{L}(2)/\mathscr{L}(1)\times\mathscr{L}(1)$ consists of a single point.*

This is far from obvious, and it is only possible because $\mathscr{L}(1)$ is a monoid but not a group. We defer its proof to XI§2. It has the following implication. Recall Lemma 4.5.

PROPOSITION 8.2. *There is an isomorphism of $\mathbb{L}$-spectra $\lambda : S\wedge_{\mathscr{L}} S\longrightarrow S$ such that $\lambda\tau=\lambda$. For based spaces X and Y, there is a natural isomorphism of $\mathbb{L}$-spectra*

$$\lambda:\Sigma^\infty X\wedge_{\mathscr{L}}\Sigma^\infty Y\cong\Sigma^\infty(X\wedge Y).$$

PROOF. The second statement follows from the first, or directly: γ induces the isomorphism

$$\mathscr{L}(2)\ltimes_{\mathscr{L}(1)\times\mathscr{L}(1)}(\mathscr{L}(0)\ltimes X)\wedge(\mathscr{L}(0)\ltimes Y)\longrightarrow\mathscr{L}(0)\ltimes X\wedge Y.$$

The relation $\lambda\tau=\lambda : S\wedge_{\mathscr{L}} S\longrightarrow S$ is clear since $\gamma\tau=\gamma$.

This formalizes our intuition that the smash product should be a stabilized generalization of the smash product of based spaces. It is natural to try to generalize the resulting isomorphism $\lambda : S\wedge_{\mathscr{L}}\Sigma^\infty X\cong\Sigma^\infty X$ to arbitrary $\mathbb{L}$-spectra, and the map does generalize.

PROPOSITION 8.3. *Let M and N be $\mathbb{L}$-spectra. There is a natural map of $\mathbb{L}$-spectra $\lambda : S\wedge_{\mathscr{L}} N\longrightarrow N$. The symmetrically defined map $M\wedge_{\mathscr{L}} S\longrightarrow M$ coincides with the composite $\lambda\tau$. Moreover, under the associativity isomorphism,*

$$\lambda\tau\wedge_{\mathscr{L}}\mathrm{id}=\mathrm{id}\wedge_{\mathscr{L}}\lambda : M\wedge_{\mathscr{L}} S\wedge_{\mathscr{L}} N\longrightarrow M\wedge_{\mathscr{L}} N,$$

and, under the commutativity isomorphism, these maps also agree with

$$\lambda : S\wedge_{\mathscr{L}}(M\wedge_{\mathscr{L}} N)\longrightarrow M\wedge_{\mathscr{L}} N.$$

PROOF. When N is the free $\mathbb{L}$-spectrum $\mathbb{L}X = \mathscr{L}(1) \ltimes X$ generated by a spectrum X, λ is given by the map

$$\begin{aligned} S \wedge_{\mathscr{L}} \mathbb{L}X &= \mathscr{L}(2) \ltimes_{\mathscr{L}(1)\times\mathscr{L}(1)} (\mathscr{L}(0) \ltimes S^0) \wedge (\mathscr{L}(1) \ltimes X) \\ &\cong (\mathscr{L}(2) \times_{\mathscr{L}(1)\times\mathscr{L}(1)} \mathscr{L}(0) \times \mathscr{L}(1)) \ltimes (S^0 \wedge X) \\ &\xrightarrow{\gamma\ltimes \mathrm{id}} \mathscr{L}(1) \ltimes X = \mathbb{L}X. \end{aligned}$$

For general N, the map just constructed induces a map of coequalizer diagrams

$$\begin{array}{ccccc} S \wedge_{\mathscr{L}} \mathbb{L}\mathbb{L}N & \rightrightarrows & S \wedge_{\mathscr{L}} \mathbb{L}N & \longrightarrow & S \wedge_{\mathscr{L}} N \\ \downarrow & & \downarrow & & \downarrow \\ \mathbb{L}\mathbb{L}N & \rightrightarrows & \mathbb{L}N & \longrightarrow & N. \end{array}$$

The symmetry is clear when M is free and follows in general by an easy comparison of coequalizer diagrams. Similarly, suppose that $M = \mathbb{L}X$ and $N = \mathbb{L}Y$ for spectra X and Y. Then, under the associativity isomorphisms of their domains given in the proof of Theorem 5.5, the two unit maps defined on $\mathbb{L}X \wedge_{\mathscr{L}} S \wedge_{\mathscr{L}} \mathbb{L}Y$ agree with the map

$$\begin{aligned} &\mathscr{L}(3) \ltimes_{\mathscr{L}(1)^3} ((\mathscr{L}(1) \ltimes X) \wedge (\mathscr{L}(0) \ltimes S^0) \wedge (\mathscr{L}(1) \ltimes Y)) \\ &\cong (\mathscr{L}(3) \times_{\mathscr{L}(1)^3} \mathscr{L}(1) \times \mathscr{L}(0) \times \mathscr{L}(1)) \ltimes (X \wedge S^0 \wedge Y) \\ &\xrightarrow{\gamma\ltimes \mathrm{id}} \mathscr{L}(2) \ltimes (X \wedge Y) \cong \mathbb{L}X \wedge_{\mathscr{L}} \mathbb{L}Y. \end{aligned}$$

The conclusion for general M and N follows by another comparison of coequalizer diagrams. The last statement can be proven similarly. □

Any attempt to show that S is a strict unit for general $\mathbb{L}$-spectra founders on the fact that Lemma 5.4 fails if $i = 0$ or $j = 0$ and $i + j > 0$. However, we shall prove the following up to homotopy version of that lemma in XI.2.2.

LEMMA 8.4. *The space*

$$\hat{\mathscr{L}}(1) \equiv \mathscr{L}(2) \times_{\mathscr{L}(1)\times\mathscr{L}(1)} \mathscr{L}(0) \times \mathscr{L}(1)$$

is contractible. Therefore $\gamma : \hat{\mathscr{L}}(1) \longrightarrow \mathscr{L}(1)$ *is a homotopy equivalence.*

Again, this assertion is far from obvious. It leads us to the following crucial result.

THEOREM 8.5. *Let M be an $\mathbb{L}$-spectrum and consider $\lambda : S \wedge_{\mathscr{L}} M \longrightarrow M$.*

(i) *If $M = \mathbb{L}X$ for a tame spectrum X, then λ is a homotopy equivalence of spectra and thus a weak equivalence of $\mathbb{L}$-spectra.*
(ii) *If M is a CW $\mathbb{L}$-spectrum, then λ is a homotopy equivalence of $\mathbb{L}$-spectra.*
(iii) *For any M, λ is a weak equivalence of $\mathbb{L}$-spectra.*

PROOF. Since $\lambda = \gamma \ltimes \mathrm{id}$ on free $\mathbb{L}$-spectra $\mathbb{L}X$, Theorem 2.5 and the lemma give (i). By Theorem 4.6(iii), (i) applies to show that $\lambda : S \wedge_{\mathscr{L}} M \longrightarrow M$ is a weak equivalence of $\mathbb{L}$-spectra when M is a CW $\mathbb{L}$-spectrum. By the Whitehead theorem for CW $\mathbb{L}$-spectra, there is a map of $\mathbb{L}$-spectra $\xi : M \longrightarrow S \wedge_{\mathscr{L}} M$ such that $\lambda \circ \xi \simeq \mathrm{id}$. To complete the proof of (ii), we must show that $\xi \circ \lambda \simeq \mathrm{id}$, and the following commutative diagram identifies this composite with $\mathrm{id} \wedge_{\mathscr{L}} (\lambda \circ \xi)$:

$$\begin{array}{ccccc} S \wedge_{\mathscr{L}} M & \xrightarrow{\mathrm{id} \wedge \xi} & S \wedge_{\mathscr{L}} S \wedge_{\mathscr{L}} M & \xrightarrow{\mathrm{id} \wedge \lambda} & S \wedge_{\mathscr{L}} M \\ \downarrow{\scriptstyle \lambda} & & \downarrow{\scriptstyle \lambda} & \swarrow\!\!\!= & \\ M & \xrightarrow[\xi]{} & S \wedge_{\mathscr{L}} M. & & \end{array}$$

The rectangle commutes by the naturality of λ and the triangle commutes by Proposition 8.3. For (iii), let M be arbitrary and consider the diagram

$$\begin{array}{ccc} \pi_n(S \wedge_{\mathscr{L}} M) \cong h\mathscr{S}[\mathbb{L}](\mathbb{L}S^n, S \wedge_{\mathscr{L}} M) & \xrightarrow{\lambda^*} & h\mathscr{S}[\mathbb{L}](S \wedge_{\mathscr{L}} \mathbb{L}S^n, S \wedge_{\mathscr{L}} M) \\ \downarrow{\scriptstyle \lambda_*} & & \downarrow{\scriptstyle \lambda_*} \\ \pi_n(M) \cong h\mathscr{S}[\mathbb{L}](\mathbb{L}S^n, M) & \xrightarrow[\lambda^*]{} & h\mathscr{S}[\mathbb{L}](S \wedge_{\mathscr{L}} \mathbb{L}S^n, M). \end{array}$$

By (ii), $\lambda : S \wedge_{\mathscr{L}} \mathbb{L}S^n \longrightarrow \mathbb{L}S^n$ is a homotopy equivalence of $\mathbb{L}$-spectra, hence the horizontal arrows are isomorphisms. The right vertical arrow is an isomorphism since, for $\mathbb{L}$-spectra K,

$$\lambda_* : \mathscr{S}[\mathbb{L}](S \wedge_{\mathscr{L}} K, S \wedge_{\mathscr{L}} M) \longrightarrow \mathscr{S}[\mathbb{L}](S \wedge_{\mathscr{L}} K, M)$$

is a natural isomorphism; its inverse sends $f : S \wedge_{\mathscr{L}} K \longrightarrow M$ to the composite

$$S \wedge_{\mathscr{L}} K \xrightarrow{\lambda^{-1} \wedge \mathrm{id}} S \wedge_{\mathscr{L}} S \wedge_{\mathscr{L}} K \xrightarrow{\mathrm{id} \wedge f} S \wedge_{\mathscr{L}} M.$$

(Compare II.1.3 below). Therefore the left vertical arrow is an isomorphism. □

REMARK 8.6. The weak equivalence $\omega : \mathbb{L}S \wedge_{\mathscr{L}} M \longrightarrow M$ of Proposition 6.2 is just the composite

$$\mathbb{L}S \wedge_{\mathscr{L}} M \xrightarrow{\xi \wedge \mathrm{id}} S \wedge_{\mathscr{L}} M \xrightarrow{\lambda} M.$$

Therefore $\xi \wedge \mathrm{id}$ is also a weak equivalence for all $\mathbb{L}$-spectra M.

COROLLARY 8.7. *For any $\mathbb{L}$-spectrum M, $\tilde{\lambda} : M \longrightarrow F_{\mathscr{L}}(S, M)$ is a weak equivalence of $\mathbb{L}$-spectra.*

PROOF. For a spectrum X, $\tilde{\lambda}_* : \mathscr{S}(X, M) \longrightarrow \mathscr{S}(X, F_{\mathscr{L}}(S, M))$ can be identified with $\tilde{\lambda}_* : \mathscr{S}[\mathbb{L}](\mathbb{L}X, M) \longrightarrow \mathscr{S}[\mathbb{L}](\mathbb{L}X, F_{\mathscr{L}}(S, M))$. In turn, by naturality and adjunction, this can be identified with

$$\lambda^* : \mathscr{S}[\mathbb{L}](\mathbb{L}X, M) \longrightarrow \mathscr{S}[\mathbb{L}](S \wedge_{\mathscr{L}} \mathbb{L}X, M) \cong \mathscr{S}[\mathbb{L}](\mathbb{L}X, F_{\mathscr{L}}(S, M)).$$

If X is a CW spectrum, then $\lambda : S \wedge_{\mathscr{L}} \mathbb{L}X \longrightarrow \mathbb{L}X$ is a homotopy equivalence of $\mathbb{L}$-spectra, hence the displayed maps all induce isomorphisms on passage to homotopy classes of maps. The conclusion follows by letting X run through the sphere spectra. □

CHAPTER II

Structured ring and module spectra

We can now define and study our basic algebraic objects. We begin with the S-modules, which we think of as analogs of modules over a fixed commutative ring k. Since the category of S-modules is symmetric monoidal under its smash product, we can define S-algebras and commutative S-algebras exactly as we define (associative and unital) k-algebras and commutative k-algebras. Intuitively, S-algebras are as close as one can get to k-algebras in stable homotopy theory, and commutative S-algebras are as close as one can get to commutative k-algebras. These basic definitions are established in the first three sections, and the material of the rest of the chapter will not be used again until Chapter VII. The reader is invited to skip directly from Section 3 to the applications in Chapters III–VI.

By analyzing free objects, we demonstrate that our new definitions are unital sharpenings of the definitions of A_∞ and E_∞ ring spectra that were first given in [48]. This allows us to use [48, 50] to supply examples and is therefore fundamentally important to the theory. We give a parallel analysis of the definitions of modules over S-algebras and over A_∞ and E_∞ ring spectra. Our new definitions drastically simplify the study of these algebraic structures. For example, in a final categorical section, we prove that the new definitions lead to elementary categorical proofs that the categories of S-algebras and of commutative S-algebras are cocomplete, as was first proven by Hopkins and McClure [32] for the categories of A_∞ and E_∞ ring spectra.

1. The category of S-modules

Here, finally, is the promised definition of S-modules.

DEFINITION 1.1. Define an S-module to be an $\mathbb{L}$-spectrum M which is unital in the sense that $\lambda : S \wedge_{\mathscr{L}} M \longrightarrow M$ is an isomorphism. Let $\mathscr{M}_S$ denote the full subcategory of $\mathscr{S}[\mathbb{L}]$ whose objects are the S-modules. For S-modules M and N, define

$$M \wedge_S N = M \wedge_{\mathscr{L}} N \quad \text{and} \quad F_S(M,N) = S \wedge_{\mathscr{L}} F_{\mathscr{L}}(M,N).$$

The justification for the name "S-module" is given by the commutative diagrams

$$\begin{array}{ccc} S \wedge_S S \wedge_S M & \xrightarrow{\lambda \wedge \mathrm{id}} & S \wedge_S M \\ {\scriptstyle \mathrm{id} \wedge \lambda}\downarrow & & \downarrow{\scriptstyle \lambda} \\ S \wedge_S M & \xrightarrow[\lambda]{} & M \end{array} \qquad \text{and} \qquad \begin{array}{ccc} M & \xrightarrow{\lambda^{-1}} & S \wedge_S M \\ & {\scriptstyle \mathrm{id}}\searrow & \downarrow{\scriptstyle \lambda} \\ & & M. \end{array}$$

For the definition to be useful, we need examples, and I.8.2 and I.8.3 provide many. We consistently retain the notation $M \wedge_{\mathscr{L}} N$ when the given $\mathbb{L}$-spectra M and N are not restricted to be S-modules.

PROPOSITION 1.2. *For any based space X, $\Sigma^\infty X$ is an S-module, and*

$$\Sigma^\infty X \wedge_S \Sigma^\infty Y \cong \Sigma^\infty (X \wedge Y).$$

For any S-module M and any $\mathbb{L}$-spectrum N, $M \wedge_{\mathscr{L}} N$ is an S-module. In particular, $S \wedge_{\mathscr{L}} N$ is an S-module for any $\mathbb{L}$-spectrum N.

PROOF. For the second statement, I.8.3 gives that λ for $M \wedge_{\mathscr{L}} N$ is determined by λ for M and is therefore an isomorphism. □

We have the following categorical relationship between $\mathscr{S}[\mathbb{L}]$ and $\mathscr{M}_S$.

LEMMA 1.3. *The functor $S \wedge_{\mathscr{L}} (-) : \mathscr{S}[\mathbb{L}] \longrightarrow \mathscr{M}_S$ is left adjoint to the functor $F_{\mathscr{L}}(S, -) : \mathscr{M}_S \longrightarrow \mathscr{S}[\mathbb{L}]$ and right adjoint to the inclusion $\ell : \mathscr{M}_S \longrightarrow \mathscr{S}[\mathbb{L}]$.*

PROOF. The first adjunction is immediate from I.7.1. For the second, let M be an S-module and N be an $\mathbb{L}$-spectrum. A map $f : M \longrightarrow S \wedge_{\mathscr{L}} N$ of S-modules determines a map $\lambda \circ f : M \longrightarrow N$ of $\mathbb{L}$-spectra, and a map $g : M \longrightarrow N$ of $\mathbb{L}$-spectra determines a map $(\mathrm{id} \wedge g) \circ \lambda^{-1} : M \longrightarrow S \wedge_{\mathscr{L}} N$ of S-modules. Using I.8.3, we see that these are inverse bijections. □

This implies that to lift right adjoint functors from $\mathscr{S}[\mathbb{L}]$ to $\mathscr{M}_S$, we must first forget down to $\mathscr{S}[\mathbb{L}]$, next apply the given functor, and then apply the functor $S \wedge_{\mathscr{L}} (-)$. For example, limits in $\mathscr{M}_S$ are created in this fashion.

PROPOSITION 1.4. *The category of S-modules is complete and cocomplete. Its colimits are created in $\mathscr{S}[\mathbb{L}]$. Its limits are created by applying the functor $S \wedge_{\mathscr{L}} (-)$ to limits in $\mathscr{S}[\mathbb{L}]$. If X is a based space and M is an S-module, then $M \wedge X$ is an S-module, and the spectrum level cofiber of a map of S-modules is an S-module. For a based space X and S-modules M and N,*

$$\mathscr{M}_S(M \wedge X, N) \cong \mathscr{M}_S(M, S \wedge_{\mathscr{L}} F(X, N)).$$

Moreover,

$$M \wedge X \cong M \wedge_S \Sigma^\infty X \quad \textit{and} \quad S \wedge_{\mathscr{L}} F(X, M) \cong F_S(\Sigma^\infty X, M).$$

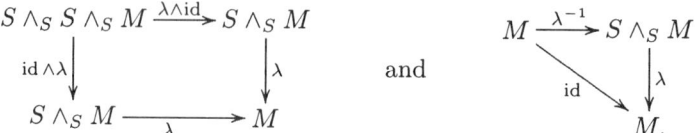

REMARK 1.5. By the path S-module of an S-module N we must understand $S \wedge_{\mathscr{L}} PN$. By the fiber of a map $f : M \longrightarrow N$ of S-modules, we must understand $S \wedge_{\mathscr{L}} Ff$. Lemma 1.3 implies that the following square of S-modules is a pullback and that its vertical arrows satisfy the CHP in the category of S-modules.

$$\begin{array}{ccc} S \wedge_{\mathscr{L}} Ff & \longrightarrow & S \wedge_{\mathscr{L}} PN \\ \downarrow & & \downarrow \\ M & \xrightarrow[f]{} & N. \end{array}$$

The resulting fiber sequences of S-modules behave in exactly the same fashion as fiber sequences of spaces or spectra.

Lemma 1.3 also explains our definition of function S-modules. Its second adjunction and the adjunction of Theorem 7.1 compose to give the adjunction displayed in the following theorem.

THEOREM 1.6. *The category $\mathscr{M}_S$ is symmetric monoidal under $\wedge_S$, and*

$$\mathscr{M}_S(M \wedge_S N, P) \cong \mathscr{M}_S(M, F_S(N, P))$$

for S-modules M, N, and P.

A homotopy in the category of S-modules is a map $M \wedge I_+ \longrightarrow N$. A map of S-modules is a weak equivalence if it is a weak equivalence as a map of spectra. The derived category $\mathscr{D}_S$ of S-modules is constructed from the homotopy category $h\mathscr{M}_S$ by adjoining formal inverses to the weak equivalences; again, the process is made rigorous by CW approximation. The free $\mathbb{L}$-spectra $\mathbb{L}X$ are not S-modules, and we define sphere S-modules by

(1.7) $$S_S^n \equiv S \wedge_{\mathscr{L}} \mathbb{L}S^n$$

and use them as the domains of attaching maps when defining cell and CW S-modules. Observe that, by I.8.7 and Lemma 1.3, we have

(1.8) $$\pi_n(M) \equiv h\mathscr{S}(S^n, M) \cong h\mathscr{S}[\mathbb{L}](\mathbb{L}S^n, F_{\mathscr{L}}(S, M)) \cong h\mathscr{M}_S(S_S^n, M)$$

for S-modules M. From here, the theory of cell and CW S-modules is exactly like the theory of cell and CW spectra and is obtained by specialization of the theory of cell R-modules to be presented in Chapter III. A weak equivalence of cell S-modules is a homotopy equivalence, any S-module is weakly equivalent to a CW S-module, and $\mathscr{D}_S$ is equivalent to the homotopy category of CW S-modules. Again, as we shall explain in Remark 1.10, the S-module S does not have the homotopy type of a CW S-module. When working homotopically, we replace it with $S_S \equiv S_S^0$.

The following comparison between CW S-modules and CW $\mathbb{L}$-spectra establishes an equivalence between $\mathscr{D}_S$ and $\bar{h}\mathscr{S}[\mathbb{L}]$ and thus between $\mathscr{D}_S$ and $\bar{h}\mathscr{S}$.

THEOREM 1.9. *The following conclusions hold.*

(i) *The functor $S \wedge_{\mathscr{L}} (-) : \mathscr{S}[\mathbb{L}] \longrightarrow \mathscr{M}_S$ carries CW $\mathbb{L}$-spectra to CW S-modules.*

(ii) *The forgetful functor $\mathscr{M}_S \longrightarrow \mathscr{S}[\mathbb{L}]$ carries S-modules of the homotopy types of CW S-modules to $\mathbb{L}$-spectra of the homotopy types of CW $\mathbb{L}$-spectra.*

(iii) *Every CW S-module M is homotopy equivalent as an S-module to $S \wedge_{\mathscr{L}} N$ for some CW $\mathbb{L}$-spectrum N.*

(iv) *The unit $\lambda : S \wedge_{\mathscr{L}} M \longrightarrow M$ is a weak equivalence for all $\mathbb{L}$-spectra M and is a homotopy equivalence of $\mathbb{L}$-spectra if M has the homotopy type of a CW $\mathbb{L}$-spectrum.*

The functors $S \wedge_{\mathscr{L}} (-)$ and the forgetful functor establish an adjoint equivalence between the stable homotopy category $\bar{h}\mathscr{S}[\mathbb{L}]$ and the derived category $\mathscr{D}_S$. This equivalence of categories preserves smash products and function spectra.

PROOF. Part (i) is immediate by induction up the sequential filtration since the functor $S \wedge_{\mathscr{L}} (-)$ preserves spheres, cones, and colimits. Part (iv) is a recapitulation of I.8.5 and, applied to sphere S-modules, it implies part (ii). Part (iii) follows from (i) and (iv) since there is a CW $\mathbb{L}$-spectrum M' and a homotopy equivalence of $\mathbb{L}$-spectra $M' \longrightarrow M$. The claimed adjoint equivalence of categories is immediate from part (iv). For smash products, the last statement is clear from (ii) and the fact that the smash product $M \wedge_S N$ of S-modules is their smash product as $\mathbb{L}$-spectra. The statement for function spectra follows formally. □

When doing classical homotopy theory, we can work interchangeably in $\bar{h}\mathscr{S}$, $\bar{h}\mathscr{S}[\mathbb{L}]$, or $\mathscr{D}_S$. These three categories are equivalent, and the equivalences preserve all structure in sight. When working on the point set level, we have reached a nearly ideal situation with our construction of $\mathscr{M}_S$. We pause to comment on Lewis' observation [37] that there is no fully ideal situation.

REMARK 1.10. Suppose given a symmetric monoidal category of spectra with a suspension spectrum functor Σ^∞ such that $S = \Sigma^\infty S^0$ is the unit for the smash product, denoted $\wedge_S$, and there is a natural isomorphism

$$\Sigma^\infty X \wedge_S \Sigma^\infty Y \cong \Sigma^\infty (X \wedge Y)$$

that is suitably compatible with the coherence isomorphisms for the unity, associativity, and commutativity of the respective smash products. Our category of S-modules satisfies all of these properties, and many other desiderata not included among Lewis's axioms. Suppose further that Σ^∞ has a right adjoint "Ω^∞" and let $QX = \operatorname{colim} \Omega^n \Sigma^n X$. Then Lewis observes that there cannot be a natural weak equivalence

$$\theta : \text{“}\Omega^\infty\text{”}\Sigma^\infty X \longrightarrow QX$$

such that $\theta \circ \eta : X \longrightarrow QX$ is the natural inclusion, where η is the unit of the adjunction. In our context, we have the two adjunction homeomorphisms

$$\mathscr{M}_S(S \wedge_{\mathscr{L}} \mathbb{L}\Sigma^\infty X, M) \cong \mathscr{T}(X, \Omega^\infty F_{\mathscr{L}}(S, M))$$

and

$$\mathscr{M}_S(\Sigma^\infty X, M) \cong \mathscr{T}(X, \mathscr{M}_S(S, M)),$$

where $\mathscr{T}$ is the category of based spaces; see VII§2 for discussion of these topologized Hom sets and of the second of these adjunctions. It is a standard property of any symmetric monoidal category that the self-maps of the unit object form a commutative monoid under composition. In our situation $\mathscr{M}_S(S,S)$ is therefore a commutative topological monoid. It cannot be weakly equivalent to QS^0, but $\Omega^\infty F_{\mathscr{L}}(S,S)$ is weakly equivalent to QS^0 Therefore the weak equivalence $S \wedge_{\mathscr{L}} \mathbb{L}S \longrightarrow S$ cannot be a homotopy equivalence of S-modules and S cannot be of the homotopy type of a CW S-module.

2. The mirror image to the category of S-modules

The categorical picture becomes clearer when we realize that the category of S-modules has a "mirror image" category to which it is naturally equivalent. We find this material quite illuminating, but it will not be used until our discussion of Quillen model categories. The reader may prefer to skip it on a first reading.

DEFINITION 2.1. Define $\mathscr{M}^S$ to be the full subcategory of $\mathscr{S}[\mathbb{L}]$ whose objects are those $\mathbb{L}$-spectra N that are counital, in the sense that $\tilde{\lambda} : N \longrightarrow F_{\mathscr{L}}(S,N)$ is an isomorphism.

Looking through the mirror at Lemma 1.3 and noting that mirrors interchange left and right, we see the following reflection.

LEMMA 2.2. *The functor* $F_{\mathscr{L}}(S,-) : \mathscr{S}[\mathbb{L}] \longrightarrow \mathscr{M}^S$ *is right adjoint to the functor* $S \wedge_{\mathscr{L}} (-) : \mathscr{M}^S \longrightarrow \mathscr{S}[\mathbb{L}]$ *and left adjoint to the inclusion* $r : \mathscr{M}^S \longrightarrow \mathscr{S}[\mathbb{L}]$.

We agree to write

(2.3) $f = F_{\mathscr{L}}(S,-) : \mathscr{S}[\mathbb{L}] \longrightarrow \mathscr{M}^S$ and $s = S \wedge_{\mathscr{L}} (-) : \mathscr{S}[\mathbb{L}] \longrightarrow \mathscr{M}_S$

in the rest of this section. With this notation, Lemmas 1.3 and 2.2 give the following mirrored pairs of adjunctions, the upper arrow being left adjoint to the lower arrow in each case.

(2.4) $\mathscr{S}[\mathbb{L}] \underset{rf\ell}{\overset{s}{\rightleftarrows}} \mathscr{M}_S \underset{s}{\overset{\ell}{\rightleftarrows}} \mathscr{S}[\mathbb{L}]$ and $\mathscr{S}[\mathbb{L}] \underset{r}{\overset{f}{\rightleftarrows}} \mathscr{M}^S \underset{f}{\overset{\ell sr}{\rightleftarrows}} \mathscr{S}[\mathbb{L}].$

The display makes new information visible. The composite of the first two left adjoints is just the functor $S \wedge_{\mathscr{L}} (-)$ and the composite of the second two right adjoints is just the functor $F_{\mathscr{L}}(S,-)$. Since these two endo-functors of $\mathscr{S}[\mathbb{L}]$ are left and right adjoint, they must be equivalent to their displayed composite adjoints.

LEMMA 2.5. *For* $\mathbb{L}$*-spectra* M*, the maps*

$$\mathrm{id} \wedge_{\mathscr{L}} \tilde{\lambda} : S \wedge_{\mathscr{L}} M \longrightarrow S \wedge_{\mathscr{L}} F_{\mathscr{L}}(S,M)$$

and

$$F_{\mathscr{L}}(\mathrm{id},\lambda) : F_{\mathscr{L}}(S, S \wedge_{\mathscr{L}} M) \longrightarrow F_{\mathscr{L}}(S,M)$$

are natural isomorphisms.

We now see that the reflection of a reflection is equivalent to the original.

PROPOSITION 2.6. *The functors*

$$f\ell : \mathscr{M}_S \longrightarrow \mathscr{M}^S \quad \text{and} \quad sr : \mathscr{M}^S \longrightarrow \mathscr{M}_S$$

are inverse equivalences of categories. More precisely,

$$\varepsilon : srf\ell M = S \wedge_{\mathscr{L}} F_{\mathscr{L}}(S, M) \longrightarrow M$$

is an isomorphism for $M \in \mathscr{M}_S$, *and*

$$\eta : N \longrightarrow F_{\mathscr{L}}(S, S \wedge_{\mathscr{L}} N) = f\ell srN$$

is an isomorphism for $N \in \mathscr{M}^S$, *where* ε *and* η *are the unit and counit of the* $(S \wedge_{\mathscr{L}} (-), F_{\mathscr{L}}(S, -))$ *adjunction.*

PROOF. The functor $s\ell : \mathscr{M}_S \longrightarrow \mathscr{M}_S$ is an equivalence, and it is left adjoint to the composite $srf\ell : \mathscr{M}_S \longrightarrow \mathscr{M}_S$. The functor $fr : \mathscr{M}^S \longrightarrow \mathscr{M}^S$ is an equivalence, and it is right adjoint to the composite $f\ell sr$. Therefore these two composites are natural equivalences. A little diagram chase from the previous lemma gives the more precise statement. □

PROPOSITION 2.7. *The category* $\mathscr{M}^S$, *hence also the category* $\mathscr{M}_S$, *is equivalent to the category of algebras over the monad* rf *in* $\mathscr{S}[\mathbb{L}]$ *determined by the adjunction* (f, r). *The category* $\mathscr{M}_S$, *hence also the category* $\mathscr{M}^S$, *is equivalent to the category of coalgebras over the comonad* ℓs *in* $\mathscr{S}[\mathbb{L}]$ *determined by the adjunction* (ℓ, s).

PROOF. The unit of the monad rf is $\tilde{\lambda} : M \longrightarrow F_{\mathscr{L}}(S, M) = rfM$ and its product is the natural isomorphism

$$\mu : rfrfM = F_{\mathscr{L}}(S, F_{\mathscr{L}}(S, M)) \cong F_{\mathscr{L}}(S, M) = rfM$$

implied by the isomorphism $S \wedge_{\mathscr{L}} S \cong S$. Clearly, if $\tilde{\lambda}$ is an isomorphism, then M is an rf-algebra with action $\tilde{\lambda}^{-1}$. Conversely if $\xi : rfM \longrightarrow M$ is an action, then $\xi \circ \tilde{\lambda} = \mathrm{id}$ and the following is a split coequalizer diagram in $\mathscr{S}[\mathbb{L}]$.

$$rfrfM \underset{\mu}{\overset{rf\xi}{\rightrightarrows}} rfM \xrightarrow{\ \xi\ } M.$$

Applying f, we obtain a split coequalizer diagram in $\mathscr{M}^S$. Since the counit $fr \longrightarrow \mathrm{id}$ of the adjunction is an isomorphism, it induces an isomorphism of diagrams

$$(frfrfM \rightrightarrows frfM) \longrightarrow (frfM \rightrightarrows fM).$$

Applying r, rfM is the (split) coequalizer of the first and M is the (split) coequalizer of the second. The resulting isomorphism $rfM \longrightarrow M$ is just the map ξ, hence is an isomorphism of rf-algebras. □

3. S-algebras and their modules

Let $\mathscr{C}$ be any symmetric monoidal category, with product $\Box$ and unit object I. Then a monoid in $\mathscr{C}$ is an object R together with maps $\eta : I \longrightarrow R$ and $\phi : R \Box R \longrightarrow R$ such that the evident associativity and unity diagrams commute; R is a commutative monoid if the evident commutativity diagram also commutes. A left R-object over a monoid R is an object M of $\mathscr{C}$ with a map $\mu : R \Box M \longrightarrow M$ such that the evident unity and associativity diagrams commute, and right R-objects are defined by symmetry. These definitions apply to our symmetric monoidal category $\mathscr{M}_S$.

DEFINITION 3.1. An S-algebra is a monoid in $\mathscr{M}_S$. A commutative S-algebra is a commutative monoid in $\mathscr{M}_S$. For an S-algebra or commutative S-algebra R, a left or right R-module is a left or right R-object in $\mathscr{M}_S$. Modules will mean left modules unless otherwise specified, and we let $\mathscr{M}_R$ denote the category of left R-modules.

Observe that if R is a commutative S-algebra, then an R-module is just a module over R regarded as an S-algebra, as in module theory in algebra. For this reason, even though our main interest is in the much richer commutative context, we work with general S-algebras wherever possible.

We insert the following lemma for later reference. It records specializations of observations that apply to monoids in any symmetric monoidal category.

LEMMA 3.2. *Let R be an S-algebra and M be an R-module. Then the following diagrams of S-modules are split coequalizers:*

$$R \wedge_S R \wedge_S R \underset{\mathrm{id} \wedge \phi}{\overset{\phi \wedge \mathrm{id}}{\rightrightarrows}} R \wedge_S R \xrightarrow{\ \phi\ } R.$$

and

$$R \wedge_S R \wedge_S M \underset{\phi \wedge \mathrm{id}}{\overset{\mathrm{id} \wedge \mu}{\rightrightarrows}} R \wedge_S M \xrightarrow{\ \mu\ } M.$$

While we have given the most conceptual form of the definitions, it is worthwhile to write out the relevant diagrams explicitly. We find that they make perfect sense for $\mathbb{L}$-spectra that might not be S-modules, and this leads us back to the earlier notions of A_∞ and E_∞ ring spectra and their modules.

DEFINITION 3.3. An A_∞ ring spectrum is an $\mathbb{L}$-spectrum R with a unit map $\eta : S \longrightarrow R$ and a product $\phi : R \wedge_{\mathscr{L}} R \longrightarrow R$ such that the following diagrams commute:

$$\begin{array}{ccccc} S \wedge_{\mathscr{L}} R & \xrightarrow{\eta \wedge \mathrm{id}} & R \wedge_{\mathscr{L}} R & \xleftarrow{\mathrm{id} \wedge \eta} & R \wedge_{\mathscr{L}} S \\ & \searrow^{\lambda} & \downarrow{\scriptstyle \phi} & \swarrow_{\lambda\tau} & \\ & & R & & \end{array}$$

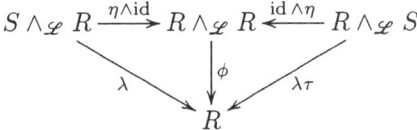

and

$$\begin{array}{ccc} R \wedge_{\mathscr{L}} R \wedge_{\mathscr{L}} R & \xrightarrow{\mathrm{id} \wedge \phi} & R \wedge_{\mathscr{L}} R \\ {\scriptstyle \phi \wedge \mathrm{id}} \downarrow & & \downarrow {\scriptstyle \phi} \\ R \wedge_{\mathscr{L}} R & \xrightarrow[\phi]{} & R; \end{array}$$

R is an E_∞ ring spectrum if the following diagram also commutes:

$$\begin{array}{ccc} R \wedge_{\mathscr{L}} R & \xrightarrow{\tau} & R \wedge_{\mathscr{L}} R \\ & {\scriptstyle \phi} \searrow \quad \swarrow {\scriptstyle \phi} & \\ & R. & \end{array}$$

A module over an A_∞ or E_∞ ring spectrum R is an $\mathbb{L}$-spectrum M with a map $\mu : R \wedge_{\mathscr{L}} M \longrightarrow M$ such that the following diagrams commute:

$$\begin{array}{ccc} S \wedge_{\mathscr{L}} M & \xrightarrow{\eta \wedge \mathrm{id}} & R \wedge_{\mathscr{L}} M \\ & {\scriptstyle \lambda} \searrow & \downarrow {\scriptstyle \mu} \\ & & M \end{array} \qquad \text{and} \qquad \begin{array}{ccc} R \wedge_{\mathscr{L}} R \wedge_{\mathscr{L}} M & \xrightarrow{\mathrm{id} \wedge \mu} & R \wedge_{\mathscr{L}} M \\ {\scriptstyle \phi \wedge \mathrm{id}} \downarrow & & \downarrow {\scriptstyle \mu} \\ R \wedge_{\mathscr{L}} M & \xrightarrow{\mu} & M. \end{array}$$

LEMMA 3.4. *An S-algebra or commutative S-algebra is an A_∞ or E_∞ ring spectrum which is also an S-module. A module over an S-algebra or commutative S-algebra R is a module over R, regarded as an A_∞ or E_∞ ring spectrum, which is also an S-module.*

In view of Proposition 1.2, this leads to the following observations.

PROPOSITION 3.5. *The following statements hold.*

(i) *S is a commutative S-algebra with unit* id *and product λ.*
(ii) *If R and R' are A_∞ or E_∞ ring spectra, then so is $R \wedge_{\mathscr{L}} R'$; if either R or R' is an S-algebra, then so is $R \wedge_{\mathscr{L}} R'$.*
(iv) *If R and R' are A_∞ ring spectra, M is an R-module and M' is an R'-module, then $M \wedge_{\mathscr{L}} M'$ is an $R \wedge_{\mathscr{L}} R'$-module.*

In particular, we have a functorial way to replace A_∞ and E_∞ ring spectra and their modules by S-algebras and commutative S-algebras and their modules.

COROLLARY 3.6. *For an A_∞ ring spectrum R, $S \wedge_{\mathscr{L}} R$ is an S-algebra and $\lambda : S \wedge_{\mathscr{L}} R \longrightarrow R$ is a weak equivalence of A_∞ ring spectra, and similarly in the E_∞ case. If M is an R-module, then $S \wedge_{\mathscr{L}} M$ is an $S \wedge_{\mathscr{L}} R$-module and $\lambda : S \wedge_{\mathscr{L}} M \longrightarrow M$ is a weak equivalence of R-modules and of modules over $S \wedge_{\mathscr{L}} R$ regarded as an A_∞ ring spectrum.*

Recall that the tensor product of commutative rings is their coproduct in the category of commutative rings. The proof consists of categorical diagram chases that apply to commutative monoids in any symmetric monoidal category.

PROPOSITION 3.7. *If R and R' are commutative S-algebras, then $R \wedge_S R'$ is the coproduct of R and R' in the category of commutative S-algebras.*

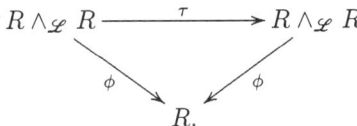

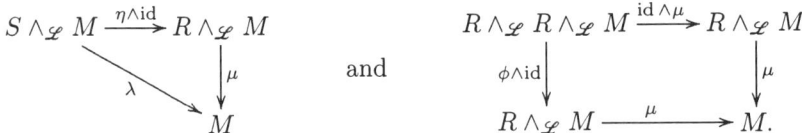

We shall construct coproducts in the category of S-algebras in Section 7, where we show more generally that the categories of S-algebras and of commutative S-algebras are cocomplete.

There is a version of the proposition that is true for E_∞ ring spectra, but this is not obvious. We shall return to this point in Chapter XIII, where we show that the category of $\mathbb{L}$-spectra under S is symmetric monoidal under a modified smash product $\star_S$ and that A_∞ and E_∞ ring spectra are exactly the monoids and commutative monoids in that symmetric monoidal category. This was the starting point for the earlier version of the present theory announced in [23].

4. Free A_∞ and E_∞ ring spectra; comparisons of definitions

We focus on A_∞ and E_∞ ring spectra here. It was proven in [48, 50] that various Thom spectra, Eilenberg-Mac Lane spectra, and connective algebraic and topological K-theory spectra are E_∞ ring spectra. Using the results stated in the previous section, we can convert these E_∞ ring spectra to weakly equivalent commutative S-algebras. However, on the face of it, the original definitions of A_∞ and E_∞ ring spectra appear to be different from those that we have given here. As in algebra, it is important to understand free A_∞ and E_∞ ring spectra, and we shall use this understanding to verify that our present definitions agree with the original ones.

There is no difficulty in constructing the relevant monads. In fact, we shall construct two pairs of monads and then relate them. The first is defined on the ground category of spectra and is transparently related to the earlier definitions. The second is defined on the ground category of S-modules and is transparently related to the present definitions. The connection between them will establish the required equivalence of definitions. In effect, our new definition of E_∞ ring spectra is obtained from the old one simply by factoring the original defining monad $\mathbb{C}$ in $\mathscr{S}$ through a new defining monad $\mathbb{P}$ in the more highly structured category $\mathscr{S}[\mathbb{L}]$.

CONSTRUCTION 4.1. Construct monads $\mathbb{B}$ and $\mathbb{C}$ in the category of spectra as follows. Let X be a spectrum and let X^j be its j-fold external smash power, with $X^0 = S^0$. Define

$$\mathbb{B}X \cong \bigvee_{j\geq 0} \mathscr{L}(j) \ltimes X^j$$

and

$$\mathbb{C}X \cong \bigvee_{j\geq 0} \mathscr{L}(j) \ltimes_{\Sigma_j} X^j,$$

where $\mathscr{L}(j) \ltimes_{\Sigma_j} X^j$ is the orbit spectrum $(\mathscr{L}(j) \ltimes X^j)/\Sigma_j$. The units of these monads are induced by the unit maps $X \cong \{1\} \ltimes X \longrightarrow \mathscr{L}(1) \ltimes X$. Their products are induced by wedge sums of maps induced by the structure maps γ of the linear isometries operad $\mathscr{L}$.

The notion of an $\mathscr{L}$-spectrum was defined in [38, VII.2.1]. The definition used permutations, and there is a corresponding notion of a non-Σ $\mathscr{L}$-spectrum. An immediate comparison of definitions gives the following result.

PROPOSITION 4.2. *The category of $\mathbb{B}$-algebras is isomorphic to the category of non-Σ $\mathscr{L}$-spectra. The category of $\mathbb{C}$-algebras is isomorphic to the category of $\mathscr{L}$-spectra.*

Actually, $\mathscr{O}$-spectra were defined in [38, VII.2.1] for any operad $\mathscr{O}$ that is augmented over $\mathscr{L}$. An E_∞ operad is one such that each $\mathscr{O}(j)$ is Σ_j-free and contractible. In earlier work, E_∞ ring spectra were understood to mean $\mathscr{O}$-spectra for any E_∞ operad $\mathscr{O}$ augmented over $\mathscr{L}$. The present theory is based on properties that are special to $\mathscr{L}$. The following result, which will be proven in XII§1, shows that restriction to $\mathscr{L}$ results in no loss of generality. There is an analogue for A_∞ ring spectra that is obtained by forgetting about permutations.

PROPOSITION 4.3. *Let $\mathscr{O}$ be an E_∞ operad over $\mathscr{L}$. There is a functor V that assigns a weakly equivalent $\mathscr{L}$-spectrum VR to an $\mathscr{O}$-spectrum R.*

CONSTRUCTION 4.4. Construct monads $\mathbb{T}$ and $\mathbb{P}$ in the category of $\mathbb{L}$-spectra as follows. Let M be an $\mathbb{L}$-spectrum and let M^j be its j-fold power with respect to $\wedge_{\mathscr{L}}$, with $M^0 = S$. Define

$$\mathbb{T}M \cong \bigvee_{j \geq 0} M^j$$

and

$$\mathbb{P}M \cong \bigvee_{j \geq 0} M^j/\Sigma_j.$$

Here passage to orbits preserves $\mathbb{L}$-spectra since it is a finite colimit. The unit is the inclusion of $M = M^1$. The product is induced by the maps

$$M^{j_1} \wedge_{\mathscr{L}} \cdots \wedge_{\mathscr{L}} M^{j_k} \longrightarrow M^{j_1 + \cdots + j_k}$$

that are given by the evident identifications if each $j_r \geq 1$ and by use of the unit map λ if any $j_r = 0$. Observe that $\mathbb{T}$ and $\mathbb{P}$ restrict to monads in the category of S-modules.

The letters $\mathbb{T}$ and $\mathbb{P}$ are mnemonic for "tensor algebra" and "polynomial" (or symmetric) algebra. As is clear for S-modules and will be made explicit in Definition 7.1, the definitions fit into a general categorical framework that includes those constructions. The following result is an easy direct consequence of our definitions.

PROPOSITION 4.5. *The categories of A_∞ ring spectra and of S-algebras are isomorphic to the categories of $\mathbb{T}$-algebras in $\mathscr{S}[\mathbb{L}]$ and of $\mathbb{T}$-algebras in $\mathscr{M}_S$. The categories of E_∞ ring spectra and of commutative S-algebras are isomorphic to the categories of $\mathbb{P}$-algebras in $\mathscr{S}[\mathbb{L}]$ and of $\mathbb{P}$-algebras in $\mathscr{M}_S$.*

To relate the monads $\mathbb{B}$ and $\mathbb{C}$ to the monads $\mathbb{T}$ and $\mathbb{P}$, recall from I.4.2 that the category of $\mathbb{L}$-spectra is the category of $\mathbb{L}$-algebras in $\mathscr{S}$. Together with Propositions 4.2 and 4.5, the following result gives the promised comparison between the old and new definitions of A_∞ and E_∞ ring spectra.

PROPOSITION 4.6. *The monads $\mathbb{B}$ and $\mathbb{T}\mathbb{L}$ are isomorphic, hence the categories of non-Σ $\mathscr{L}$-spectra and of A_∞ ring spectra are isomorphic. The monads $\mathbb{C}$ and $\mathbb{P}\mathbb{L}$ are isomorphic, hence the categories of $\mathscr{L}$-spectra and of E_∞ ring spectra are isomorphic.*

PROOF. The isomorphisms on objects are immediate from I.5.6 applied to $\mathbb{L}$-spectra $M_i = \mathbb{L}X_i$. Since these isomorphisms are induced from the structure maps γ of $\mathscr{L}$, the comparison of monad structures is immediate. In both statements, the second clause is a categorical consequence of the first, as we shall show in Lemma 6.1 below. □

REMARK 4.7. Observe that we have quotient maps of monads $\mathbb{B} \longrightarrow \mathbb{C}$ and $\mathbb{T} \longrightarrow \mathbb{P}$. In Section 6, we shall give categorical definitions that show how to exploit these maps to construct an E_∞ ring spectrum $\mathbb{C} \otimes_{\mathbb{B}} R$ (or $\mathbb{P} \otimes_{\mathbb{T}} R$) from an A_∞ ring spectrum R by "passage to quotients", just as we construct commutative algebras as quotients of associative algebras; see Lemma 6.7 and Corollary 7.3. Formally, $\mathbb{C} \otimes_{\mathbb{B}} R$ is a coequalizer of a right action of $\mathbb{B}$ on $\mathbb{C}$ and the given action of $\mathbb{B}$ on R.

REMARK 4.8. Passage to orbits and passage to coequalizers are often hard to analyze homotopically. We show how to deal with the first difficulty in III§5, where we show that symmetric powers and extended powers of S-modules (and, more generally, R-modules) are essentially equivalent. One often circumvents the second difficulty by replacing a construction like $\mathbb{C} \otimes_{\mathbb{B}} R$ with its associated bar construction $B(\mathbb{C}, \mathbb{B}, R)$, which we shall introduce in XII§1.

REMARK 4.9. There are reduced monads $\tilde{\mathbb{B}}$ and $\tilde{\mathbb{C}}$ in the category $\mathscr{S}\backslash S$ of spectra under S and $\tilde{\mathbb{T}}$ and $\tilde{\mathbb{P}}$ in the category $\mathscr{S}[\mathbb{L}]\backslash S$ of $\mathbb{L}$-spectra under S. They are constructed from the unreduced monads by unit map identifications similar to the basepoint identifications in the James construction or the infinite symmetric product. Observe that $\mathscr{S}\backslash S$ is the category of algebras over the monad $\mathbb{U}$ that is specified by $\mathbb{U}X = X \vee S$, with product given by the folding map $S \vee S \longrightarrow S$, and similarly for $\mathscr{S}[\mathbb{L}]\backslash S$. In all four cases, the unreduced monad is the composite of the reduced monad with $\mathbb{U}$, hence, by Lemma 6.1 below, the reduced and unreduced monads have the same algebras. The difference is that, when considering the reduced monad, one is considering the unit map $S \longrightarrow R$ as preassigned and then ensuring that the unit map created by the monad action coincides with it. It follows that the monad $\mathbb{U}$ acts from the right on the unreduced monads, and it is easy to write down this action directly. The reduced monad $\tilde{\mathbb{C}}$ can then be constructed from $\mathbb{C}$ by setting $\tilde{\mathbb{C}}X = \mathbb{C} \otimes_{\mathbb{U}} X$ for a spectrum X under S, with structure maps induced by passage to coequalizers, and similarly for our other monads. A more explicit description is given in [38, VII§3], where $\tilde{\mathbb{C}}$ is denoted by $\mathbb{C}$. While the monad $\mathbb{C}$ is more convenient for formal work, the monad $\tilde{\mathbb{C}}$ is of far greater homotopical interest.

5. Free modules over A_∞ and E_∞ ring spectra

There is an analogue for modules of the original explicit definition of A_∞ and E_∞ ring spectra in terms of twisted half-smash products, and there is an analogous comparison of definitions.

PROPOSITION 5.1. *The category of modules over an $\mathscr{L}$-spectrum R is isomorphic to the category of spectra M together with associative, unital, and, in the E_∞ context, equivariant systems of action maps*

$$\mathscr{L}(j) \ltimes (R^{j-1} \wedge M) \longrightarrow M.$$

Since we shall not need the details, we shall not write out the relevant diagrams. They make sense for any operad $\mathscr{O}$ augmented over $\mathscr{L}$, and they are exact analogs of diagrams that are written out in the context of algebraic operads in [35, I.4.1]. Remarkably, with this alternative form of the definition, it is far from obvious that a module over an E_∞ ring spectrum R is the same thing as a module over R regarded as an A_∞ ring spectrum. In fact, this appears to be false in the context of modules over an $\mathscr{O}$-spectrum R for a general E_∞ operad $\mathscr{O}$ augmented over $\mathscr{L}$. However, we have the following analogue of Proposition 4.3, which will be proven in XII§1. Again, there is an analogue for A_∞ ring spectra and modules that is obtained by forgetting about permutations.

PROPOSITION 5.2. *Let $\mathscr{O}$ be an E_∞ operad over $\mathscr{L}$ and R be an $\mathscr{O}$-spectrum. There is a functor V that assigns a weakly equivalent VR-module to an R-module M, where VR is the $\mathscr{L}$-spectrum of Proposition 4.3.*

There is a conceptual monadic proof of Proposition 5.1 that is based on analogs of Propositions 4.2, 4.5, and 4.6. To carry out this argument, we need to know that there is a free R-module functor. This is obvious enough when we are considering S-modules: $R \wedge_S M$ is then the free R-module generated by an S-module M. For a general A_∞ ring spectrum R and an $\mathbb{L}$-spectrum M, $R \wedge_{\mathscr{L}} M$ is an R-module but, since M need not be isomorphic to $S \wedge_{\mathscr{L}} M$, it is not the free R-module generated by M.

DEFINITION 5.3. For an A_∞ ring spectrum R and an $\mathbb{L}$-spectrum M, define an $\mathbb{L}$-spectrum $\mathbb{R}M$ and maps of $\mathbb{L}$-spectra $\pi : R \wedge_{\mathscr{L}} M \longrightarrow \mathbb{R}M$ and $\eta : M \longrightarrow \mathbb{R}M$ by the pushout diagram

$$\begin{array}{ccc} S \wedge_{\mathscr{L}} M & \xrightarrow{\eta \wedge \mathrm{id}} & R \wedge_{\mathscr{L}} M \\ {\scriptstyle \lambda}\downarrow & & \downarrow{\scriptstyle \pi} \\ M & \xrightarrow[\eta]{} & \mathbb{R}M. \end{array}$$

Dually, define an $\mathbb{L}$-spectrum $\mathbb{R}^{\#}M$ by the pullback diagram

$$\begin{array}{ccc} \mathbb{R}^{\#}M & \longrightarrow & M \\ \downarrow & & \downarrow{\scriptstyle \tilde{\lambda}} \\ F_{\mathscr{L}}(R, M) & \xrightarrow[F(\eta,\mathrm{id})]{} & F_{\mathscr{L}}(S, M). \end{array}$$

These are special cases of general constructions to be studied in Chapter XIII. Such constructions permeated earlier versions of the present theory. Of course, π is an isomorphism if M is an S-module. As will be generalized in XIII.1.4, we deduce the following homotopical property by applying the functor $S \wedge_{\mathscr{L}} (-)$ to the defining pushout diagram.

PROPOSITION 5.4. *The map* $\pi : R \wedge_{\mathscr{L}} M \longrightarrow \mathbb{R}M$ *is a weak equivalence for any* $\mathbb{L}$*-spectrum* M.

The unit diagram of an R-module M ensures that its product factors through a map $\mathbb{R}M \longrightarrow M$. More formally, elementary inspections of definitions give the following result.

PROPOSITION 5.5. *Let* R *be an* A_∞ *ring spectrum. Then* $\mathbb{R}$ *is a monad in* $\mathscr{S}[\mathbb{L}]$ *with unit* $\eta : M \longrightarrow \mathbb{R}M$ *and with product* $\mu : \mathbb{R}\mathbb{R} \longrightarrow \mathbb{R}$ *induced from the product* $\phi : R \wedge_{\mathscr{L}} R \longrightarrow R$. *A left* R*-module is an algebra over the monad* $\mathbb{R}$ *and, for an* $\mathbb{L}$*-spectrum* M, $\mathbb{R}M$ *is the free* R*-module generated by* M. *The functor* $\mathbb{R}^{\#}$ *is right adjoint to* $\mathbb{R}$ *and is therefore a comonad in* $\mathscr{S}[\mathbb{L}]$ *such that an* R*-module is a coalgebra over* $\mathbb{R}^{\#}$.

It is logical to denote the category of R-modules by $\mathscr{S}[\mathbb{L}][\mathbb{R}]$, reserving the notation $\mathscr{M}_R$ for the case when R is an S-algebra and R-modules are required to be S-modules. We have freeness and cofreeness adjunctions

$$\mathscr{S}[\mathbb{L}][\mathbb{R}](\mathbb{R}M, N) \cong \mathscr{S}[\mathbb{L}](M, N)$$

and

$$\mathscr{S}[\mathbb{L}][\mathbb{R}](N, \mathbb{R}^{\#}M) \cong \mathscr{S}[\mathbb{L}](N, M)$$

for $\mathbb{L}$-spectra M and R-modules N.

Clearly there results a composite adjunction that starts with spectra.

PROPOSITION 5.6. *For a spectrum* X, *define* $\mathbb{F}X = \mathbb{R}\mathbb{L}X$. *Then* $\mathbb{F}X$ *is the free* R*-module generated by* X. *Thus*

$$\mathscr{S}[\mathbb{L}][\mathbb{R}](\mathbb{F}X, N) \cong \mathscr{S}(X, N)$$

for an R*-module* N. *Dually, define* $\mathbb{F}^{\#}X = \mathbb{R}^{\#}\mathbb{L}^{\#}X$. *Then* $\mathbb{F}^{\#}X$ *is the cofree* R*-module generated by* X, *so that*

$$\mathscr{S}[\mathbb{L}][\mathbb{R}](N, \mathbb{F}^{\#}X) \cong \mathscr{S}(N, X).$$

In Construction 6.2, we shall show how to combine the monads of the previous section with these free module constructions to obtain monads $\mathbb{B}[1]$ and $\mathbb{C}[1]$ in the category of pairs of spectra such that a $\mathbb{B}[1]$-algebra or $\mathbb{C}[1]$-algebra (R, M) is an A_∞ or E_∞ ring spectrum R together with an R-module M in the alternative operad action sense described in Proposition 5.1. The construction will also give monads $\mathbb{T}[1]$ and $\mathbb{P}[1]$ in the category of pairs of $\mathbb{L}$-spectra such that a $\mathbb{T}[1]$-algebra or $\mathbb{P}[1]$-algebra $(R; M)$ is an A_∞ or E_∞ ring spectrum R together with an R-module M in the sense of Definition 3.3. The monad $\mathbb{B}[1]$ has the general form

$$\mathbb{B}[1](X; Y) = (\mathbb{B}X; \mathbb{B}(X; Y)),$$

and similarly in the other three cases. Propositions 4.6 and 5.5, together with inspection of the cited construction, directly imply the following analogue of Proposition 4.6. By Lemma 6.1, this in turn implies Proposition 5.1.

PROPOSITION 5.7. *The monads $\mathbb{B}[1]$ and $\mathbb{T}[1] \circ (\mathbb{L}, \mathbb{L})$ are isomorphic. The monads $\mathbb{C}[1]$ and $\mathbb{P}[1] \circ (\mathbb{L}, \mathbb{L})$ are isomorphic. The second coordinates of the four monads are given explicitly as follows. Applied to a pair of spectra $(X; Y)$,*

$$\mathbb{B}(X;Y) = \bigvee_{j \geq 1} \mathscr{L}(j) \ltimes (X^{j-1} \wedge Y)$$

and

$$\mathbb{C}(X;Y) = \bigvee_{j \geq 1} \mathscr{L}(j) \ltimes_{\Sigma_{j-1}} (X^{j-1} \wedge Y).$$

Applied to a pair of $\mathbb{L}$-spectra $(M; N)$,

$$\mathbb{T}(M;N) = \bigvee_{j \geq 1} M^{j-1} \wedge_S N$$

and

$$\mathbb{P}(M;N) = \bigvee_{j \geq 1} (M^{j-1}/\Sigma_{j-1}) \wedge_S N.$$

If N is an S-module, then so are $\mathbb{T}(M;N)$ and $\mathbb{P}(M;N)$.

REMARK 5.8. Construction 6.2 applies equally well to give reduced versions of our four monads, giving monads in the category of pairs (of spectra or $\mathbb{L}$-spectra), the first coordinate of which lies under S. The monad $\tilde{\mathbb{B}}[1]$ has the form

$$\tilde{\mathbb{B}}[1](X;Y) = (\tilde{\mathbb{B}}X; \tilde{\mathbb{B}}(X;Y))$$

and similarly in the other three cases. Inspection of definitions shows that

$$\tilde{\mathbb{B}}S = \tilde{\mathbb{C}}S = S \quad \text{and} \quad \tilde{\mathbb{B}}(S;Y) = \tilde{\mathbb{C}}(S;Y) = \mathscr{L}(1) \ltimes Y.$$

This fact dictates our original definition of $\mathbb{L}$-spectra and is thus the conceptual starting point of our entire theory.

6. Composites of monads and monadic tensor products

In this section and the next, we collect a number of purely categorical observations and constructions that are needed in our work. We shall return to these topics in Chapter VII, but we shall make no further use of this material until then. The reader may prefer to skip these sections on a first reading. We here give the description of algebras over composite monads that was at the heart of our comparisons of definitions and formalize the tensor product construction that appeared briefly in Section 4.

LEMMA 6.1. *Let $\mathbb{S}$ be a monad in a category $\mathscr{C}$ and let $\mathbb{T}$ be a monad in the category $\mathscr{C}[\mathbb{S}]$ of $\mathbb{S}$-algebras. Then the category $\mathscr{C}[\mathbb{S}][\mathbb{T}]$ of $\mathbb{T}$-algebras in $\mathscr{C}[\mathbb{S}]$ is isomorphic to the category $\mathscr{C}[\mathbb{TS}]$ of algebras over the composite monad $\mathbb{TS}$ in $\mathscr{C}$. Moreover, the unit of $\mathbb{T}$ defines a map $\mathbb{S} \longrightarrow \mathbb{TS}$ of monads in $\mathscr{C}$. An analogous assertion holds for comonads.*

PROOF. Strictly speaking, in constructing $\mathbb{TS}$, we are regarding $\mathbb{S}$ as the free $\mathbb{S}$-algebra functor $\mathscr{C} \longrightarrow \mathscr{C}[\mathbb{S}]$, applying the functor $\mathbb{T}$, and then applying the forgetful functor back to $\mathscr{C}$. We continue to neglect notation for forgetful functors and to write $\mathbb{S}$ and $\mathbb{T}$ ambiguously for both the given monads and the resulting free functors. The unit of $\mathbb{TS}$ is given by the composite of unit maps

$$X \longrightarrow \mathbb{S}X \longrightarrow \mathbb{TS}X.$$

The product of $\mathbb{TS}$ is given by the composite maps

$$\mathbb{TSTS}X \longrightarrow \mathbb{TTS}X \longrightarrow \mathbb{TS}X,$$

where the second arrow is given by the product of $\mathbb{T}$ and the first is obtained by application of $\mathbb{T}$ to the action $\mathbb{STS}X \longrightarrow \mathbb{TS}X$ given by the fact that $\mathbb{T}$ takes $\mathbb{S}$-algebras to $\mathbb{S}$-algebras. If R is a $\mathbb{T}$-algebra in $\mathscr{C}[\mathbb{S}]$, with action ξ by $\mathbb{S}$ and action χ by $\mathbb{T}$, then it is a $\mathbb{TS}$-algebra with action the composite

$$\mathbb{TS}R \xrightarrow{\mathbb{T}\xi} \mathbb{T}R \xrightarrow{\chi} R.$$

If Q is a $\mathbb{TS}$-algebra with action ω, then Q is a $\mathbb{T}$-algebra in $\mathscr{C}[\mathbb{S}]$ with actions the composites

$$\mathbb{S}Q \xrightarrow{\eta} \mathbb{TS}Q \xrightarrow{\omega} Q \qquad \text{and} \qquad \mathbb{T}Q \xrightarrow{\mathbb{T}\eta} \mathbb{TS}Q \xrightarrow{\omega} Q.$$

These correspondences establish the required isomorphism of categories. Easy diagram chases show that $\mathbb{S} \longrightarrow \mathbb{TS}$ is a map of monads. □

When applying this to modules, we used the following construction.

CONSTRUCTION 6.2. For a category $\mathscr{C}$, let $\mathscr{C}[1]$ be the category of pairs $(X;Y)$ in $\mathscr{C}$ and pairs of maps. Let $\mathbb{S}$ be any of the monads $\mathbb{B}$, $\mathbb{C}$, $\mathbb{T}$, or $\mathbb{P}$, and let $\mathscr{C}$ be its ground category $\mathscr{S}$, $\mathscr{S}[\mathbb{L}]$, or $\mathscr{M}_S$. Construct a monad $\mathbb{S}[1]$ in $\mathscr{C}[1]$ as follows. On a pair $(X;Y)$, the functor $\mathbb{S}[1]$ is given by

$$\mathbb{S}[1](X;Y) = (\mathbb{S}X;\mathbb{S}(X;Y)),$$

where $\mathbb{S}(X;Y)$ is the free $\mathbb{S}X$-module generated by Y. This functor factors through the evident category of pairs

$$(\mathbb{S}\text{-algebra; object of } \mathscr{C})$$

as the composite of $(\mathbb{S};\text{id})$ and (id; free module), where the free module functor is that associated to the algebra in the first variable. Since the identity functor is a monad in a trivial way, each of these functors is a monad. Therefore, by Lemma 6.1, their composite $\mathbb{S}[1]$ is a monad such that an $\mathbb{S}[1]$-algebra $(R;M)$ is an $\mathbb{S}$-algebra R together with an R-module M.

We used the following definition in our construction of E_∞ ring spectra from A_∞ ring spectra.

DEFINITION 6.3. Let $(\mathbb{S},\mu,\eta)$ be a monad in a cocomplete category $\mathscr{C}$. A (right) $\mathbb{S}$-functor in a category $\mathscr{C}'$ is a functor $F:\mathscr{C}\longrightarrow\mathscr{C}'$ together with a natural transformation $\nu: F\mathbb{S}\longrightarrow F$ such that the following diagrams commute:

$$\begin{array}{ccc} F\mathbb{S} & \xleftarrow{F\eta} & F \\ {\scriptstyle\nu}\downarrow & \swarrow{\scriptstyle \mathrm{id}} & \\ F & & \end{array} \qquad\text{and}\qquad \begin{array}{ccc} F\mathbb{S}\mathbb{S} & \xrightarrow{\nu S} & F\mathbb{S} \\ {\scriptstyle F\mu}\downarrow & & \downarrow{\scriptstyle\nu} \\ F\mathbb{S} & \xrightarrow{\nu} & F. \end{array}$$

Given an $\mathbb{S}$-algebra (R,ξ), define $F\otimes_{\mathbb{S}}R$ to be the coequalizer displayed in the diagram

$$F\mathbb{S}R \underset{F\xi}{\overset{\nu}{\rightrightarrows}} FR \longrightarrow F\otimes_{\mathbb{S}}R.$$

Given a monad $\mathbb{S}'$ in $\mathscr{C}'$ and a left action $\lambda:\mathbb{S}'F\longrightarrow F$, we say that F is an $(\mathbb{S}',\mathbb{S})$-bifunctor if the following diagram commutes:

$$\begin{array}{ccc} \mathbb{S}'F\mathbb{S} & \xrightarrow{\lambda\mathbb{S}} & F\mathbb{S} \\ {\scriptstyle \mathbb{S}'\nu}\downarrow & & \downarrow{\scriptstyle\nu} \\ \mathbb{S}'F & \xrightarrow{\lambda} & F. \end{array}$$

EXAMPLE 6.4. The functor $\mathbb{S}$ is an $(\mathbb{S},\mathbb{S})$-bifunctor, with both left and right action μ. If $\pi:\mathbb{S}\longrightarrow\mathbb{S}'$ is a map of monads in $\mathscr{C}$, then $\mathbb{S}'$ is an $(\mathbb{S}',\mathbb{S})$-bifunctor with right action $\nu=\mu'\circ\mathbb{S}'\pi:\mathbb{S}'\mathbb{S}\longrightarrow\mathbb{S}'$. Observe that, for $X\in\mathscr{C}$, $\mathbb{S}'\otimes_{\mathbb{S}}\mathbb{S}X\cong\mathbb{S}'X$.

When $\mathscr{C}'$ in Definition 6.3 has a forgetful functor to the category of spectra, we shall construct a bar construction $B(F,\mathbb{S},R)$ that will give the appropriate homotopical version of $F\otimes_{\mathbb{S}}R$ in XII§1. Assuming that F is an $(\mathbb{S}',\mathbb{S})$-bifunctor for one of the monads constructed earlier in this chapter, we will find that $B(F,\mathbb{S},R)$ is an $\mathbb{S}'$-algebra. It is natural to ask whether or not $F\otimes_{\mathbb{S}}R$ is itself an $\mathbb{S}'$-algebra. To answer this, we need another categorical definition.

DEFINITION 6.5. In any category $\mathscr{C}$, a coequalizer diagram

$$A \underset{f}{\overset{e}{\rightrightarrows}} B \xrightarrow{g} C.$$

is said to be a reflexive coequalizer if there is a map $h:B\longrightarrow A$ such that $e\circ h=\mathrm{id}$ and $f\circ h=\mathrm{id}$.

The following categorical observation is standard and easy. Although their stated hypotheses are different, the proofs of similar results in [43, p. 147] and [4, pp. 106-108] apply to give the first statement, and the second statement follows.

LEMMA 6.6. *Let $\mathbb{S}$ be a monad in $\mathscr{C}$ such that $\mathbb{S}$ preserves reflexive coequalizers. If*

$$A \underset{f}{\overset{e}{\rightrightarrows}} B \xrightarrow{g} C$$

is a reflexive coequalizer in $\mathscr{C}$ such that A and B are $\mathbb{S}$-algebras and e and f are maps of $\mathbb{S}$-algebras, then C has a unique structure of $\mathbb{S}$-algebra such that g is a *map of $\mathbb{S}$-algebras, and g is the coequalizer of e and f in the category $\mathscr{C}[\mathbb{S}]$. If, further, $\mathbb{T}$ is a monad in $\mathscr{C}[\mathbb{S}]$ such that $\mathbb{T}$ preserves reflexive coequalizers, then $\mathbb{T}\circ\mathbb{S}$ also preserves reflexive coequalizers.*

Since the coequalizer diagram used to define $F\otimes_{\mathbb{S}} R$ is reflexive, via the map $F\eta : FR \longrightarrow F\mathbb{S}R$, the first statement implies an answer to the question we asked originally.

LEMMA 6.7. *Let $\mathbb{S}$ be a monad in $\mathscr{C}$, $\mathbb{S}'$ be a monad in $\mathscr{C}'$, R be an $\mathbb{S}$-algebra, and $F : \mathscr{C} \longrightarrow \mathscr{C}'$ be an $(\mathbb{S}',\mathbb{S})$-bifunctor. If $\mathbb{S}'$ preserves reflexive coequalizers, then $F\otimes_{\mathbb{S}} R$ is an $\mathbb{S}'$-algebra.*

7. Limits and colimits of S-algebras

We here prove that the categories of A_∞ and E_∞ ring spectra and of S-algebras and commutative S-algebras are complete and cocomplete. In fact, completeness follows immediately from Proposition 4.5. All four of our categories are categories of algebras over a monad in a complete category, and it follows that they are complete, with their limits created in their respective ground categories [43, VI.2, ex. 2]. The first statement of Lemma 6.6 applies to construct colimits, but to explain this properly we need some preliminary definitions that put our definitions of A_∞ and E_∞ ring spectra in perspective.

DEFINITION 7.1. A weak symmetric monoidal category $\mathscr{C}$ with product $\Box$ and unit object I is defined in exactly the same way as a symmetric monoidal category [43, p. 180], except that its unit map $\lambda : I\,\Box\, X \longrightarrow X$ is not required to be an isomorphism; $\mathscr{C}$ is said to be closed if the functor $(-)\,\Box\, Y$ has a right adjoint $\mathrm{Hom}(Y,-)$ for each $Y\in\mathscr{C}$. Monoids and commutative monoids in $\mathscr{C}$ are defined in terms of diagrams of the form displayed in Definition 3.3. As in Construction 4.4 and Proposition 4.5, if $\mathscr{C}$ is cocomplete, then there are monads $\mathbb{T}$ and $\mathbb{P}$ in $\mathscr{C}$ whose algebras are the monoids and commutative monoids in $\mathscr{C}$. For $X\in\mathscr{C}$,

$$\mathbb{T}X \cong \coprod_{j\geq 0} X^j \quad \text{and} \quad \mathbb{P}X \cong \coprod_{j\geq 0} X^j/\Sigma_j.$$

The proof of the following result is abstracted from an argument that Hopkins gave for the monad $\mathbb{C}$ [32]. He proceeded by reduction to a proof that the j-fold symmetric powers of based spaces preserve reflexive coequalizers. With our new associative smash products, an abstraction of the latter proof makes the reduction unnecessary.

PROPOSITION 7.2. *Let $\mathscr{C}$ be any cocomplete closed weak symmetric monoidal category. Then the monads $\mathbb{T}$ and $\mathbb{P}$ in $\mathscr{C}$ preserve reflexive coequalizers.*

PROOF. For $\mathbb{T}$, it suffices to prove that the j-fold product $X_1\,\Box\cdots\Box\, X_j$ preserves reflexive coequalizers. Thus let

$$X_i \underset{f_i}{\overset{e_i}{\rightrightarrows}} Y_i \xrightarrow{g_i} Z_i$$

be reflexive coequalizer diagrams in $\mathscr{C}$, $1 \leq i \leq j$, and let $h_i : Y_i \longrightarrow X_i$ satisfy $e_i \circ h_i = \mathrm{id}$ and $f_i \circ h_i = \mathrm{id}$. Let

$$\varepsilon = e_1 \,\square \cdots \square\, e_j, \quad \phi = f_1 \,\square \cdots \square\, f_j, \quad \text{and} \quad \gamma = g_1 \,\square \cdots \square\, g_j.$$

Let $\beta : Y_1 \,\square \cdots \square\, Y_j \longrightarrow Z$ be the coequalizer of ε and ϕ. Since $\gamma\varepsilon = \gamma\phi$, there is a unique map $\xi : Z \longrightarrow Z_1 \,\square \cdots \square\, Z_j$ such that $\xi \circ \beta = \gamma$. We claim that ξ is an isomorphism, and we proceed by induction on j. Let

$$\varepsilon_i = (\mathrm{id})^{i-1} \square e_i \square (\mathrm{id})^{j-i} : Y_1 \square \cdots \square Y_{i-1} \square X_i \square Y_{i+1} \square \cdots \square Y_j \longrightarrow Y_1 \square \cdots \square Y_j$$

and, similarly, define $\phi_i = (\mathrm{id})^{i-1} \square f_i \square (\mathrm{id})^{j-i}$. We observe first that $Z_1 \square \cdots \square Z_j$ is the colimit of the diagram given by the j pairs of maps $\{\varepsilon_i, \phi_i\}$. Indeed, for any map $\alpha : Y_1 \,\square \cdots \square\, Y_j \longrightarrow W$ such that $\alpha \circ \varepsilon_i = \alpha \circ \phi_i$ for $1 \leq i \leq j$, we obtain unique maps $\hat{\alpha}$ and $\tilde{\alpha}$ that make the following diagram commute by the induction hypothesis and the fact that the $\square$-product preserves colimits and epimorphisms:

$$\begin{array}{ccc}
Y_1 \square \cdots \square Y_{j-1} \square X_j & \xrightarrow{g_1 \square \cdots \square g_{j-1} \square \mathrm{id}} & Z_1 \square \cdots \square Z_{j-1} \square X_j \\
{\scriptstyle \mathrm{id} \square e_j} \downdownarrows {\scriptstyle \mathrm{id} \square f_j} & & {\scriptstyle \mathrm{id} \square e_j} \downdownarrows {\scriptstyle \mathrm{id} \square f_j} \\
Y_1 \square \cdots \square Y_{j-1} \square Y_j & \xrightarrow{g_1 \square \cdots \square g_{j-1} \square \mathrm{id}} & Z_1 \square \cdots \square Z_{j-1} \square Y_j \\
{\scriptstyle \alpha} \downarrow & \swarrow {\scriptstyle \hat{\alpha}} & \downarrow {\scriptstyle \mathrm{id} \square g_j} \\
W & \xleftarrow[\tilde{\alpha}]{} & Z_1 \square \cdots \square Z_{j-1} \square Z_j .
\end{array}$$

Now let $k_i = h_1 \,\square \cdots \square\, h_{i-1} \,\square\, \mathrm{id} \square h_{i+1} \,\square \cdots \square\, h_j$. Visibly

$$\varepsilon_i = \varepsilon \circ k_i \quad \text{and} \quad \phi_i = \phi \circ k_i.$$

Since $\beta\varepsilon = \beta\phi$, $\beta\varepsilon_i = \beta\phi_i$ for $1 \leq i \leq j$ and the universal property gives a map $\zeta : Z_1 \,\square \cdots \square\, Z_j \longrightarrow Z$. It is easy to check from the universal properties that ζ and ξ are inverse isomorphisms. In the symmetric case, we may take our j given coequalizer diagrams to be the same and compose the j-fold power, regarded as a functor to the category of Σ_j-objects in $\mathscr{C}$, with the orbit functor. The latter is constructed as a coequalizer in $\mathscr{C}$ and is a left adjoint, so preserves coequalizers. $\square$

COROLLARY 7.3. *The functors $\mathbb{T}$ and $\mathbb{P}$ on $\mathscr{S}[\mathbb{L}]$, their restrictions to functors $\mathbb{T}$ and $\mathbb{P}$ on $\mathscr{M}_S$, and the functors $\mathbb{B}$ and $\mathbb{C}$ on $\mathscr{S}$ preserve reflexive coequalizers.*

PROOF. This is immediate since $\mathbb{B} = \mathbb{TL}$, $\mathbb{C} = \mathbb{PL}$, the functor $\mathbb{L} : \mathscr{S} \longrightarrow \mathscr{S}[\mathbb{L}]$ preserves colimits, and colimits in $\mathscr{S}[\mathbb{L}]$ and in $\mathscr{M}_S$ are created in $\mathscr{S}$. $\square$

Our claim that the categories of A_∞ and E_∞ ring spectra and of S-algebras and commutative S-algebras are cocomplete is now an immediate corollary of the following known result, which we also learned from Hopkins.

PROPOSITION 7.4. *Let $\mathbb{S}$ be a monad in a cocomplete category $\mathscr{C}$. If $\mathbb{S}$ preserves reflexive coequalizers, then $\mathscr{C}[\mathbb{S}]$ is cocomplete.*

PROOF. Consider a diagram $\{R_i\}$ of $\mathbb{S}$-algebras. Let $\operatorname{colim} R_i$ be its colimit in $\mathscr{C}$ and let $\iota_i : R_i \longrightarrow \operatorname{colim} R_i$ be the natural maps. Let

$$\alpha : \operatorname{colim} \mathbb{S}R_i \longrightarrow \mathbb{S} \operatorname{colim} R_i$$

be the unique map in $\mathscr{C}$ whose composite with the natural map $\mathbb{S}R_i \longrightarrow \operatorname{colim} \mathbb{S}R_i$ is $\mathbb{S}\iota_i$ for each i. Define $\operatorname{colim}_{\mathbb{S}} R_i$ by the following coequalizer diagram in $\mathscr{C}$:

$$\mathbb{S}(\operatorname{colim} \mathbb{S}R_i) \underset{\mu \circ \mathbb{S}\alpha}{\overset{\mathbb{S}(\operatorname{colim} \xi_i)}{\rightrightarrows}} \mathbb{S}(\operatorname{colim} R_i) \longrightarrow \operatorname{colim}_{\mathbb{S}} R_i.$$

This is a reflexive coequalizer, via $\mathbb{S}(\operatorname{colim} \eta_i)$. Thus, by Lemma 6.6, $\operatorname{colim}_{\mathbb{S}} R_i$ is an $\mathbb{S}$-algebra such that the displayed diagram is a coequalizer in $\mathscr{C}[\mathbb{S}]$. It follows easily that $\operatorname{colim}_{\mathbb{S}} R_i$ is the colimit of $\{R_i\}$ in $\mathscr{C}[\mathbb{S}]$. $\square$

This result is closely related to the following standard result of Linton [41] (see also [4, Thm 2, p. 319]).

THEOREM 7.5 (LINTON). *Let $\mathbb{S}$ be a monad in a cocomplete category $\mathscr{C}$. If $\mathscr{C}[\mathbb{S}]$ has coequalizers, then $\mathscr{C}[\mathbb{S}]$ is cocomplete.*

CHAPTER III

The homotopy theory of R-modules

We here develop the homotopy theory of modules over an S-algebra R. The classical theory of cell spectra generalizes to give a theory of cell modules over R. The derived category $\mathscr{D}_R$ of R-modules is constructed from the category of R-modules by formally adjoining inverses to the weak equivalences, and it is equivalent to the homotopy category of cell R-modules. We define the smash product over R, $\wedge_R$, and the function R-module functor, F_R, by direct mimicry of the definitions of tensor product and Hom functors for modules over an algebra. When specialized to commutative S-algebras, our smash product of R-modules is again an R-module, and similarly for F_R. Here the category of R-modules has structure precisely like the category of S-modules, and duality theory works exactly as it does for spectra. We assume familiarity with II§§1,3 and work in the ground category $\mathscr{M}_S$ of S-modules.

1. The category of R-modules; free and cofree R-modules

Fix an S-algebra R. We understand R-modules to be left R-modules unless otherwise specified. We first observe that the category $\mathscr{M}_R$ of R-modules is closed under various constructions in the underlying categories of spectra and S-modules. As in algebra, an R-module is the same thing as an algebra over the monad $R \wedge_S (-)$ in $\mathscr{M}_S$ or, equivalently, a coalgebra over the adjoint comonad $F_S(R, -)$ in $\mathscr{M}_S$. The functors $R \wedge_S (-)$ and $F_S(R, -)$ from $\mathscr{M}_S$ to $\mathscr{M}_R$ are left and right adjoint to the forgetful functor. That is, $R \wedge_S (-)$ and $F_S(R, -)$ are the free and cofree functors from S-modules to R-modules. Together with II.1.4 and formal arguments exactly like those in algebra, this leads to the following result.

THEOREM 1.1. *The category of R-modules is complete and cocomplete, with both limits and colimits created in the underlying category $\mathscr{M}_S$. Let X be a based space, K be an S-module, and M and N be R-modules. Then the following conclusions hold, where the displayed isomorphisms are obtained by restriction of the corresponding isomorphisms for S-modules.*

(i) *$M \wedge X$ is an R-module and the spectrum level cofiber of a map of R-modules is an R-module.*

(ii) $S \wedge_{\mathscr{L}} F(X,N)$ *is an* R*-module and*

$$\mathscr{M}_R(M \wedge X, N) \cong \mathscr{M}_R(M, S \wedge_{\mathscr{L}} F(X,N)).$$

(iii) $M \wedge_S K$ *and* $F_S(K,N)$ *are* R*-modules and*

$$\mathscr{M}_R(M \wedge_S K, N) \cong \mathscr{M}_R(M, F_S(K,N)).$$

(iv) $F_S(M,K)$ *is a right* R*-module.*

(v) *As* R*-modules,*

$$M \wedge X \cong M \wedge_S \Sigma^\infty X \quad \text{and} \quad S \wedge_{\mathscr{L}} F(X,N) \cong F_S(\Sigma^\infty X, N).$$

The cofiber and fiber of a map of R*-modules are* R*-modules, where the fiber is understood to be obtained by application of the functor* $S \wedge_{\mathscr{L}} (-)$ *to the fiber constructed in the category of spectra.*

PROOF. The only point that might need comment is the R-module structure on $S \wedge_{\mathscr{L}} F(X,N)$. The evaluation map $\varepsilon : F(X,N) \wedge X \longrightarrow N$ is a map of $\mathbb{L}$-spectra. The adjoint of $R \wedge_{\mathscr{L}} \varepsilon$ is a map of $\mathbb{L}$-spectra

$$\tilde{\varepsilon} : R \wedge_{\mathscr{L}} F(X,N) \longrightarrow F(X, R \wedge_S N),$$

and we obtain the desired action upon applying $S \wedge_{\mathscr{L}} \tilde{\varepsilon}$ and using the given action of R on N. This leads to the R-module structure on the specified fiber of a map of R-modules; compare II.1.5. □

The free R-module functor on spectra is the starting point of cellular theory.

DEFINITION 1.2. Define the free R-module generated by a spectrum X to be

$$\mathbb{F}_R X = R \wedge_S \mathbb{F}_S X,$$

where $\mathbb{F}_S X = S \wedge_{\mathscr{L}} \mathbb{L}X$. Equivalently, since $R \wedge_S S \cong R$,

$$\mathbb{F}_R X = R \wedge_{\mathscr{L}} \mathbb{L}X.$$

We abbreviate $\mathbb{F}X = \mathbb{F}_R X$ when R is clear from the context.

The term "free" is technically a misnomer, since $\mathbb{F}$ is not left adjoint to the forgetful functor. However, it is nearly so.

PROPOSITION 1.3. *The functor* $\mathbb{F} : \mathscr{S} \longrightarrow \mathscr{M}_R$ *is left adjoint to the functor that sends an* R*-module* M *to the spectrum* $F_{\mathscr{L}}(S,M)$*, and there is a natural map of* R*-modules* $\xi : \mathbb{F}M \longrightarrow M$ *whose adjoint* $M \longrightarrow F_{\mathscr{L}}(S,M)$ *is a weak equivalence of spectra. Therefore*

$$\pi_n(M) \cong h\mathscr{M}_R(\mathbb{F}S^n, M).$$

PROOF. In view of II.1.3, we have the chain of isomorphisms

$$\mathscr{M}_R(\mathbb{F}_R X, N) \cong \mathscr{M}_S(\mathbb{F}_S X, N) \cong \mathscr{S}[\mathbb{L}](\mathbb{L}X, F_{\mathscr{L}}(S,N)) \cong \mathscr{S}(X, F_{\mathscr{L}}(S,N)).$$

By I.8.7, we have a natural weak equivalence $\tilde{\lambda} : M \longrightarrow F_{\mathscr{L}}(S,M)$ of $\mathscr{S}[\mathbb{L}]$-spectra. Thought of as a map of spectra, its adjoint is the required R-map ξ. The statement about the homotopy groups $\pi_n(M) = h\mathscr{S}(S^n, M)$ is clear; compare II.1.8. □

The following central theorem shows that we have homotopical control on $\mathbb{F}X$ without any hypotheses (such as tameness or CW homotopy type) on R.

THEOREM 1.4. *In the stable homotopy category $\bar{h}\mathscr{S}$, $\mathbb{F}X$ is naturally isomorphic to the internal smash product $R \wedge X$. Moreover, the composite*

$$\zeta : \mathbb{F}S \xrightarrow{\mathbb{F}\eta} \mathbb{F}R \xrightarrow{\xi} R$$

is a weak equivalence of R-modules.

PROOF. The first statement is clear from II.1.9 and I.6.7, but we point out a variant proof that makes clear that the weak equivalence is one of R-module spectra (in the homotopical sense). In X§5, we shall construct a tame A_∞ ring spectrum KR and a weak equivalence of A_∞ ring spectra $r : KR \longrightarrow R$. Since we are working in the stable homotopy category, we may take X to be a CW spectrum. Then, by I.4.6 and I.6.6,

$$r \wedge_{\mathscr{L}} \mathrm{id} : KR \wedge_{\mathscr{L}} \mathbb{L}X \longrightarrow R \wedge_{\mathscr{L}} \mathbb{L}X = \mathbb{F}X$$

is a weak equivalence. By I.4.6 and I.6.7, there are natural weak equivalences

$$KR \wedge X \longrightarrow KR \wedge \mathbb{L}X \longrightarrow KR \wedge_{\mathscr{L}} \mathbb{L}X.$$

For the second statement, observe that ζ is the common composite in the diagram

$$\begin{array}{ccccc} R \wedge_{\mathscr{L}} \mathbb{L}S & \xrightarrow{\mathrm{id} \wedge \xi} & R \wedge_{\mathscr{L}} S & & \\ \Big\downarrow{\scriptstyle \mathrm{id} \wedge \mathbb{L}\eta} & & \Big\downarrow{\scriptstyle \mathrm{id} \wedge \eta} & \searrow{\scriptstyle \cong} & \\ R \wedge_{\mathscr{L}} \mathbb{L}R & \xrightarrow[\mathrm{id} \wedge \xi]{} & R \wedge_{\mathscr{L}} R & \xrightarrow[\phi]{} & R. \end{array}$$

By I.8.6, the top map $\mathrm{id} \wedge \xi$ is a weak equivalence. □

COROLLARY 1.5. *If X is a wedge of sphere spectra, then $\pi_*(\mathbb{F}X)$ is the free $\pi_*(R)$-module with one generator of degree n for each wedge summand S^n.*

We shall need one further fundamental property of free R-modules.

DEFINITION 1.6. A compact spectrum is one of the form $\Sigma^\infty_V X$ for a compact space X and an indexing space $V \subset U$. A compact R-module is one of the form $\mathbb{F}K$ for a compact spectrum K.

PROPOSITION 1.7. *Let L be a finite colimit of compact R-modules and let $\{M_i\}$ be a sequence of R-modules and (spacewise) inclusions $M_i \longrightarrow M_{i+1}$. Then*

$$\mathscr{M}_R(L, \operatorname{colim} M_i) \cong \operatorname{colim} \mathscr{M}_R(L, M_i).$$

The generalization from compact R-modules to their finite colimits is immediate. The compact case would be elementary if the free functor were left adjoint to the forgetful functor, and we shall show in XI§2 that this is near enough to being true to give the conclusion.

While they play a less central role, we shall also make use of cofree R-modules. Recall from I§4 that $\mathbb{L}^\# : \mathscr{S} \longrightarrow \mathscr{S}$ is the right adjoint of $\mathbb{L}$ and gives a comonad whose coalgebras are the $\mathbb{L}$-spectra. In particular, $\mathbb{L}^{\#}X$ is an $\mathbb{L}$-spectrum for any spectrum X.

DEFINITION 1.8. Define the cofree S-module generated by a spectrum X to be $\mathbb{F}^{\#}_S X = S \wedge_{\mathscr{L}} \mathbb{L}^{\#}X$. Then define the cofree R-module generated by X to be

$$\mathbb{F}^{\#}_R X = F_S(R, \mathbb{F}^{\#}_S X)$$

with left action of R induced by the right action of R on itself. We abbreviate $\mathbb{F}^{\#}X = \mathbb{F}^{\#}_R X$ when R is clear from the context.

The term "cofree" is not a misnomer, since here we do have the expected adjunction.

PROPOSITION 1.9. *The functor* $\mathbb{F}^{\#}_R : \mathscr{S} \longrightarrow \mathscr{M}_R$ *is right adjoint to the forgetful functor* $\mathscr{M}_R \longrightarrow \mathscr{S}$.

PROOF. Let M be an R-module and X be a spectrum. Lemma 5.5(ii) below gives the first of the following isomorphisms, and II.1.3 and I.4.7 give the others:

$$\mathscr{M}_R(M, \mathbb{F}^{\#}_R X) \cong \mathscr{M}_S(M, \mathbb{F}^{\#}_S X) \cong \mathscr{S}[\mathbb{L}](M, \mathbb{L}^{\#}X) \cong \mathscr{S}(M, X). \quad \square$$

THEOREM 1.10. *In the stable homotopy category* $\bar{h}\mathscr{S}$, $\mathbb{F}^{\#}X$ *is naturally isomorphic to the internal function spectrum* $F(R, X)$.

PROOF. This is immediate from II.1.9, I.7.3, and I.4.8. $\square$

2. Cell and CW R-modules; the derived category of R-modules

To develop cell and CW theories for R-modules, we think of the free R-modules $S^n_R \equiv \mathbb{F}S^n$ as "sphere R-modules". This is consistent with the sphere S-modules of II.1.7. For cells, we note that the cone functor $CX = X \wedge I$ commutes with $\mathbb{F}$, so that $C\mathbb{F}S^n \cong \mathbb{F}CS^n$. Since $\mathbb{F}$ has a right adjoint, maps out of sphere R-modules and their cones are induced by maps on the spectrum level; the fact that the right adjoint is not the obvious forgetful functor will create no difficulties. In fact, we can simply parrot the cell theory of spectra from [38, I§5], reducing proofs to those given there via adjunction.

DEFINITIONS 2.1. We define cell and relative cell R-modules.

(i) A cell R-module M is the union of an expanding sequence of sub R-modules M_n such that $M_0 = *$ and M_{n+1} is the cofiber of a map $\phi_n : F_n \longrightarrow M_n$, where F_n is a (possibly empty) wedge of sphere modules S^q_R (of varying dimensions). The restriction of ϕ_n to a wedge summand S^q_R is called an attaching map. The induced map

$$CS^q_R \longrightarrow M_{n+1} \subset M$$

is called a cell. The sequence $\{M_n\}$ is called the sequential filtration of M.

(ii) For an R-module L, a relative cell R-module (M, L) is an R-module M specified as in (i), but with $M_0 = L$.

(iii) A map $f : M \longrightarrow N$ between cell R-modules is sequentially cellular if $f(M_n) \subset N_n$ for all n.

(iv) A submodule L of a cell R-module M is a cell submodule if L is a cell R-module such that $L_n \subset M_n$ and the composite of each attaching map $S^q_R \longrightarrow L_n$ of L with the inclusion $L_n \longrightarrow M_n$ is an attaching map of M. Thus every cell of L is a cell of M. Observe that (M, L) may be viewed as a relative cell R-module.

(v) A cell R-module is finite dimensional if it has cells in finitely many dimensions. It is finite if it has finitely many cells.

The sequential filtration is essential for inductive arguments, but it should be regarded as flexible and subject to change whenever convenient. It merely records the order in which cells are attached and, as long as the cells to which new cells are attached are already present, it doesn't matter in what order cells are attached.

LEMMA 2.2. *Let $f : M \longrightarrow N$ be an R-map between cell R-modules. Then M admits a new sequential filtration with respect to which f is sequentially cellular.*

PROOF. Assume inductively that M_n has been given a filtration as a cell R-module $M_n = \cup M'_q$ such that $f(M'_q) \subset N_q$ for all q. Let $\chi : S^r_R \longrightarrow M_n$ be an attaching map for the construction of M_{n+1} from M_n and let $\tilde{\chi} : CS^r_R \longrightarrow M_{n+1}$ be the corresponding cell. By Proposition 1.7, there is a minimal q such that both

$$\mathrm{Im}(\chi) \subset M'_q \quad \text{and} \quad \mathrm{Im}(f \circ \tilde{\chi}) \subset N_{q+1}.$$

Extend the filtration of M_n to M_{n+1} by taking χ to be a typical attaching map of a cell of M'_{q+1}. $\square$

We shall occasionally need the following two reassuring results. Their proofs are similar to those of their spectrum level analogs [38, pp 494–495] and depend on Proposition 1.7 and its proof.

LEMMA 2.3. *A map from a compact R-module to a cell R-module has image contained in a finite subcomplex, and a cell R-module is the colimit of its finite subcomplexes.*

If K and L are subcomplexes of a cell R-module M, then we understand their intersection and union in the combinatorial sense. That is, $K \cap L$ is the cell R-module constructed from the attaching maps and cells that are in both K and L and $K \cup L$ is the cell R-module constructed from the attaching maps and cells that are in either K or L. However, we also have their categorical intersection and union, namely the pullback of the inclusions of K and L in M and the pushout of the resulting maps from the categorical intersection to K and to L.

LEMMA 2.4. *For subcomplexes K and L of a cell R-module M, the canonical map from the combinatorial intersection to the categorical intersection and from the categorical union to the combinatorial union of K and L are isomorphisms of R-modules.*

DEFINITION 2.5. A cell R-module M is said to be a CW R-module if each cell is attached only to cells of lower dimension. The n-skeleton M^n of a CW R-module is the union of its cells of dimension at most n. A map $f : M \longrightarrow N$ between CW R-modules is cellular if $f(M^n) \subset N^n$ for all n. We do not require that f also be sequentially cellular but, by Lemma 2.2, that can always be arranged by changing the order in which cells are attached. Relative CW R-modules (M, L) are defined similarly, with each cell attached only to the union of L and the cells of lower dimension.

PROPOSITION 2.6. *The collection of cell R-modules enjoys the following closure properties.*

(i) *A wedge of cell R-modules is a cell R-module.*
(ii) *The pushout of a map along the inclusion of a cell submodule is a cell R-module.*
(iii) *The union of a sequence of inclusions of cell submodules is a cell R-module.*
(iv) *The smash product of a cell R-module and a based cell space (with based attaching maps) is a cell R-module.*
(v) *The smash product over S of a cell R-module and a cell S-module is a cell R-module.*

The same statements hold with "cell" replaced by "CW", provided that, in (ii), the given map is cellular.

PROOF. In (ii), we apply Lemma 2.2 to ensure that the given map is sequentially cellular. Part (v) follows from I.6.1, which implies that the smash product of a sphere R-module and a sphere S-module is a sphere R-module. Otherwise the proofs are the same as for cell and CW spectra [38, I§5]. □

The following result is the "Homotopy Extension and Lifting Property".

THEOREM 2.7 (HELP). *Let (M, L) be a relative cell R-module and let $e : N \longrightarrow P$ be a weak equivalence of R-modules. Then, given maps $f : M \longrightarrow P$, $g : L \longrightarrow N$, and $h : L \wedge I_+ \longrightarrow P$ such that $f|_L = hi_0$ and $eg = hi_1$ in the following diagram, there are maps $\tilde{g}$ and $\tilde{h}$ that make the entire diagram commute.*

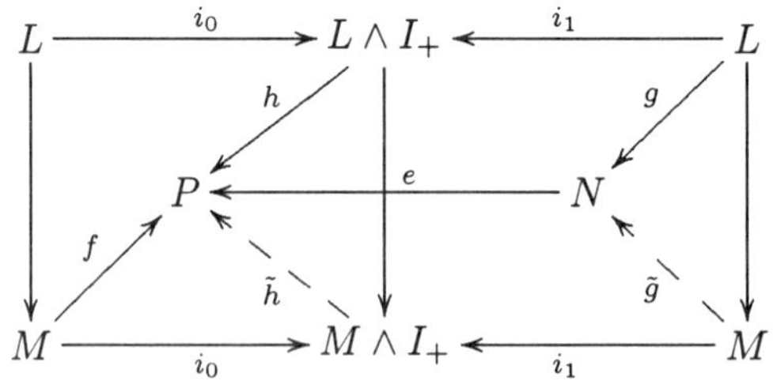

PROOF. This is proven for $(M, L) = (CS^q_R, S^q_R)$ by reduction to the spectrum level analog. Technically, we use that the fact that our spheres are obtained from sphere spectra by applying a functor that is left adjoint to a functor that preserves weak equivalences (even though it is not the obvious forgetful functor). The general case follows by induction up the sequential filtration, and the inductive step reduces directly to the case of (CS_R^q, S_R^q) already handled. □

The Whitehead theorem is a formal consequence.

THEOREM 2.8 (WHITEHEAD). *If M is a cell R-module and $e : N \longrightarrow P$ is a weak equivalence of R-modules, then $e_* : h\mathscr{M}_R(M,N) \longrightarrow h\mathscr{M}_R(M,P)$ is an isomorphism. Therefore a weak equivalence between cell R-modules is a homotopy equivalence.*

Recall that a spectrum is "connective" if it is (-1)-connected. When R is connective, $\pi_q(N/N^q) = 0$ for any CW R-module and we can prove the following cellular approximation theorem exactly as in [38, I.5.8]. For non-connective R, this result fails and we must content ourselves with cell R-modules. For connective R, there is no significant loss of information if we restrict attention to CW R-modules.

THEOREM 2.9 (CELLULAR APPROXIMATION). *Assume that R is connective and let (M, L) and (M', L') be relative CW R-modules. Then any map $f : (M, L) \longrightarrow (M', L')$ is homotopic relative to L to a cellular map. Therefore, for cell R-modules M and M', any map $M \longrightarrow M'$ is homotopic to a cellular map, and any two homotopic cellular maps are cellularly homotopic.*

THEOREM 2.10 (APPROXIMATION BY CELL MODULES). *For any R-module M, there is a cell R-module ΓM and a weak equivalence $\gamma : \Gamma M \longrightarrow M$. If R is connective, then ΓM can be chosen to be a CW R-module.*

PROOF. Choose a wedge of sphere R-modules N_0 and a map $\gamma_0 : N_0 \longrightarrow M$ that induces an epimorphism on homotopy groups. Given $\gamma_n : N_n \longrightarrow M$, we construct N_{n+1} from N_n as a homotopy coequalizer of pairs of representative maps for all pairs of unequal elements of any $\pi_q(N_n)$ that map to the same element in $\pi_q(M)$. We have homotopies that allow us to extend γ_n to γ_{n+1}. We let ΓM be the union of the N_n, and the γ_n give a map $\gamma : \Gamma M \longrightarrow M$. We deduce from Proposition 1.7 that γ is a weak equivalence, and we deduce from Proposition 2.6 that ΓM is a cell R-module. If R is connective, we may take our representative maps to be cellular, and ΓN is then a CW R-module. □

CONSTRUCTION 2.11. For each R-module M, choose a cell R-module ΓM and a weak equivalence $\gamma : \Gamma M \longrightarrow M$. By the Whitehead theorem, for a map $f : M \longrightarrow N$, there is a map $\Gamma f : \Gamma M \longrightarrow \Gamma N$, unique up to homotopy, such that the following diagram is homotopy commutative:

$$\begin{array}{ccc} \Gamma M & \xrightarrow{\Gamma f} & \Gamma N \\ {\scriptstyle\gamma}\downarrow & & \downarrow{\scriptstyle\gamma} \\ M & \xrightarrow[f]{} & N. \end{array}$$

Thus Γ is a functor $h\mathscr{M}_R \longrightarrow h\mathscr{M}_R$, and γ is a natural transformation from Γ to the identity. The derived category $\mathscr{D}_R$ can be described as the category whose objects are the R-modules and whose morphisms are specified by

$$\mathscr{D}_R(M,N) = h\mathscr{M}_R(\Gamma M, \Gamma N),$$

with the evident composition. When M is a cell R-module,

$$\mathscr{D}_R(M,N) \cong h\mathscr{M}_R(M,N).$$

Using the identity function on objects and Γ on morphisms, we obtain a functor $i : h\mathscr{M}_R \longrightarrow \mathscr{D}_R$ that sends weak equivalences to isomorphisms and is universal with this property. Let $\mathscr{C}_R$ be the full subcategory of $\mathscr{M}_R$ whose objects are the cell R-modules. Then the functor Γ induces an equivalence of categories $\mathscr{D}_R \longrightarrow h\mathscr{C}_R$ with inverse the composite of i and the inclusion of $h\mathscr{C}_R$ in $h\mathscr{M}_R$.

Therefore the derived category and the homotopy category of cell R-modules can be used interchangeably. Homotopy-preserving functors on R-modules that do not preserve weak equivalences are transported to the derived category by first applying Γ, then the given functor.

The category $\mathscr{D}_R$ has all homotopy limits and colimits. They are created as the corresponding constructions on the underlying diagrams of S-modules; equivalently, homotopy colimits are created on the spectrum level and homotopy limits are created from spectrum level homotopy limits, which are $\mathscr{S}[\mathbb{L}]$-spectra, by applying the functor $S\wedge_{\mathscr{L}}(-)$. Explicit functorial constructions will be given in X§3. We have enough information to quote the categorical form of Brown's representability theorem given in [13].

THEOREM 2.12 (BROWN). *A contravariant functor $k : \mathscr{D}_R \longrightarrow$ Sets is representable in the form $k(M) \cong \mathscr{D}_R(M,N)$ for some R-module N if and only if k converts wedges to products and converts homotopy pushouts to weak pullbacks.*

REMARK 2.13. There is a variant of Brown's theorem, due to Adams [3], that applies to functors that are defined only on finite CW spectra. Working in an algebraic context, Neeman [56] observed that Adams' variant does not generalize so readily. Rather, it requires a countability hypothesis that is satisfied automatically in the classical context of finite CW spectra. In our context, Adams' variant applies provided that each homotopy group $\pi_n(R)$ is countable.

3. The smash product of R-modules

We mimic the definition of tensor products of modules over algebras.

DEFINITION 3.1. Let R be an S-algebra and let M be a right and N be a left R-module. Define $M\wedge_R N$ to be the coequalizer displayed in the following diagram of S-modules:

$$M\wedge_S R\wedge_S N \underset{\mathrm{id}\wedge_S \nu}{\overset{\mu\wedge_S \mathrm{id}}{\rightrightarrows}} M\wedge_S N \longrightarrow M\wedge_R N,$$

where μ and ν are the given actions of R on M and N.

When $R = S$, we are coequalizing the same isomorphism (see I.8.3). Therefore our new $M \wedge_S N$ coincides with our old $M \wedge_S N$.

We shall shortly construct function R-modules satisfying the usual adjunction. It will follow that the functor $\wedge_R$ preserves colimits in each of its variables. It is clear that smash products with spaces commute with $\wedge_R$, in the sense that

$$(X \wedge M) \wedge_R N \cong X \wedge (M \wedge_R N) \cong (M \wedge_R N) \wedge X \cong M \wedge_R (N \wedge X).$$

Therefore the functor $\wedge_R$ commutes with cofiber sequences in each of its variables. We also have the following adjunction, which complements Theorem 1.1(iv).

LEMMA 3.2. *For an S-module K,*

$$\mathscr{M}_S(M \wedge_R N, K) \cong \mathscr{M}_R(N, F_S(M, K)).$$

The commutativity, associativity, and unity properties of the smash product over S and comparisons of coequalizer diagrams give commutativity, associativity, and unity properties of the smash product over R, exactly as in algebra. We state these properties in the generality of their algebraic counterparts.

An S-algebra R with product $\phi : R \wedge_S R \longrightarrow R$ has an opposite S-algebra R^{op} with product $\phi \circ \tau$, and a left R-module with action μ is a right R^{op}-module with action $\mu \circ \tau$.

LEMMA 3.3. *For a right R-module M and left R-module N,*

$$M \wedge_R N \cong N \wedge_{R^{op}} M.$$

For S-algebras R and R', we define an (R, R')-bimodule to be a left R and right R'-module M such that the evident diagram commutes:

$$\begin{array}{ccc} R \wedge_S M \wedge_S R' & \longrightarrow & M \wedge_S R' \\ \downarrow & & \downarrow \\ R \wedge_S M & \longrightarrow & M. \end{array}$$

As in algebra, an (R, R')-bimodule is the same thing as an $(R \wedge_S (R')^{op})$-module.

PROPOSITION 3.4. *Let M be an (R, R')-bimodule, N be an (R', R'')-bimodule, and P be an (R'', R''')-bimodule. Then $M \wedge_{R'} N$ is an (R, R'')-bimodule and*

$$(M \wedge_{R'} N) \wedge_{R''} P \cong M \wedge_{R'} (N \wedge_{R''} P)$$

as (R, R''')-bimodules.

The unity isomorphism has already been displayed, in the guise of a split coequalizer diagram, in II.3.2. We restate the conclusion.

LEMMA 3.5. *The action $\nu : R \wedge_S N \longrightarrow N$ of an R-module N factors through an isomorphism of R-modules $\lambda : R \wedge_R N \longrightarrow N$.*

For an S-module K, $R \wedge_S K \cong K \wedge_S R$ is an (R,R)-bimodule. In particular, this applies to the free left R-module $\mathbb{F}_R X = R \wedge_S \mathbb{F}_S X$ generated by a spectrum X, which may be identified with the free right R-module generated by X. The following instances of the isomorphisms above will be used in conjunction with the weak equivalences of I.6.7 and II.1.9. They allow us to deduce homotopical properties of $\wedge_R$ from corresponding properties of $\wedge_S$.

PROPOSITION 3.6. *Let K and L be S-modules and let N be an R-module. There is a natural isomorphism of R-modules*

$$(K \wedge_S R) \wedge_R N \cong K \wedge_S N.$$

There is also a natural isomorphism of (R,R)-bimodules

$$(K \wedge_S R) \wedge_R (R \wedge_S L) \cong R \wedge_S (K \wedge_S L).$$

Using I.6.1, we obtain the following consequence, in which we use an isomorphism of universes $f : U \oplus U \longrightarrow U$ to define the internal smash product $f_*(X \wedge Y)$.

COROLLARY 3.7. *Let X and Y be spectra and let N be an R-module. There is a natural isomorphism of R-modules*

$$\mathbb{F}_R X \wedge_R N \cong \mathbb{F}_S X \wedge_S N.$$

There is also a natural isomorphism of (R,R)-bimodules

$$\mathbb{F}_R X \wedge_R \mathbb{F}_R Y \cong \mathbb{F}_R f_*(X \wedge Y).$$

THEOREM 3.8. *If M is a cell R-module and $\phi : N \longrightarrow N'$ is a weak equivalence of R-modules, then*

$$\mathrm{id} \wedge_R \phi : M \wedge_R N \longrightarrow M \wedge_R N'$$

is a weak equivalence of S-modules.

PROOF. When $M = \mathbb{F}_R X$ for a CW spectrum X, the conclusion is immediate from the corollary and I.6.6. The general case follows from the case of sphere R-modules by induction up the sequential filtration and passage to colimits. □

We construct $\wedge_R$ as a functor

$$r\mathscr{D}_R \times \ell\mathscr{D}_R \longrightarrow \mathscr{D}_S$$

by approximating one of the variables by a cell R-module; here "r" and "ℓ" indicate right and left R-modules. That is, the derived smash product of M and N is represented by $\Gamma M \wedge_R N$.

The following technical sharpening of Corollary 1.5 will be the starting point for our later construction of a spectral sequence for the calculation of $\pi_*(M \wedge_R N)$.

PROPOSITION 3.9. *Let X be a wedge of sphere spectra and let N be a cell R-module. Then there is an isomorphism*

$$\pi_*(\mathbb{F}_R X \wedge_R N) \cong (\pi_*(R) \otimes H_*(X)) \otimes_{\pi_*(R)} \pi_*(N)$$

that is natural in the R-modules $\mathbb{F}_R X$ and N.

PROOF. The point is that naturality on general maps $g : \mathbb{F}_R X \longrightarrow \mathbb{F}_R X'$, with their induced maps $\pi_*(R) \otimes H_*(X) \longrightarrow \pi_*(R) \otimes H_*(X')$, and not just on maps of the form $g = \mathbb{F}_R f$, $f : X \longrightarrow X'$, will be essential in the cited application. The diagram

$$\mathbb{F}_R X \wedge_S R \wedge_S N \rightrightarrows \mathbb{F}_R X \wedge_S N \longrightarrow \mathbb{F}_R X \wedge_R N$$

is a split coequalizer in $\mathscr{M}_S$ and thus in $\mathscr{S}$, and it is visibly natural in both $\mathbb{F}_R X$ and N. It remains a coequalizer on applying π_*, and the required naturality follows. □

Finally, we record an analogue of the behavior of tensor products of modules with respect to tensor products of algebras.

PROPOSITION 3.10. *Let R and R' be S-algebras, M and N be right and left R-modules, and M' and N' be right and left R'-modules. Then there is a natural isomorphism of S-modules*

$$(M \wedge_S M') \wedge_{R \wedge_S R'} (N \wedge_S N') \cong (M \wedge_R N) \wedge_S (M' \wedge_{R'} N').$$

If M is a cell R-module and N' is a cell R'-module, then $M \wedge_S N'$ is a cell $R \wedge_S R'$-module.

PROOF. The first statement is a comparison of coequalizer diagrams. The second statement holds since, on spheres, I.6.1 implies isomorphisms

$$(\mathbb{L}S^q \wedge_{\mathscr{L}} R) \wedge_S (R' \wedge_{\mathscr{L}} \mathbb{L}S^r) \cong (R \wedge_S R') \wedge_{\mathscr{L}} \mathbb{L}S^{q+r}. \ \square$$

4. Change of S-algebras; q-cofibrant S-algebras

In this section, we assume given a map of S-algebras

$$\phi : R \longrightarrow R',$$

and we study the relationship between the categories of R-modules and of R'-modules. By pullback along ϕ, we obtain a functor $\phi^* : \mathscr{M}_{R'} \longrightarrow \mathscr{M}_R$. It preserves weak equivalences and thus induces a functor $\phi^* : \mathscr{D}_{R'} \longrightarrow \mathscr{D}_R$. It is vital to the theory that this functor is an equivalence of categories when ϕ is a weak equivalence. As we explain, this allows us to replace general S-algebras by better behaved "q-cofibrant" ones whenever convenient, without changing the derived category.

Regard R' as a right R-module via the composite

$$R' \wedge_S R \xrightarrow{\mathrm{id} \wedge \phi} R' \wedge_S R' \xrightarrow{\mu'} R'.$$

Observe that R' is an (R', R)-bimodule with the evident left action by R' and that, for an R-module M, $R' \wedge_R M$ is an R'-module.

PROPOSITION 4.1. *Define $\phi_* : \mathscr{M}_R \longrightarrow \mathscr{M}_{R'}$ by $\phi_* M = R' \wedge_R M$. Then ϕ_* is left adjoint to ϕ^*, and the adjunction induces a derived adjunction*

$$\mathscr{D}_{R'}(\phi_* M, M') \cong \mathscr{D}_R(M, \phi^* M').$$

Moreover, the functor ϕ_* preserves cell modules.

PROOF. The required isomorphism

$$\mathscr{M}_{R'}(\phi_*M, M') \cong \mathscr{M}_R(M, \phi^*M')$$

is proven exactly as in algebra. It sends an R-map $M \longrightarrow M'$ to the induced composite

$$R' \wedge_R M \longrightarrow R' \wedge_R M' \longrightarrow R' \wedge_{R'} M' \cong M',$$

and it sends an R'-map $R' \wedge_R M \longrightarrow M'$ to its restriction along the canonical map $M \longrightarrow R' \wedge_R M$. Since the functor ϕ^* preserves weak equivalences, it is formal that the functor ϕ_* carries R-modules of the homotopy types of cell modules to R'-modules of the homotopy types of cell modules and induces an adjunction on derived categories [38, I.5.13]. Clearly

$$R' \wedge_R (R \wedge_S L) \cong R' \wedge_S L$$

for an S-module L. Therefore the functor ϕ_* carries sphere R-modules to sphere R'-modules. Since, as a left adjoint, ϕ_* preserves colimits, this implies that ϕ_* preserves cell modules and not just homotopy types of cell modules. □

THEOREM 4.2. *Let $\phi : R \longrightarrow R'$ be a weak equivalence of S-algebras. Then $\phi_* : \mathscr{D}_R \longrightarrow \mathscr{D}_{R'}$ and $\phi^* : \mathscr{D}_{R'} \longrightarrow \mathscr{D}_R$ are inverse adjoint equivalences of categories.*

PROOF. If M is a cell R-module, then the unit

$$\phi \wedge_R \mathrm{id} : M \cong R \wedge_R M \longrightarrow R' \wedge_R M$$

of the adjunction is a weak equivalence by Theorem 3.8. Now let M' be an R'-module. In the derived category, the composite $\phi_*\phi^*M'$ means $R' \wedge_R \Gamma M'$, where $\Gamma M'$ is a cell R-module for which there is a weak equivalence of R-modules $\gamma : \Gamma M' \longrightarrow \phi^*M'$. The counit of the adjunction is given by

$$\mathrm{id} \wedge_\phi \gamma : R' \wedge_R \Gamma M' \longrightarrow R' \wedge_{R'} M' \cong M'.$$

An easy diagram chase shows that the composite map of R-modules

$$\Gamma M' \cong R \wedge_R \Gamma M' \xrightarrow{\phi \wedge_R \mathrm{id}} R' \wedge_R \Gamma M' \xrightarrow{\mathrm{id} \wedge_\phi \gamma} R' \wedge_{R'} M' \cong M'$$

coincides with γ. Since $\phi \wedge_R \mathrm{id}$ is a weak equivalence, so is $\mathrm{id} \wedge_\phi \gamma$. □

We shall give the category of S-algebras a Quillen (closed) model category structure in Chapter VII. We will then have the notion of a "q-cofibrant S-algebra", which is a retract of a "cell S-algebra". For any S-algebra R, there is a weak equivalence $\lambda : \Lambda R \longrightarrow R$, where ΛR is a cell S-algebra. By the previous result, λ induces an adjoint equivalence between the categories $\mathscr{D}_R$ and $\mathscr{D}_{\Lambda R}$. Actually, we will have two quite different model categories, one for S-algebras and another for commutative S-algebras. The comments that we have just made apply in either context. As we shall explain in VII§6, the forgetful functor from R-algebras to R-modules is better behaved homotopically in the non-commutative case than in the commutative case. In fact, VII.6.2 will give the following result.

THEOREM 4.3. *If R is a q-cofibrant S-algebra, then (R, S) is a retract of a relative cell S-module, the inclusion $S \longrightarrow R$ being the unit of R. Therefore (R, S) has the homotopy type of a relative CW S-module.*

Since we can approximate a commutative S-algebra by a non-commutative cell S-algebra without changing the derived category of modules (up to equivalence), we can use the previous result to obtain homotopical information about the derived categories of commutative S-algebras.

We illustrate the force of these ideas by using them to obtain a complementary adjunction to the case of Proposition 4.1 that is obtained by specializing to the unit $\eta : S \longrightarrow R$ of an S-algebra R:

$$\mathscr{D}_R(R \wedge_S M, N) \cong \mathscr{D}_S(M, \eta^* N)$$

for S-modules M and R-modules N.

PROPOSITION 4.4. *The forgetful functor $\eta^* : \mathscr{D}_R \longrightarrow \mathscr{D}_S$ has a right adjoint $\eta^\# : \mathscr{D}_S \longrightarrow \mathscr{D}_R$, so that*

$$\mathscr{D}_R(N, \eta^\# M) \cong \mathscr{D}_S(\eta^* N, M)$$

for S-modules M and R-modules N.

PROOF. On the point set level, we have the adjunction

$$\mathscr{M}_S(\eta^* N, M) \cong \mathscr{M}_R(N, F_S(R, M)).$$

Here we regard R as an (S, R)-bimodule, and the right action of R on itself induces a left action of R on $F_S(R, M)$ (as with Hom functors in algebra). However, there is no reason to believe that the functor $F_S(R, M)$ of M preserves weak equivalences, so that it is not clear how to pass to derived categories. Let $\lambda : \Lambda R \longrightarrow R$ be a weak equivalence of S-algebras, where ΛR is a cell S-algebra. It follows easily from the previous theorem that the functor

$$F_S(\Lambda R, M) : \mathscr{M}_S \longrightarrow \mathscr{M}_{\Lambda R}$$

of M does preserve weak equivalences. We therefore have an adjunction

$$\mathscr{D}_S((\eta')^* N', M) \cong \mathscr{D}_{\Lambda R}(N', F_S(\Lambda R, M))$$

for ΛR-modules N' and S-modules M, where η' is the unit of ΛR. Theorem 4.2 implies that we also have an adjunction

$$\mathscr{D}_{\Lambda R}(\lambda^* N, N') \cong \mathscr{D}_R(N, \lambda_* N').$$

Since $\eta = \lambda \circ \eta' : S \longrightarrow R$ and these forgetful functors all preserve weak equivalences, $\eta^* = (\eta')^* \circ \lambda^* : \mathscr{D}_R \longrightarrow \mathscr{D}_S$. We define $\eta^\#(M) = \lambda_* F_S(\Lambda R, M)$ and obtain the desired adjunction as the composite of the adjunctions just given. □

5. Symmetric and extended powers of R-modules

Let R be a commutative S-algebra and M be an R-module. The jth symmetric power of M is defined to be M^j/Σ_j and the jth extended power of M is defined to be

$$D_jM = (E\Sigma_j)_+ \wedge_{\Sigma_j} M^j.$$

In both notions, M^j denotes the j-th power of M with respect to $\wedge_R$. One of the most striking features of our smash product of R-modules is that, in the derived category $\mathscr{D}_R$, these are essentially equivalent notions. This fact will give us homotopical control on free commutative R-algebras and will play an important role in our study of Bousfield localizations of commutative R-algebras in Chapter VIII; it will not be needed before then.

To explain this fact, observe that I.5.6 implies that, for a spectrum K, we have an equivariant isomorphism

$$(\mathbb{L}K)^j \cong \mathscr{L}(j) \ltimes K^j,$$

where the j-th power is taken with respect to $\wedge_{\mathscr{L}}$ on the left and with respect to the external smash product on the right. Therefore

$$(\mathbb{L}K)^j/\Sigma_j \cong \mathscr{L}(j) \ltimes_{\Sigma_j} K^j.$$

This is the core of the claimed equivalence between symmetric and extended powers of R-modules. However, to retain suffficient homotopical control on our constructions to prove the equivalence, we must assume that R is a q-cofibrant commutative S-algebra and apply results to be proven in VII§6. Note that, as the initial object in the category of commutative S-algebras, S itself is q-cofibrant.

THEOREM 5.1. *Let R be a q-cofibrant commutative S-algebra. If M is a cell R-module, then the projection*

$$\pi : (E\Sigma_j)_+ \wedge_{\Sigma_j} M^j \longrightarrow M^j/\Sigma_j$$

is a homotopy equivalence of spectra.

PROOF. The conclusion is trivial for $j = 1$ and we may assume inductively that it holds for $i < j$.

We first prove the result for any j when M is the free R-module generated by a CW-spectrum X. Expanding definitions and commuting the smash product with $(E\Sigma_j)_+$ through our constructions, we find that

$$M^j \cong R \wedge_S S \wedge_{\mathscr{L}} (\mathscr{L}(j) \ltimes X^j),$$

$$(E\Sigma_j)_+ \wedge M^j \cong R \wedge_S S \wedge_{\mathscr{L}} ((E\Sigma_j \times \mathscr{L}(j)) \ltimes X^j),$$

and π is induced from the Σ_j-equivalence $E\Sigma_j \times \mathscr{L}(j) \longrightarrow \mathscr{L}(j)$ by passage to orbits. When $R = S$, π is a homotopy equivalence of spectra by I.8.5 and the equivariant version of I.2.5. For general R, VII.6.5 and VII.6.7 imply that the functor $R \wedge_S (-)$ carries this homotopy equivalence to a weak equivalence. However, the domain and target have the homotopy types of CW spectra, by VII.6.6.

Next, let M be a subcomplex of a cell R-module N and assume that the conclusion holds for M and N/M. As explained for the (external) smash power of spectra in [14, pp 37-38] and works equally well for the (internal) smash power of R-modules, we have a filtration of N^j by Σ_j-cofibrations of R-modules

$$M^j = F_j N^j \subset F_{j-1} N^j \subset \cdots \subset F_1 N^j \subset F_0 N^j = N^j.$$

Here $F_i N^j$ is the union of the subcomplexes $M_1 \wedge_R \cdots \wedge_R M_j$, where each M_k is either M or N and i of the M_k are M. The subquotients can be identified equivariantly as

$$F_i N^j / F_{i+1} N^j \cong \Sigma_j \times_{\Sigma_i \times \Sigma_{j-i}} (M^i \wedge_R (N/M)^{j-i}).$$

As a $(\Sigma_i \times \Sigma_{j-i})$-space, $E\Sigma_j$ is homotopy equivalent to $E\Sigma_i \times E\Sigma_{j-i}$, and there result homotopy equivalences

$$(E\Sigma_j)_+ \wedge_{\Sigma_j} F_i N^j / F_{i+1} N^j \simeq ((E\Sigma_i)_+ \wedge_{\Sigma_i} M^i) \wedge_R ((E\Sigma_{j-i})_+ \wedge_{\Sigma_{j-i}} (N/M)^{j-i}).$$

Applying the original induction hypothesis on j and inducting up the filtration, we deduce the conclusion for N.

Finally, turning to the general case, let $\{M_n\}$ be the sequential filtration of M, with $M_0 = *$. By the first step, the conclusion holds for each M_{n+1}/M_n. By the second step, the conclusion for M_n implies the conclusion for M_{n+1}. Since M^j is the colimit of the sequence of Σ_j-cofibrations of R-modules $(M_n)^j \longrightarrow (M_{n+1})^j$, the conclusion for M follows. □

6. Function R-modules

Let R be an S-algebra. We have a function R-module functor F_R to go with our smash product. Its definition is dictated by the expected adjunction.

DEFINITION 6.1. Let M and N be (left) R-modules. Define $F_R(M, N)$ to be the equalizer displayed in the following diagram of S-modules:

$$F_R(M,N) \longrightarrow F_S(M,N) \underset{\omega}{\overset{\mu^*}{\rightrightarrows}} F_S(R \wedge_S M, N).$$

Here $\mu^* = F_S(\mu, \mathrm{id})$ and ω is the adjoint of the composite

$$R \wedge_S (M \wedge_S F_S(M,N)) \xrightarrow{\mathrm{id} \wedge \varepsilon} R \wedge_S N \xrightarrow{\nu} N.$$

When $R = S$, our new and old function S-modules $F_S(M, N)$ are identical. We state the expected adjunction in a general form, but we are most interested in the case $R' = S$.

LEMMA 6.2. *Let M be an (R, R')-bimodule, N be an R'-module, and P be an R-module. Then*

$$\mathscr{M}_R(M \wedge_{R'} N, P) \cong \mathscr{M}_{R'}(N, F_R(M, P)).$$

PROOF. The general case follows from the case $R = S$ of Lemma 3.2 by use of the coequalizer definition of $\wedge_{R'}$ and the equalizer definition of F_R. □

As in algebra, this leads to a function module analogue of Proposition 3.4.

PROPOSITION 6.3. *Let M be an (R,R')-bimodule, N be an (R',R'')-bimodule, and P be an (R,R''')-bimodule. Then $F_R(M,P)$ is an (R',R''')-bimodule, and*

$$F_R(M \wedge_{R'} N, P) \cong F_{R'}(N, F_R(M,P))$$

as (R'',R''')-bimodules.

Similarly, the unit isomorphism of Lemma 3.5 implies a counit isomorphism.

LEMMA 6.4. *The adjoint $\tilde{\lambda} : M \longrightarrow F_R(R,M)$ is an isomorphism.*

We also have analogs of Proposition 3.6 and Corollary 3.7. While we are interested primarily in the versions relating F_R to the functor $\wedge_S$, there are also versions relating F_R to the functor F_S. The following lemma is needed for the latter versions. Its algebraic analogue is proven by a formal argument that applies equally well in topology.

LEMMA 6.5. *Let R and R' be S-algebras.*

(i) *Let M be an R-module, M' be an R'-module and P be an $R \wedge_S R'$-module. Then there is a natural bijection*

$$\mathscr{M}_R(M, F_{R'}(M',P)) \cong \mathscr{M}_{R\wedge_S R'}(M \wedge_S M', P).$$

(ii) *Let M be a left R-module, N be a right R-module, and K be an S-module. Then there is a natural bijection*

$$\mathscr{M}_R(M, F_S(N,K)) \cong \mathscr{M}_S(N \wedge_R M, K).$$

PROOF. It suffices to check (i) when $M = R \wedge_S L$ and $M' = R' \wedge_S L'$ are the free modules generated by S-modules L and L'. Similarly, it suffices to check (ii) when $M = R \wedge_S L$. These cases are easy consequences of our adjunctions. □

PROPOSITION 6.6. *Let K be an S-module and M be a left R-module. There is a natural isomorphism of left R-modules*

$$F_R(K \wedge_S R, M) \cong F_S(K,M)$$

and a natural isomorphism of right R-modules

$$F_R(M, F_S(R,K)) \cong F_S(M,K).$$

PROOF. The first isomorphism is immediate from the following chain of isomorphisms of represented functors on left R-modules N, which result from Proposition 6.3, Proposition 3.6, and Theorem 1.1(iii), respectively.

$$\begin{aligned}\mathscr{M}_R(N, F_R(K \wedge_S R, M)) &\cong \mathscr{M}_R((K \wedge_S R) \wedge_R N, M)\\ &\cong \mathscr{M}_R(K \wedge_S N, M)\\ &\cong \mathscr{M}_R(N, F_S(K,M)).\end{aligned}$$

The second isomorphism results from the following chain of isomorphisms of represented functors on right R-modules N:

$$\begin{aligned}\mathscr{M}_{R^{op}}(N, F_R(M, F_S(R,K))) &\cong \mathscr{M}_{R\wedge_S R^{op}}(M \wedge_S N, F_S(R,K)) \\ &\cong \mathscr{M}_S(R \wedge_{R\wedge_S R^{op}} (M \wedge_S N), K) \\ &\cong \mathscr{M}_S(M \wedge_{R^{op}} N, K) \cong \mathscr{M}_{R^{op}}(N, F_S(M,K)).\end{aligned}$$

The first two isomorphisms are instances of isomorphisms of the lemma. The third follows from the fact that there is a natural isomorphism

$$R \wedge_{R\wedge_S R^{op}} (M \wedge_S N) \cong M \wedge_{R^{op}} N,$$

as is easily checked when M and N are free R-modules and follows in general. □

COROLLARY 6.7. *Let X be a spectrum and M be an R-module. There is a natural isomorphism of left R-modules*

$$F_R(\mathbb{F}_R X, M) \cong F_S(\mathbb{F}_S X, M)$$

and a natural isomorphism of right R-modules

$$F_R(M, \mathbb{F}_R^{\#} X) \cong F_S(M, \mathbb{F}_S^{\#} X)).$$

The functor $F_R(M,N)$ converts colimits and cofiber sequences in M to limits and fiber sequences and it preserves limits and fiber sequences in N, as we see formally on the spectrum level (compare [38, III.2.5]) and deduce in order on the levels of $\mathbb{L}$-spectra, S-modules, and R-modules (compare II.1.5 and Theorem 1.1). Using the previous corollary to deal with sphere R-modules and proceeding by induction up the sequential filtration of M, we obtain the analogue of Theorem 3.8.

THEOREM 6.8. *If M is a cell R-module and $\phi : N \longrightarrow N'$ is a weak equivalence of R-modules, then*

$$F_R(\mathrm{id}, \phi) : F_R(M, N) \longrightarrow F_R(M, N')$$

is a weak equivalence.

In the derived category $\mathscr{D}_R$, $F_R(M,N)$ means $F_R(\Gamma M, N)$, where ΓM is a cell approximation of M. We are entitled to conclude that

$$\mathscr{D}_R(M \wedge_S N, P) \cong \mathscr{D}_S(N, F_R(M, P)).$$

As in Proposition 3.9, we have the following calculational sharpening of Corollary 6.7. It will be the starting point for our later construction of a spectral sequence for the calculation of $\pi_*(F_R(M,N))$.

COROLLARY 6.9. *Let X be a wedge of sphere spectra and N be an R-module. Then there is an isomorphism*

$$\pi_*(F_R(\mathbb{F}X, N)) \cong \mathrm{Hom}_{\pi_*(R)}(\pi_*(R) \otimes H_*(X), \pi_*(N))$$

that is natural in the R-modules $\mathbb{F}X$ and N.

We end this section by recording a composition pairing that is a formal implication of Lemma 6.2 and Proposition 6.3. This works exactly as with tensor products and Hom in algebra and, as there, it is convenient for this purpose to use the commutativity of the smash product over S to rewrite our adjunctions and isomorphisms with their variables occurring in the same order on both sides, returning to our original conventions of I§7. Thus, for an S-module L and for R-modules M and N, we have the natural isomorphism of S-modules

$$F_R(L \wedge_S M, N) \cong F_S(L, F_R(M, N)). \tag{6.10}$$

Let P be another R-module. Using the evaluation R-map

$$\varepsilon : F_R(M, N) \wedge_S M \longrightarrow N,$$

we obtain a composite R-map

$$F_R(N, P) \wedge_S F_R(M, N) \wedge_S M \xrightarrow{\mathrm{id} \wedge_S \varepsilon} F_R(N, P) \wedge_S N \xrightarrow{\varepsilon} P.$$

Its adjoint is a composition pairing of S-modules

$$\pi : F_R(N, P) \wedge_S F_R(M, N) \longrightarrow F_R(M, P). \tag{6.11}$$

This pairing is unital and associative in the sense that the following diagrams commute; let $\eta : S \longrightarrow F_R(M, M)$ be the adjoint of $\lambda : S \wedge_S M \longrightarrow M$:

$$\begin{array}{ccc} F_R(N, P) \wedge_S S & & \\ {\scriptstyle \mathrm{id} \wedge_S \eta} \downarrow & \searrow {\scriptstyle \lambda\tau} & \\ F_R(N, P) \wedge_S F_R(N, N) & \xrightarrow[\pi]{} & F_R(N, P), \end{array}$$

$$\begin{array}{ccc} S \wedge_S F_R(M, N) & & \\ {\scriptstyle \eta \wedge_S \mathrm{id}} \downarrow & \searrow {\scriptstyle \lambda} & \\ F_R(N, N) \wedge_S F_R(M, N) & \xrightarrow[\pi]{} & F_R(M, N), \end{array}$$

and, for another R-module L,

$$\begin{array}{ccc} F_R(N, P) \wedge_S F_R(M, N) \wedge_S F_R(L, M) & \xrightarrow{\mathrm{id} \wedge_S \pi} & F_R(N, P) \wedge_S F_R(L, N) \\ {\scriptstyle \pi \wedge_S \mathrm{id}} \downarrow & & \downarrow {\scriptstyle \pi} \\ F_R(M, P) \wedge_S F_R(L, M) & \xrightarrow[\pi]{} & F_R(L, P). \end{array}$$

This leads to a host of examples of S-algebras and their modules.

PROPOSITION 6.12. *Let R be an S-algebra and let M and N be (left) R-modules. Then $F_R(N, N)$ is an S-algebra with product π and unit η. Moreover, $F_R(M, N)$ is an $(F_R(N, N), F_R(M, M))$-bimodule with left and right actions given by π.*

7. Commutative S-algebras and duality theory

We assume that R is a commutative S-algebra in this section, and we show that the study of modules over R works in exactly the same way as the study of modules over commutative algebras. If $\mu : R \wedge_S M \longrightarrow M$ gives M a left R-module structure, then $\mu \circ \tau : M \wedge_S R \longrightarrow M$ gives M a right R-module structure such that M is an (R, R)-bimodule. As in the study of modules over commutative algebras, this leads to the following important conclusion.

THEOREM 7.1. *If M and N are R-modules, then $M \wedge_R N$ and $F_R(M, N)$ have canonical R-module structures induced from the R-module structure of M or, equivalently, N. The smash product over R is commutative, associative and unital. There is an adjunction*

$$\mathscr{M}_R(L \wedge_R M, N) \cong \mathscr{M}_R(L, F_R(M, N)). \tag{7.2}$$

Moreover, the adjunction passes to derived categories.

We have the following consequence of Corollary 3.7.

PROPOSITION 7.3. *If M and M' are cell R-modules, then $M \wedge_R M'$ is a cell R-module with one $(p+q)$-cell for each p-cell of M and q-cell of M'.*

For R-modules L, M and N, we have a natural isomorphism of R-modules

$$F_R(L \wedge_R M, N) \cong F_R(L, F_R(M, N)) \tag{7.4}$$

because both sides represent the same functor. Exactly as in the previous section, but working entirely with R-modules, we obtain a natural associative and R-unital composition pairing

$$\pi : F_R(M, N) \wedge_R F_R(L, M) \longrightarrow F_R(L, N). \tag{7.5}$$

The formal duality theory explained in [38, Ch. III] applies to the stable category of R-modules. Define the dual of M to be $D_R M = F_R(M, R)$. We have an evaluation map $\varepsilon : D_R M \wedge_R M \longrightarrow R$ and a map $\eta : R \longrightarrow F_R(M, M)$, namely the adjoint of $\lambda : R \wedge_R M \longrightarrow M$. There is also a natural map

$$\nu : F_R(L, M) \wedge_R N \longrightarrow F_R(L, M \wedge_R N). \tag{7.6}$$

By composition with the isomorphism $F_R(\mathrm{id}, \lambda)$, ν specializes to a map

$$\nu : D_R M \wedge_R M \longrightarrow F_R(M, M). \tag{7.7}$$

We say that M is "strongly dualizable", if it has a coevaluation map $\bar{\eta} : R \longrightarrow M \wedge_R D_R M$ such that the following diagram commutes in $\mathscr{D}_R$:

$$\begin{array}{ccc} R & \xrightarrow{\bar{\eta}} & M \wedge_R D_R M \\ {\scriptstyle \eta}\downarrow & & \downarrow{\scriptstyle \tau} \\ F_R(M, M) & \xleftarrow[\nu]{} & D_R M \wedge_R M. \end{array} \tag{7.8}$$

The definition has many purely formal implications. The map ν of (7.6) is an isomorphism in $\mathscr{D}_R$ if either L or N is strongly dualizable. The map ν of

(7.7) is an isomorphism in $\mathscr{D}_R$ if and only if M is strongly dualizable, and the coevaluation map $\bar{\eta}$ is then the composite $\tau\nu^{-1}\eta$ in (7.8). The natural map

$$\rho : M \longrightarrow D_R D_R M$$

is an isomorphism in $\mathscr{D}_R$ if M is strongly dualizable. The natural map

$$\wedge : F_R(M, N) \wedge_R F_R(M', N') \longrightarrow F_R(M \wedge_R M', N \wedge_R N')$$

is an isomorphism in $\mathscr{D}_R$ if M and M' are strongly dualizable or if M is strongly dualizable and $N = R$.

Say that a cell R-module N is a wedge summand up to homotopy of a cell R-module M if there is a homotopy equivalence of R-modules between M and $N \vee N'$ for some cell R-module N'. In contrast with the usual stable homotopy category, if M is finite it does not follow that N must have the homotopy type of a finite cell R-module. Via Eilenberg-Mac Lane spectra, finitely generated projective modules that are not free give rise to explicit counterexamples. Define a semi-finite R-module to be an R-module that is a wedge summand up to homotopy of a finite cell R-module, and note for use in Chapter VI that this notion makes sense even when R is not commutative.

THEOREM 7.9. *A cell R-module is strongly dualizable if and only if it is semi-finite.*

PROOF. Observe first that S_R^q is strongly dualizable with dual S_R^{-q}, hence any finite wedge of sphere R-modules is strongly dualizable. Observe next that the cofiber of a map between strongly dualizable R-modules is strongly dualizable. In fact, the evaluation map ε induces a natural map

$$\varepsilon_{\#} : \mathscr{D}_R(L, N \wedge_R D_R M) \longrightarrow \mathscr{D}_R(L \wedge_R M, N),$$

and M is strongly dualizable if and only if $\varepsilon_{\#}$ is an isomorphism for all L and N [38, III.1.6]. Since both sides convert cofiber sequences in the variable M into long exact sequences, the five lemma gives the observation. We conclude by induction on the number of cells that a finite cell R-module is strongly dualizable. It is formal that a wedge summand in $\mathscr{D}_R$ of a strongly dualizable cell R-module is strongly dualizable. For the converse, let N be a cell R-module that is strongly dualizable with coevaluation map $\bar{\eta} : R \longrightarrow N \wedge_R D_R N$. Since $\bar{\eta}$ is determined by its restriction to S and S is compact, $\bar{\eta}$ factors through $M \wedge_R D_R N$ for some finite cell subcomplex M of N. By [38, III.1.2], the bottom composite in the following commutative diagram is the identity (in $\mathscr{D}_R$):

$$\begin{array}{ccccccc}
 & & M \wedge_R D_R N \wedge_R N & \xrightarrow{\mathrm{id} \wedge \varepsilon} & M \wedge_R R & \xrightarrow{\cong} & M \\
 & \nearrow & \downarrow & & \downarrow & & \downarrow \\
N \cong R \wedge_R N & \xrightarrow[\bar{\eta} \wedge \mathrm{id}]{} & N \wedge_R D_R N \wedge_R N & \xrightarrow[\mathrm{id} \wedge \varepsilon]{} & N \wedge_R R & \xrightarrow[\cong]{} & N.
\end{array}$$

Therefore N is a retract up to homotopy and thus, by a comparison of exact triangles, a wedge summand up to homotopy of M: retractions split in triangulated categories. $\square$

CHAPTER IV

The algebraic theory of R-modules

We define generalized Tor and Ext groups as the homotopy groups of derived smash product and function modules, and we interpret these groups in terms of generalized homology and cohomology theories on R-modules. Specializing to Eilenberg-Mac Lane spectra, these groups give the classical Tor and Ext groups, and we show how to topologically realize classical algebraic derived categories of complexes of modules over a ring. Starting with a connective S-algebra R, rather than an Eilenberg-Mac Lane spectrum, the discussion generalizes to give ordinary homology and cohomology theories on R-modules, together with Atiyah-Hirzebruch spectral sequences for the computation of generalized homology and cohomology theories on R-modules.

In Sections 4 and 5, we construct "hyperhomology" spectral sequences for the calculation of our generalized Tor and Ext groups in terms of ordinary Tor and Ext groups, and we show that these specialize to give universal coefficient and Künneth spectral sequences for homology and cohomology theories defined on spectra. In Sections 6 and 7, we generalize to Eilenberg-Moore spectral sequences for the computation of $E_*(M \wedge_R N)$ under varying hypotheses on R and E. In particular, we give a bar construction approximation to $M \wedge_R N$ that allows us to view the classical space level Eilenberg-Moore-Rothenberg-Steenrod spectral sequence as a special case.

Except that his theory was intrinsically restricted to the A_∞ context, Robinson's series of papers [59, 60, 61, 62, 63] gave earlier versions of many of the results of this chapter. Of course, with the earlier technology, the proofs were substantially more difficult.

As usual, for a spectrum E, we shall often abbreviate notations by setting

$$E_n = \pi_n(E) = E^{-n}.$$

1. Tor and Ext; homology and cohomology; duality

DEFINITION 1.1. Let R be an S-algebra. For a right R-module M and a left R-module N, define

$$\mathrm{Tor}_n^R(M, N) = \pi_n(M \wedge_R N).$$

For left R-modules M and N, define

$$\mathrm{Ext}^n_R(M,N) = \pi_{-n}(F_R(M,N)).$$

Here the smash product and function modules are understood to be taken in the derived category $\mathscr{D}_R$. For Tor, this means that M or N must be replaced by a weakly equivalent cell R-module before applying the module level functor $\wedge_R$. For Ext, this means that M must be approximated by a cell R-module before applying F_R. At this point in our work, however, we act as traditional topologists, taking it for granted that all spectra and modules are to be approximated as cell modules, without change of notation, whenever necessary. We will point out explicitly any places where this gives rise to mathematical issues.

Clearly $\mathrm{Tor}^R_*(M,N)$ and $\mathrm{Ext}^*_R(M,N)$ are R_*-modules when R is commutative. Various properties reminiscent of those of the classical Tor and Ext functors follow directly from the definition and the results of the previous chapters. The intuition is that the definition gives an analogue of the differential Tor and Ext functors (alias hyperhomology and cohomology functors) in the context of differential graded modules over differential graded algebras. In particular, the grading should not be thought of as the resolution grading of the classical torsion product, but rather as a total grading that sums a resolution degree and an internal degree; this idea will be made precise by the grading of the spectral sequences that we shall construct for the calculation of these functors.

PROPOSITION 1.2. *$\mathrm{Tor}^R_*(M,N)$ satisfies the following properties.*

(i) *If R, M, and N are connective, then $\mathrm{Tor}^R_n(M,N) = 0$ for $n < 0$.*

(ii) *A cofiber sequence $N' \longrightarrow N \longrightarrow N''$ gives rise to a long exact sequence*

$$\cdots \longrightarrow \mathrm{Tor}^R_n(M,N') \longrightarrow \mathrm{Tor}^R_n(M,N) \longrightarrow \mathrm{Tor}^R_n(M,N'') \longrightarrow \mathrm{Tor}^R_{n-1}(M,N') \longrightarrow \cdots.$$

(iii) $\mathrm{Tor}^R_*(M,R) \cong \pi_*(M)$ *and, for a spectrum X,*

$$\mathrm{Tor}^R_*(M,\mathbb{F}X) \cong \pi_*(M \wedge X).$$

(iv) *The functor $\mathrm{Tor}^R_*(M,-)$ carries wedges to direct sums.*

PROOF. In (i), M and N can be approximated by CW R-modules with cells of non-negative dimension, hence it suffices to check the conclusion for $N = S^r_R$, $r \geq 0$, in which case it is immediate from (iii). Part (iii) follows from III.1.4 and III.3.7. □

The commutativity and associativity relations for the smash product imply various further properties. We content ourselves with the following specialization.

PROPOSITION 1.3. *If R is commutative, then*

$$\mathrm{Tor}^R_*(M,N) \cong \mathrm{Tor}^R_*(N,M)$$

and

$$\mathrm{Tor}^R_*(M \wedge_R N, P) \cong \mathrm{Tor}^R_*(M, N \wedge_R P).$$

Say that a spectrum N is coconnective if $\pi_q N = 0$ for $q > 0$.

PROPOSITION 1.4. $\operatorname{Ext}^*_R(M, N)$ *satisfies the following properties.*

(i) *If* R *and* M *are connective and* N *is coconnective, then* $\operatorname{Ext}^n_R(M, N) = 0$ *for* $n < 0$.

(ii) *Fiber sequences* $N' \longrightarrow N \longrightarrow N''$ *give rise to long exact sequences*

$$\cdots \longrightarrow \operatorname{Ext}^n_R(M, N') \longrightarrow \operatorname{Ext}^n_R(M, N) \longrightarrow \operatorname{Ext}^n_R(M, N'') \longrightarrow \operatorname{Ext}^{n+1}_R(M, N') \longrightarrow \cdots;$$

cofiber sequences $M' \longrightarrow M \longrightarrow M''$ *give rise to long exact sequences*

$$\cdots \longrightarrow \operatorname{Ext}^n_R(M'', N) \longrightarrow \operatorname{Ext}^n_R(M, N) \longrightarrow \operatorname{Ext}^n_R(M', N) \longrightarrow \operatorname{Ext}^{n+1}_R(M'', N) \longrightarrow \cdots.$$

(iii) $\operatorname{Ext}^*_R(R, N) \cong \pi_{-*}(N)$ *and, for a spectrum* X,

$$\operatorname{Ext}^*_R(\mathbb{F}X, N) \cong \pi_{-*}(F(X, N))$$

and

$$\operatorname{Ext}^*_R(M, \mathbb{F}^\# X) \cong \pi_{-*}(F(M, X)).$$

(iv) *The functor* $\operatorname{Ext}^*_R(-, N)$ *carries wedges to products and the functor* $\operatorname{Ext}^*_R(M, -)$ *carries products to products.*

PROOF. It suffices to check (i) for $M = S^r_R$, $r \geq 0$, in which case the conclusion is immediate from (iii). Part (iii) follows from III.1.4, III.1.10, and III.6.7. □

Passing to homotopy groups from the pairings of III.6.11 and III.7.5, we obtain the following further property.

PROPOSITION 1.5. *There is a natural, associative, and unital system of pairings*

$$\pi : \operatorname{Ext}^*_R(M, N) \otimes_{\pi^*(S)} \operatorname{Ext}^*_R(L, M) \longrightarrow \operatorname{Ext}^*_R(L, N).$$

If R *is commutative, then these are pairings of* R_**-modules, and the tensor product may be taken over* R_*.

PROOF. The first statement is clear. The second uses the the fact that $\pi_q(M) = \mathscr{D}_R(S^q_R, M)$, together with the equivalences of R-modules

$$S^q_R \wedge_R S^r_R \cong S^{q+r}_R$$

given by III.3.7; the system becomes associative and unital on passage to $\mathscr{D}_R$. □

The formal duality theory of III§7 implies the following result, together with various other such isomorphisms.

PROPOSITION 1.6. *Let* R *be commutative. For a finite cell* R*-module* M *and any* R*-module* N,

$$\operatorname{Tor}^R_n(D_R M, N) \cong \operatorname{Ext}^{-n}_R(M, N).$$

We think of the derived category $\mathscr{D}_R$ as a stable homotopy category. Changing notations, we may reinterpret the functors Tor and Ext as prescribing homology and cohomology theories in this category.

DEFINITION 1.7. Let E' be a right and E a left R-module. For left R-modules M and N, define

$$E'^R_n(M) = \pi_n(E' \wedge_R M) \quad \text{and} \quad E^n_R(M) = \pi_{-n}(F_R(M, E)).$$

The properties of Tor and Ext translate directly to statements about homology and cohomology. All of the standard homotopical machinery is available to us, and the previous result now takes the form of Spanier-Whitehead duality.

COROLLARY 1.8. *Let R be commutative. For a finite cell R-module M and any R-module E,*

$$E^R_n(D_R M) \cong E^{-n}_R(M).$$

Since the equivalence between the classical stable homotopy category and the derived category of S-modules preserves smash products and function spectra, we obtain versions of all of the usual homology and cohomology theories on spectra by taking $R = S$. Moreover the following reinterpretation of Propositions 1.2(iii) and 1.4(iii) shows that the specializations to R-modules of all of the usual homology and cohomology theories on spectra are given by instances of our new homology and cohomology theories on R-modules.

COROLLARY 1.9. *For a spectrum E and a (left) R-module M,*

$$E_*(M) \cong (\mathbb{F}E)^R_*(M) \quad \text{and} \quad E^*(M) \cong (\mathbb{F}^\# E)^*_R(M).$$

2. Eilenberg-Mac Lane spectra and derived categories

In this section, we change notation and let R denote a discrete ring. Applying multiplicative infinite loop space theory [50] to obtain an A_∞ ring spectrum and then applying the functor $S \wedge_{\mathscr{L}} (-)$, we obtain an Eilenberg-Mac Lane spectrum $HR = K(R, 0)$ that is an S-algebra and is a commutative S-algebra if R is commutative. An elaboration of multiplicative infinite loop space theory, followed by application of the functor $S \wedge_{\mathscr{L}} (-)$, can be used to realize passage to Eilenberg-Mac Lane spectra as a point-set level functor H from R-modules in the sense of algebra to HR-modules. We shall shortly use the present theory to give two different homotopical constructions of such Eilenberg-Mac Lane HR-modules. Granting this for the moment, we have the following result.

THEOREM 2.1. *For a ring R and R-modules M and N,*

$$\mathrm{Tor}^R_*(M, N) \cong \mathrm{Tor}^{HR}_*(HM, HN)$$

and

$$\mathrm{Ext}^*_R(M, N) \cong \mathrm{Ext}^*_{HR}(HM, HN).$$

If R is commutative, then these are isomorphisms of R-modules. Under the second isomorphism, the topologically defined pairing

$$\mathrm{Ext}^*_{HR}(HM, HN) \otimes \mathrm{Ext}^*_{HR}(HL, HM) \longrightarrow \mathrm{Ext}^*_{HR}(HL, HN)$$

coincides with the algebraic Yoneda product.

PROOF. If $0 \longrightarrow N' \longrightarrow N \longrightarrow N'' \longrightarrow 0$ is a short exact sequence of R-modules, then $HN' \longrightarrow HN \longrightarrow HN''$ is equivalent to a cofiber sequence. The conclusion is now immediate from Propositions 1.2 and 1.4, together with the axioms for algebraic Tor and Ext functors. It should be noted that right exactness and proper behavior on free modules together imply algebraically that

$$\mathrm{Tor}_0^R(M,N) \cong M \otimes_R N \quad \text{and} \quad \mathrm{Ext}_R^0(M,N) \cong \mathrm{Hom}_R(M,N).$$

It is important to remember that the axioms for Ext require verifications about free or injective modules, but not both. The last statement follows from Yoneda's axiomatization [73], which only requires proper behavior in degree zero and proper behavior relating connecting homomorphisms to products. The last follows topologically from commutation with cofiber sequences, which is easily derived from the adjoint construction of our pairings in III§6. □

We can elaborate this result to an equivalence of derived categories. We shall restrict attention to morphisms of degree zero since the extension to graded morphisms is formal. Recall from [69] or [35, Ch.III] that the derived category $\mathscr{D}_R$ is obtained from the homotopy category of chain complexes over R by localizing at the quasi-isomorphisms, exactly as we obtained the category $\mathscr{D}_{HR}$ from the homotopy category of HR-modules by localizing at the weak equivalences. The algebraic theory of cell and CW chain complexes over R in [35, Ch.III] makes the analogy especially close. The proof of the equivalence is quite simple. The category $\mathscr{D}_{HR}$ is equivalent to the homotopy category of CW HR-modules and cellular maps. We will see that CW HR-modules have associated chain complexes. This gives a functor $\mathscr{D}_{HR} \longrightarrow \mathscr{D}_R$, and we will obtain an inverse functor directly from Brown's representability theorem.

DEFINITION 2.2. Let M be a CW HR-module. Define the associated chain complex $C_*(M)$ of R-modules by letting $C_n(M) = \pi_n(M^n, M^{n-1})$ and letting the differential $d_n : C_n(M) \longrightarrow C_{n-1}(M)$ be the connecting homorphism of the triple (M^n, M^{n-1}, M^{n-2}). Observe that a cellular map of HR-modules induces a map of chain complexes and that a cellular homotopy induces a chain homotopy. Observe too that, since M^n/M^{n-1} is a wedge of free modules $S^n_{HR} \simeq HR \wedge S^n$, $C_n(M)$ is a free R-module.

LEMMA 2.3. *For CW HR-modules M, the homology groups $H_*(C_*(M))$ are naturally isomorphic to the homotopy groups of M.*

PROOF. Since HR is connective, the inclusion $M^n \longrightarrow M$ induces a bijection on π_q for $q < n$ and a surjection on π_n. By induction up the sequential filtration of M^{n-1}, $\pi_q(M^{n-1}) = 0$ for $q \geq n$. Therefore the quotient map $M \longrightarrow M/M^{n-1}$ induces a monomorphism on π_n. The conclusion follows by a simple diagram chase. □

THEOREM 2.4. *The cellular chain functor C_* on HR-modules induces an equivalence of categories $\mathscr{D}_{HR} \longrightarrow \mathscr{D}_R$. The inverse equivalence Φ satisfies*

$$H_*(X) \cong \pi_*(\Phi(X)).$$

PROOF. The functor C_* carries wedges to direct sums and carries homotopy colimits of cellular diagrams to chain level homotopy colimits. For a fixed chain complex X, the functor k on $\mathscr{D}_{HR}$ specified by $k(M) = \mathscr{D}_R(C_*(M), X)$ therefore satisfies the wedge and Mayer-Vietoris axioms. By Brown's representability theorem, III.2.12, k is represented by an HR-module spectrum $\Phi(X)$. By the functoriality of the representation, this gives a functor $\Phi : \mathscr{D}_R \longrightarrow \mathscr{D}_{HR}$ and an adjunction

$$\mathscr{D}_R(C_*(M), X) \cong \mathscr{D}_{HR}(M, \Phi(X)).$$

Since $H_n(X) \cong \mathscr{D}_R(\Sigma^n R, X)$, where $\Sigma^n R$ is the free R-module on one generator of degree n and $C_*(S^n_{HR}) = \Sigma^n R$, this implies that $H_*(X) \cong \pi_*(\Phi(X))$. We claim that the unit $\eta : M \longrightarrow \Phi(C_*(M))$ and counit $\varepsilon : C_*(\Phi(X)) \longrightarrow X$ of the adjunction are natural isomorphisms. On hom sets, the functor C_* coincides with

$$\eta_* : \mathscr{D}_{HR}(L, M) \longrightarrow \mathscr{D}_{HR}(L, \Phi(C_*(M))) \cong \mathscr{D}_R(C_*(L), C_*(M)).$$

As L runs through the S^n_{HR}, η_* runs through the isomorphisms

$$\pi_n(M) \longrightarrow H_n(C_*(M))$$

of the previous lemma. Therefore η is an isomorphism in $\mathscr{D}_{HR}$ for all M. Since the composite

$$\Phi X \xrightarrow{\eta} \Phi C_*(\Phi X) \xrightarrow{\Phi\varepsilon} \Phi X$$

is the identity, it follows that $\Phi\varepsilon$ is an isomorphism in $\mathscr{D}_{HR}$ for all X. The following natural diagram commutes:

$$\begin{array}{ccc} \mathscr{D}_{HR}(L, \Phi C_*(\Phi X)) & \cong & \mathscr{D}_R(C_*(L), C_*(\Phi X)) \\ \downarrow{\scriptstyle (\Phi\varepsilon)_*} & & \downarrow{\scriptstyle \varepsilon_*} \\ \mathscr{D}_{HR}(L, \Phi X) & \cong & \mathscr{D}_R(C_*(L), X). \end{array}$$

As L runs through the sphere modules S^n_{HR}, the resulting isomorphisms ε_* show that ε induces an isomorphism on all homology groups and is therefore an isomorphism in $\mathscr{D}_R$. □

In the commutative case, we have the following important addendum to the theorem. See [35, III] for a discussion of tensor product and Hom functors in the derived category $\mathscr{D}_R$. As in topology, they are constructed by first applying CW approximation of R-modules and then taking the point-set level functor.

PROPOSITION 2.5. *Assume that R is commutative. If M and N are CW HR-modules, then $M \wedge_{HR} N$ is a CW HR-module such that*

$$C_*(M \wedge_{HR} N) \cong C_*(M) \otimes_R C_*(N).$$

Therefore such an isomorphism holds in the derived category $\mathscr{D}_R$ for general HR-modules M and N. Moreover, in $\mathscr{D}_R$,

$$C_*(F_{HR}(M, N)) \cong \mathrm{Hom}_R(C_*(M), C_*(N)).$$

If X and Y are chain complexes, then

$$\Phi(X \otimes_R Y) \cong \Phi X \wedge_{HR} \Phi Y$$

and

$$\Phi \operatorname{Hom}_R(X, Y) \cong F_{HR}(\Phi X, \Phi Y)$$

in $\mathscr{D}_{HR}$.

PROOF. The first statement is implied by III.7.3, and the last three derived category level isomorphisms are all formal consequences of the first. □

Regarding R-modules as chain complexes concentrated in degree zero, we see that the functor Φ restricts to a functor H that assigns an Eilenberg-Mac Lane HR-module spectrum HM to an R-module M. We give a more explicit construction.

CONSTRUCTION 2.6. (i) For an R-module M, we construct $HM = K(M, 0)$ as a CW module L with sequential filtration $\{L_n\}$ and skeletal filtration $\{L^n\}$ related by $L^{n-1} = L_n$. Choose a free resolution

$$\cdots \longrightarrow F_n \xrightarrow{d_n} F_{n-1} \longrightarrow \cdots \longrightarrow F_0 \xrightarrow{\varepsilon} M \longrightarrow 0$$

of M. Let K_0 be a wedge of 0-spheres, with one sphere for each basis element of F_0. For $n \geq 1$, let K_n be a wedge of $(n-1)$-spheres, with one sphere for each basis element of F_n. Define $L_1 = \mathbb{F}K_0$. For $n \geq 2$, L_n will have two nonvanishing homotopy groups, namely $\pi_0(L_n) = M$ and $\pi_{n-1}(L_n) = \operatorname{Im} d_n$, and the inclusion $i_n : L_n \longrightarrow L_{n+1}$ will induce an isomorphism on π_0. By freeness, we can realize d_1 by a map of HR-modules $\mathbb{F}K_1 \longrightarrow \mathbb{F}K_0$. Let L_2 be its cofiber. Then the resulting map $\mathbb{F}K_0 \longrightarrow L_2$ realizes ε on π_0 and the resulting map $L_2 \longrightarrow \Sigma\mathbb{F}K_1$ realizes the inclusion $\operatorname{Im} d_2 \subset F_1$ on π_1. Inductively, given L_n, we can realize $d_n : F_n \longrightarrow \operatorname{Im} d_n$ on the $(n-1)$st homotopy group by a map of HR-modules $\mathbb{F}K_n \longrightarrow L_n$. We let L_{n+1} be its cofiber. The claimed properties follow immediately. The union $L = \cup L_n$ is the desired CW HR-module HM.
(ii) Given a map $f : M \longrightarrow M'$ of R-modules, we construct a cellular map $Hf : HM \longrightarrow HM'$ of CW HR-modules that realizes f on π_0. Construct $L' = HM'$ as above, writing F'_n, etc. As usual, we can construct a sequence of R-maps $f_n : F_n \longrightarrow F'_n$ that give a map of resolutions. We can realize f_n on homotopy groups by an HR-map $\mathbb{F}K_n \longrightarrow \mathbb{F}K'_n$. Starting with $L_1 = \mathbb{F}K_0$ and proceeding inductively, we can use a standard cofibration sequence argument, carried out in the category of HR-modules, to construct HR-maps $L_n \longrightarrow L'_n$ such that the middle squares commute and the left and right squares commute up to homotopy in the following diagrams of HR-modules:

$$\begin{array}{ccccccc}
\mathbb{F}K_{n+1} & \longrightarrow & L_n & \longrightarrow & L_{n+1} & \longrightarrow & \Sigma\mathbb{F}K_{n+1} \\
\downarrow & & \downarrow & & \downarrow & & \downarrow \\
\mathbb{F}K'_{n+1} & \longrightarrow & L'_n & \longrightarrow & L'_{n+1} & \longrightarrow & \Sigma\mathbb{F}K'_{n+1}.
\end{array}$$

On passage to unions, we obtain the desired cellular map $Hf : HM \longrightarrow HM'$. A similar argument works to show that if we choose another map $g_* : F_* \longrightarrow F'_*$

of resolutions over f and repeat the construction, then the resulting HR-maps are homotopic.

REMARK 2.7. Since they are $H\mathbb{Z}$-module spectra, the underlying spectra of the HR-module spectra studied in this section all have the homotopy types of Eilenberg-Mac Lane spectra.

3. The Atiyah-Hirzebruch spectral sequence

We assume given a connective S-algebra R in this section, and we let $k = \pi_0(R)$. Since R is connective, its derived category $\mathscr{D}_R$ is equivalent to the homotopy category $h\mathscr{C}\mathscr{W}_R$ of CW R-modules and cellular maps. We shall see that the Eilenberg-Mac Lane spectrum Hk is an R-module that plays a role in the study of R-modules analogous to the role played by $H\mathbb{Z}$ in the category of spectra. We use this insight to construct Atiyah-Hirzebruch spectral sequences and prove a Hurewicz theorem in the category of R-modules. Although we shall not assume this, the theory is most useful when R is commutative; of course, k may well be commutative even when R is not. Remember that modules mean left modules unless otherwise specified.

PROPOSITION 3.1. *There is a map of S-algebras $\pi : R \longrightarrow Hk$ that realizes the identity homomorphism on $\pi_0(R) = k$.*

PROOF. We sketch two proofs. The first is an application of multiplicative infinite loop space theory. By [48, VII.2.4], the zeroth space R_0 of R is an A_∞ ring space. Modulo some point-set care to ensure continuity (e.g, we could replace R by a weakly equivalent "q-cofibrant" S-algebra, which is of the homotopy type of a CW spectrum by VII.6.5 and VII.6.6), we obtain a discretization map $\delta : R_0 \longrightarrow k$, and it is immediate from the definitions that it is an A_∞ ring map. By [48, VII§4], there is a functor E from A_∞ ring spaces to A_∞ ring spectra, hence there results a map of A_∞ ring spectra $ER_0 \longrightarrow Ek$. By [48, VII.3.2 and 4.3] and the connectivity of R, there is a natural weak equivalence of A_∞ ring spectra between ER_0 and R, and the homotopical properties of E immediately imply that Ek is an Hk. Now apply the functor $S \wedge_{\mathscr{L}} (-)$ to replace A_∞ ring spectra by S-algebras, and replace R by the weakly equivalent S-algebra $S \wedge_{\mathscr{L}} ER_0$.

The second proof makes more serious use of the Quillen model category structure on the category of S-algebras that we construct in VII§§4,5. Using it, we can mimic the classical space level argument of killing higher homotopy groups, successively attaching cell S-algebras to kill the higher homotopy groups of R. □

It follows that Hk is an (R, R)-bimodule. If R and therefore also Hk are commutative S-algebras, then Hk is a commutative R-algebra in the sense of VII§1 below. If j is a k-module, then the Hk-module Hj is an R-module by pullback along π. We consider the homology and cohomology theories represented by the Hj as ordinary homology and cohomology theories defined on the derived category of R-modules: they clearly satisfy the analogs of the Eilenberg–Steenrod axioms for ordinary homology and cohomology theories; here the R-module R

plays the role of a point in the dimension axiom. We agree to alter the notations of Definition 1.7 by setting

$$(3.2)\qquad H^R_*(M;j') = (Hj')^R_*(M) \quad \text{and} \quad H^*_R(M;j) = (Hj)^*_R(M)$$

for a left R-module M, a right k-module j' and a left k-module j. We have symmetric definitions with left and right reversed.

These theories can be calculated as the homology and cohomology of the cellular chain complexes of CW R-modules. In fact, the definition of the associated chain complex of a CW R-module M is formally identical to Definition 2.2.

DEFINITION 3.3. Let M be a CW R-module. Define the associated chain complex $C^R_*(M)$ of k-modules by letting $C^R_n(M) = \pi_n(M^n, M^{n-1})$ and letting the differential $d_n : C^R_n(M) \longrightarrow C^R_{n-1}(M)$ be the connecting homorphism of the triple (M^n, M^{n-1}, M^{n-2}). Observe that a cellular map of R-modules induces a map of chain complexes and that a cellular homotopy induces a chain homotopy. Observe too that, since M^n/M^{n-1} is a wedge of free modules $S^n_R \simeq R \wedge S^n$, $C^R_n(M)$ is a free k-module. For right and left k-modules j' and j, define chain and cochain complexes of abelian groups

$$C^R_*(M;j') = j' \otimes_k C^R_*(M) \quad \text{and} \quad C^*_R(M;j) = \mathrm{Hom}_k(C^R_*(M), j).$$

Clearly these chain and cochain functors induce covariant and contravariant functors from the derived category $\mathscr{D}_R$ to the derived category $\mathscr{D}_{\mathbb{Z}}$ of chain complexes over $\mathbb{Z}$, interpreted as homologically or cohomologically graded, respectively. When k is commutative, these functors take values in $\mathscr{D}_k$. We have the following analogue of Proposition 2.5.

PROPOSITION 3.4. *If R is a commutative S-algebra and M and N are CW R-modules, then $M \wedge_R N$ is a CW R-module such that*

$$C^R_*(M \wedge_R N) \cong C^R_*(M) \otimes_k C^R_*(N).$$

Therefore such an isomorphism holds in the derived category $\mathscr{D}_k$ for general R-modules M and N. Moreover, in $\mathscr{D}_k$, there is a natural map

$$\tilde{\varepsilon} : C^R_*(F_R(M,N)) \longrightarrow \mathrm{Hom}_k(C^R_*(M), C^R_*(N)),$$

and $\tilde{\varepsilon}$ is an isomorphism if M is a finite CW R-module.

PROOF. The first statement is implied by III.7.3. For the second, the evaluation map $F_R(M,N) \wedge_R M \longrightarrow N$ induces a map

$$C^R_*(F_R(M,N)) \otimes_k C^R_*(M) \cong C^R_*(F_R(M,N) \wedge_R M) \longrightarrow C_*(N)$$

in $\mathscr{D}_k$, and its adjoint is the required map $\tilde{\varepsilon}$. Clearly $\tilde{\varepsilon}$ is an isomorphism when M is a sphere R-module. It is therefore an isomorphism for all finite CW R-modules since the functors F_R and Hom_k both convert cofibration sequences in the first variable to fibration sequences. □

We cannot expect the derived chain complex functor to preserve function objects in general, as the case $R = S$ makes clear.

By checking the Eilenberg-Steenrod axioms, as in the classical case $R = S$, we reach the following conclusion. Alternatively, we could use the Atiyah-Hirzebruch spectral sequence below.

THEOREM 3.5. *For R-modules M and right and left k-modules j' and j, there are natural isomorphisms*

$$H^R_*(M; j') = H(C^R_*(M; j')) \quad \text{and} \quad H^*_R(M; j) = H(C^*_R(M; j)).$$

The map $\pi \wedge \mathrm{id} : M \cong R \wedge_R M \longrightarrow Hk \wedge_R M$ induces the Hurewicz homomorphism $h : \pi_*(M) \longrightarrow H^R_*(M; k)$, and the proof of the Hurewicz theorem is exactly the same as in the classical case.

THEOREM 3.6 (HUREWICZ). *Let M be an $(n-1)$-connected R-module. Then $H^R_i(M; k) = 0$ for $i < n$ and $h : \pi_n(M) \longrightarrow H^R_n(M; k)$ is an isomorphism.*

PROOF. We may replace M by a weakly equivalent cell R-module with no q-cells for $q < n$. Then the n-skeleton of M is a wedge of sphere R-modules S^n_R and, for $q > n$, the quotients M^q/M^{q-1} are wedges of sphere R-modules S^q_R. The proof is an easy inductive comparison of the long exact homotopy and homology exact sequences of the pairs (M^q, M^{q-1}). □

Applying a generalized homology or cohomology theory to the skeletal filtration of a CW R-module M, we obtain an exact couple and thus a spectral sequence that generalizes the chain and cochain description of the ordinary represented homology and cohomology of M.

THEOREM 3.7 (ATIYAH-HIRZEBRUCH SPECTRAL SEQUENCE). *For a homology theory E^R_* and a cohomology theory E^*_R on R-modules, there are natural spectral sequences of the form*

$$E^2_{p,q} = H^R_p(M; E^R_q) \Longrightarrow E^R_{p+q}(M)$$

and

$$E_2^{p,q} = H^p_R(M; E^q_R) \Longrightarrow E^{p+q}_R(M).$$

Convergence is as in the classical case, and we refer the reader to Boardman [7, §14] (see also [25, App B]) for discussion. If M is bounded below, then the homology spectral sequence converges strongly to $E^R_*(M)$ and the cohomology spectral sequence converges conditionally to $E^*_R(M)$. If, further, for each fixed (p, q) there are only finitely many r such that d^r is non-zero on $E^r_{p,q}$, then the cohomology spectral sequence converges strongly.

The multiplicative properties of the spectral sequences are as one would expect from Proposition 3.4.

4. Universal coefficient and Künneth spectral sequences

There are spectral sequences for the calculation of our Tor and Ext groups that are analogous to the Eilenberg-Moore (or hyperhomology) spectral sequences in differential homological algebra. Compare [19, 29, 35]. They specialize to give universal coefficient and Künneth spectral sequences in the homology and cohomology theory of spectra. We state our results in this section and give the construction in the next. Fix an S-algebra R.

THEOREM 4.1. *For right R-modules M and left R-modules N, there is a natural spectral sequence of the form*

$$E^2_{p,q} = \mathrm{Tor}^{R_*}_{p,q}(M_*, N_*) \Longrightarrow \mathrm{Tor}^R_{p+q}(M, N). \tag{4.2}$$

For left R-modules M and N, there is a natural spectral sequence of the form

$$E_2^{p,q} = \mathrm{Ext}^{p,q}_{R^*}(M^*, N^*) \Longrightarrow \mathrm{Ext}^{p+q}_R(M, N). \tag{4.3}$$

If R is commutative, then these are spectral sequences of differential R_*-modules.

The Tor spectral sequence is of standard homological type, with

$$d^r_{p,q} : E^r_{p,q} \longrightarrow E^r_{p-r,q+r-1}.$$

It lies in the right half-plane and converges strongly. The Ext spectral sequence is of standard cohomological type, with

$$d_r : E_r^{p,q} \longrightarrow E_r^{p+r,q-r+1}.$$

It lies in the right half plane and converges conditionally. We have the following addendum.

PROPOSITION 4.4. *The pairing $F_R(M,N) \wedge_S F_R(L,M) \longrightarrow F_R(L,N)$ induces a pairing of spectral sequences that coincides with the algebraic Yoneda pairing*

$$\mathrm{Ext}^{*;*}_{R^*}(M^*, N^*) \otimes_{S^*} \mathrm{Ext}^{*;*}_{R^*}(L^*, M^*) \longrightarrow \mathrm{Ext}^{*;*}_{R^*}(L^*, N^*)$$

on the E_2-level and that converges to the induced pairing of Ext groups.

The rest of the results of this section are corollaries of the results already stated. With the specializations of variables that we cite, the conclusions are immediate from the properties of our free and cofree functors and properties of Tor and Ext recorded in Section 1. Recalling Definition 1.7, we see that our spectral sequences can be viewed as universal coefficient spectral sequences for the computation of homology and cohomology theories on R-modules. Via Corollary 1.9, they specialize to give universal coefficient spectral sequences for the computation of homology and cohomology theories on spectra. Thus, setting $M = \mathbb{F}X$ in the two spectral sequences of Theorem 4.1, we obtain the following result. We have written the stars to indicate the way the grading is usually thought of in cohomology.

THEOREM 4.5 (UNIVERSAL COEFFICIENT). *For an R-module N and any spectrum X, there are spectral sequences of the form*

$$\mathrm{Tor}^{R_*}_{*,*}(R_*(X), N_*) \Longrightarrow N_*(X)$$

and

$$\mathrm{Ext}^{*,*}_{R^*}(R_{-*}(X), N^*) \Longrightarrow N^*(X).$$

Of course, replacing R and N by Eilenberg-Mac Lane spectra HR and HN for a ring R and R-module N, we obtain the classical universal coefficient theorem. Here we are thinking of the module N as defining theories acting on general spectra. By instead taking $N = \mathbb{F}E$ and $N = \mathbb{F}^{\#}E$ in the two spectral sequences of Theorem 4.1, we obtain spectral sequences that are suitable for calculating the E-homology and cohomology of M.

THEOREM 4.6. *For an R-module M and any spectrum E, there are spectral sequences of the form*

$$\mathrm{Tor}^{R_*}_{*,*}(M_*, E_*(R)) \Longrightarrow E_*(M)$$

and

$$\mathrm{Ext}^{*,*}_{R^*}(M^*, E^*(R)) \Longrightarrow E^*(M).$$

When E is also an R-module, we can take $M = E$ and so obtain spectral sequences that converge to the E-Steenrod algebra $E^*(E)$ and its dual $E_*(E)$. For example, when $R = S$ and $M = E = H\mathbb{Z}_p$, the cohomology spectral sequence is a backwards Adams spectral sequence that converges from $\mathrm{Ext}^{*,*}_{S^*}(\mathbb{Z}_p, \mathbb{Z}_p)$ to the mod p Steenrod algebra A. Such a spectral sequence was first studied in [40].

Replacing N by $\mathbb{F}Y$ and by $F_R(\mathbb{F}Y, R)$ in the two universal coefficient spectral sequences, we arrive at Künneth spectral sequences.

THEOREM 4.7 (KÜNNETH). *For any spectra X and Y, there are spectral sequences of the form*

$$\mathrm{Tor}^{R_*}_{*,*}(R_*(X), R_*(Y)) \Longrightarrow R_*(X \wedge Y)$$

and

$$\mathrm{Ext}^{*,*}_{R^*}(R_{-*}(X), R^*(Y)) \Longrightarrow R^*(X \wedge Y).$$

A reference to Adams [1] is mandatory. He was the first to observe that one can derive Künneth spectral sequences from universal coefficient spectral sequences, and he observed that, by duality, the four spectral sequences of Theorems 4.5 and 4.7 imply two more universal coefficient and two more Künneth spectral sequences. He derived spectral sequences of this sort under the hypothesis that his given ring spectrum E is the colimit of finite subspectra E_α such that $H^*(E_\alpha; E^*)$ is E^*-projective and the Atiyah-Hirzebruch spectral sequence converging from $H^*(E_\alpha; E^*)$ to $E^*(E_\alpha)$ satisfies $E_2 = E_\infty$. Of course, this is an ad hoc calculational hypothesis that requires case-by-case verification. It covers some cases that are not covered by our results, and conversely.

5. The construction of the spectral sequences

The construction is similar to the construction of Eilenberg-Mac Lane spectra at the end of Section 2. For a right R-module M, we choose a $\pi_*(R)$-free resolution

$$(5.1)\qquad \cdots \longrightarrow F_p \xrightarrow{d_p} F_{p-1} \longrightarrow \cdots \longrightarrow F_0 \xrightarrow{\varepsilon} \pi_*(M) \longrightarrow 0.$$

Let $Q_0 = \ker \varepsilon$ and $Q_p = \ker d_p$ for $p \geq 1$, so that d_p defines an epimorphism $F_p \longrightarrow Q_{p-1}$. For $p \geq 0$, let K_p be the wedge of one $(p+s)$-sphere for each basis element of F_p of degree s and let $M_0 = M$. Proceeding inductively, we can use freeness to construct cofiber sequences of right R-modules

$$(5.2)\qquad \mathbb{F}K_p \xrightarrow{k_p} M_p \xrightarrow{i_p} M_{p+1} \xrightarrow{j_{p+1}} \Sigma\mathbb{F}K_p$$

for $p \geq 0$ that satisfy the following properties:

(i) k_0 realizes ε on π_*.
(ii) $\pi_*(M_p) = \Sigma^p Q_{p-1}$ for $p \geq 1$.
(iii) k_p realizes $\Sigma^p d_p : \Sigma^p F_p \longrightarrow \Sigma^p Q_{p-1}$ on π_* for $p \geq 1$.
(iv) i_p induces the zero homomorphism on π_* for $p \geq 0$.
(v) j_{p+1} realizes the inclusion $\Sigma^{p+1} Q_p \longrightarrow \Sigma^{p+1} F_p$ on π_* for $p \geq 0$.

Observe that (iii) implies the case $p+1$ of (ii) together with (iv) and (v).

To obtain the spectral sequence for Tor, we define

$$(5.3)\qquad D^1_{p,q} = \pi_{p+q+1}(M_{p+1} \wedge_R N) \quad\text{and}\quad E^1_{p,q} = \pi_{p+q}(\mathbb{F}K_p \wedge_R N).$$

The maps displayed in (5.2) give maps

$$i \equiv (i_p)_* : D^1_{p-1,q+1} \longrightarrow D^1_{p,q}$$

$$j \equiv (j_{p+1})_* : D^1_{p,q} \longrightarrow E^1_{p,q}$$

$$k \equiv (k_p)_* : E^1_{p,q} \longrightarrow D^1_{p-1,q}.$$

These display an exact couple in standard homological form. We see from III.3.9 that $E^1_{p,q} \cong (F_p \otimes_{R_*} N_*)_q$ and that d_1 agrees under the isomorphism with $d \otimes 1$. This proves that

$$E^2_{p,q} = \mathrm{Tor}^{R_*}_{p,q}(M_*, N_*).$$

Observe that $k : E^1_{0,q} \longrightarrow D^1_{-1,q}$ can and must be interpreted as

$$\pi_q(\mathbb{F}K_0 \wedge_R N) \longrightarrow \pi_q(M \wedge_R N).$$

On passage to E^2, it induces the edge homomorphism

$$(5.4)\qquad E^2_{0,q} = M_* \otimes_{R_*} N_* \longrightarrow \pi_*(M \wedge_R N).$$

The convergence is standard, although it appears a bit differently than in most spectral sequences in current use. Write $i_{0,p}$ for both the evident composite map $M \longrightarrow M_p$ and its smash product with N. We filter $\pi_*(M \wedge_R N)$ by letting $F_p\pi_*(M \wedge_R N)$ be the kernel of

$$(i_{0,p+1})_* : \pi_*(M \wedge_R N) \longrightarrow \pi_*(M_{p+1} \wedge_R N).$$

By (iv) above, we see that the telescope $\operatorname{tel} M_p$ is trivial. Since the functor $(-) \wedge_R N$ commutes with telescopes, $\operatorname{tel}(M_p \wedge_R N)$ is also trivial. This implies that the filtration is exhaustive. Consider the (p,q)th term of the associated bigraded group of the filtration. It is defined as usual by

$$E^0_{p,q}\pi_*(M \wedge_R N) = F_p\pi_{p+q}(M \wedge_R N)/F_{p-1}\pi_{p+q}(M \wedge_R N),$$

and the definition of the filtration immediately implies that this group is isomorphic to the image of

$$(i_{0,p})_* : \pi_{p+q}(M \wedge_R N) \longrightarrow \pi_{p+q}(M_p \wedge_R N).$$

The target of $(i_{0,p})_*$ is $D^1_{p-1,q}$, and of course $E^1_{p,q} = \pi_{p+q}(\mathbb{F}K_p \wedge_R N)$ also maps into $D^1_{p-1,q}$, via k. It is a routine exercise in the definition of a spectral sequence to check that k induces an isomorphism

$$E^\infty_{p,q} \longrightarrow \operatorname{Im}(i_{0,p})_*.$$

(We know of no published source, but this verification is given in [7, §6].)

To see the functoriality of the spectral sequence, suppose given a map $f : M \longrightarrow M'$ of R-modules and apply the constructions above to M', writing F'_p, etc. Construct a sequence of maps of R_*-modules $f_p : F_p \longrightarrow F'_p$ that give a map of resolutions. We can realize the maps f_p on homotopy groups by R-module maps $\mathbb{F}K_p \longrightarrow \mathbb{F}K'_p$. Starting with $f = f_0$ and proceeding inductively, a standard cofiber sequence argument allows us to construct a map $M_{p+1} \longrightarrow M'_{p+1}$ such that the following diagram of R-modules commutes up to homotopy:

$$\begin{array}{ccccccc} \mathbb{F}K_p & \longrightarrow & M_p & \longrightarrow & M_{p+1} & \longrightarrow & \Sigma\mathbb{F}K_p \\ \downarrow & & \downarrow & & \downarrow & & \downarrow \\ \mathbb{F}K'_p & \longrightarrow & M'_p & \longrightarrow & M'_{p+1} & \longrightarrow & \Sigma\mathbb{F}K'_p. \end{array}$$

There results a map of spectral sequences that realizes the induced map

$$\operatorname{Tor}^{R_*}_{*,*}(M_*, N_*) \longrightarrow \operatorname{Tor}^{R_*}_{*,*}(M'_*, N_*)$$

on E^2 and converges to $(f \wedge_R \operatorname{id})_*$. Functoriality in N is obvious.

To obtain the analogous spectral sequence for Ext, we switch from right to left modules in our resolution (5.1) of M_* and its realization by R-modules. We define

(5.5) $\quad D^{p,q}_1 = \pi_{-p-q}(F_R(M_p, N))$ and $E^{p,q}_1 = \pi_{-p-q}(F_R(\mathbb{F}K_p, N)).$

The maps displayed in (5.1) give maps

$$i \equiv (i_p)^* : D^{p+1,q-1}_1 \longrightarrow D^{p,q}_1$$

$$j \equiv (k_p)^* : D^{p,q}_1 \longrightarrow E^{p,q}_1$$

$$k \equiv (j_{p+1})^* : E^{p,q}_1 \longrightarrow D^{p+1,q}_1.$$

These display an exact couple in standard cohomological form. We see by III.6.9 that $E_1^{p,q} \cong \mathrm{Hom}^q_{R^*}(F_p, N^*)$, where F_p is regraded cohomologically, and that d_1 agrees with $\mathrm{Hom}(d,1)$ under the isomorphism. This proves that

$$E_2^{p,q} = \mathrm{Ext}^{p,q}_{R^*}(M^*, N^*).$$

Observe that $j : D_1^{0,q} \longrightarrow E_1^{0,q}$ can and must be interpreted as

$$\pi_{-q}(F_R(M,N)) \longrightarrow \pi_{-q}(F_R(\mathbb{F}K_0, N)).$$

On passage to E_2, it induces the edge homomorphism

$$\pi_{-q}(F_R(M,N)) \longrightarrow \mathrm{Hom}^q_{R^*}(M^*, N^*) = E_2^{0,q}. \tag{5.6}$$

To see the convergence, let

$$\iota^{0,p} : F_R(M_p, N) \longrightarrow F_R(M,N)$$

be the map induced by the evident iterate $M \longrightarrow M_p$ and filter $\pi_*(F_R(M,N))$ by letting $F^p\pi_*(F_R(M,N))$ be the image of

$$(\iota^{0,p})_* : \pi_*(F_R(M_p,N)) \longrightarrow \pi_*(F_R(M,N)).$$

The (p,q)th term of the associated bigraded group of the filtration is

$$E_0^{p,q}\pi_*(F_R(M,N)) = F^p\pi_{-p-q}(F_R(M,N))/F^{p+1}\pi_{-p-q}(F_R(M,N)).$$

The group $E_\infty^{p,q}$ is defined as the subquotient $Z_\infty^{p,q}/B_\infty^{p,q}$ of $E_1^{p,q}$, where

$$B_\infty^{p,q} = j(\ker(\iota^{0,p})_*),$$

and a routine exercise in the definition of a spectral sequence shows that the additive relation $(\iota^{0,p})_* \circ j^{-1}$ induces an isomorphism

$$E_\infty^{p,q} \cong E_0^{p,q}\pi_*(F_R(M,N)).$$

Since $\mathrm{tel}\, M_p$ is trivial, so is the homotopy limit, or "microscope",

$$\mathrm{mic}\, F_R(M_p, N) \cong F_R(\mathrm{tel}\, M_p, N).$$

By the $\lim^1$ exact sequence for the computation of $\pi_*(\mathrm{mic}\, F_R(M_p,N))$, we conclude that

$$\lim \pi_*(F_R(M_p,N)) = 0 \quad \text{and} \quad \lim{}^1 \pi_*(F_R(M_p,N)) = 0.$$

This means that the spectral sequence $\{E_r^{p,q}\}$ is conditionally convergent. The functoriality of the spectral sequence is clear from the argument for torsion products already given.

Finally, turning to the proof of Proposition 4.4, consider the pairing

$$F_R(M,N) \wedge_R F_R(L,M) \longrightarrow F_R(L,N).$$

Construct a sequence $\{L_p\}$ as in (5.2). Then the maps $M \longrightarrow M_p$ induce a compatible system of pairings

$$F_R(M_p,N) \wedge_R F_R(L_{p'},M) \longrightarrow F_R(M,N) \wedge_R F_R(L_{p'},M) \longrightarrow F_R(L_{p'},N).$$

These induce the required pairing of spectral sequences. The convergence is clear, and the behavior on E_2 terms is correct by comparison with the axioms or by comparison with the usual construction of Yoneda products.

6. Eilenberg-Moore type spectral sequences

Let R be an S-algebra and let M be a right and N a left R-module. Let E be an associative ring spectrum in the sense of classical stable homotopy theory. By I.6.7 and II.1.9, we may assume without loss of generality that E is an associative S-ring spectrum (in the sense to be defined formally in V§2). Under several different further hypotheses, we shall construct a spectral sequence of the form

$$\mathrm{Tor}^{E_*(R)}_{p,q}(E_*(M), E_*(N)) \Longrightarrow E_{p+q}(M \wedge_R N). \tag{6.1}$$

The simplest version of this spectral sequence is the following one.

THEOREM 6.2. *A spectral sequence of the form (6.1) exists if $E_*(R)$ is a flat right R_*-module.*

PROOF. By a standard comparison of homology theories argument, the flatness hypothesis implies that, for left R-modules N, the natural map

$$E_*(R) \otimes_{R_*} N_* \longrightarrow \pi_*((E \wedge_S R) \wedge_R N) \cong \pi_*(E \wedge_S N) = E_*(N)$$

is an isomorphism. It also ensures that the functor $E_*(R) \otimes_{R_*} (-)$ carries the exact sequence (5.1) to an exact sequence of $E_*(R)$-modules. We now apply the functor $E_*(-) = \pi_*(E \wedge_S -)$, rather than the functor π_*, to the sequence of cofibrations obtained from (5.2) by smashing over R with N and find that the rest of the proof of Theorem 4.1 carries over verbatim. In fact, if R is commutative, then the spectral sequence (6.1) results from the first spectral sequence of Theorem 4.1 by applying the exact functor $E_*(R) \otimes_{R_*} (-)$. □

This flatness hypothesis is generally unrealistic. By assuming that E is also an S-algebra and exploiting the S-algebra $E \wedge_S R$, we can obtain a theorem like this without flatness hypotheses. We need a lemma.

LEMMA 6.3. *Let R be an S-algebra such that (R, S) has the homotopy type of a relative CW S-module and let M and N be right and left cell R-modules. Then M, N, and $M \wedge_R N$ have the homotopy types of cell S-modules.*

PROOF. Up to homotopy, $S \longrightarrow R$ is a cofibration of S-modules and R/S is a CW S-module. Since $\mathbb{F}_R X = R \wedge_S \mathbb{F}_S X$, it follows from the cofiber sequence

$$\mathbb{F}_S X \longrightarrow R \wedge_S \mathbb{F}_S X \longrightarrow R/S \wedge_S \mathbb{F}_S X$$

that $\mathbb{F}_R X$ has the homotopy type of a CW S-module if X has the homotopy type of a CW spectrum. Therefore sphere R-modules and, by III.3.7, their smash products are of the homotopy types of CW S-modules. The conclusion follows. □

THEOREM 6.4. *Let E and R be S-algebras and assume that (R, S) is of the homotopy type of a relative CW S-module. Let M be a right and N a left R-module. Then there is a spectral sequence of the form*

$$\mathrm{Tor}_{p,q}^{E_*(R)}(E_*(M), E_*(N)) \Longrightarrow E_{p+q}(M \wedge_R N).$$

PROOF. Replace the triple $(R; M, N)$ in Theorem 4.1 by the triple

$$(E \wedge_S R; E \wedge_S M, E \wedge_S N).$$

The E^2 term of the resulting spectral sequence is

$$\mathrm{Tor}_{*,*}^{(E\wedge_S R)_*}((E \wedge_S M)_*, (E \wedge_S N)_*).$$

It converges to $\pi_*((E \wedge_S M) \wedge_{E\wedge_S R} (E \wedge_S N))$ and, by III.3.10, we have

$$(E \wedge_S M) \wedge_{E\wedge_S R} (E \wedge_S N) \cong E \wedge_S (M \wedge_R N).$$

Since we are working in derived categories, we may assume that M and N are cell R-modules. Then M, N, and $M \wedge_R N$ are of the homotopy types of CW S-modules, and I.6.7 and II.1.9 imply that their smash products over S with E are isomorphic in $\bar{h}\mathscr{S}$ to the corresponding internal smash products. This is also true for R/S, and of course $E \wedge_S S \cong E \simeq E \wedge S$. We conclude that

$$(E \wedge_S R)_* \cong E_*(R), \quad (E \wedge_S M)_* \cong E_*(M) \quad \text{and} \quad (E \wedge_S N)_* \cong E_*(N),$$

so that the E^2 term of the spectral sequence is as stated, and

$$\pi_*((E \wedge_S M) \wedge_{E\wedge_S R} (E \wedge_S N)) \cong E_*(M \wedge_R N),$$

so that the target of the spectral sequence is also as stated. □

The hypothesis that (R, S) is of the homotopy type of a relative CW S-module results in no loss of generality since, as discussed in III§4, the model category theory of Chapter VII implies that, for any S-algebra R, there is a q-cofibrant S-algebra ΛR and a weak equivalence $\lambda : \Lambda R \longrightarrow R$. The map λ induces an equivalence of categories $\mathscr{D}_R \approx \mathscr{D}_{\Lambda R}$, and $(\Lambda R, S)$ is of the homotopy type of a relative CW S-module.

REMARK 6.5. To deal with multiplicative structures, it is important to work with commutative S-algebras. As we shall see in Chapter VII, the category of commutative S-algebas also admits a model category structure. However, we do not believe that its q-cofibrant objects are of the homotopy types of relative CW S-modules. This is a significant technical difference between the theories of S-algebras and of commutative S-algebras. One way of getting around this difficulty is to approximate commutative S-algebras by q-cofibrant non-commutative S-algebras. We shall find a more satisfactory solution in VII§6, where we examine the homotopical properties of q-cofibrant commutative S-algebras. The results there show that the proofs of Theorem 6.4 and of Theorem 7.7 below work in the commutative context provided that we assume that our given commutative S-algebras are q-cofibrant.

7. The bar constructions $B(M, R, N)$ and $B(X, G, Y)$

The spectral sequence (6.1) is reminiscent of the Rothenberg-Steenrod-Eilenberg-Moore spectral sequence

$$\mathrm{Tor}_{*,*}^{E_*(G)}(E_*(X), E_*(Y)) \Longrightarrow E_*B(X, G, Y), \tag{7.1}$$

where G is a topological monoid, X is a right G-space, Y is a left G-space, and $B(X, G, Y)$ is the two-sided bar construction [51]. We here describe a spectrum level two-sided bar construction $B(M, R, N)$ that explains the analogy. We will use the bar construction to derive a version of (6.1) for general commutative ring spectra E that applies under different, and more realistic, flatness hypotheses than those of Theorem 6.2, and we will show that the classical spectral sequence (7.1) is a special case.

Definition 7.2. For an S-algebra (R, ϕ, η), a right R-module (M, μ), and a left R-module (N, ν), define a simplicial S-module $B_*(M, R, N)$ by setting

$$B_p(M, R, N) = M \wedge_S R^p \wedge_S N,$$

where R^p is the p-fold $\wedge_S$-power, interpreted as S if $p = 0$. The face and degeneracy operators on $B_p(M, R, N)$ are

$$d_i = \begin{cases} \mu \wedge (\mathrm{id}_R)^{p-1} \wedge \mathrm{id}_N & \text{if } i = 0 \\ \mathrm{id}_M \wedge (\mathrm{id}_R)^{i-1} \wedge \phi \wedge (\mathrm{id}_R)^{p-i-1} \wedge \mathrm{id}_N & \text{if } 0 < i < p \\ \mathrm{id}_M \wedge (\mathrm{id}_R)^{p-1} \wedge \nu & \text{if } i = p \end{cases}$$

and $s_i = \mathrm{id}_M \wedge (\mathrm{id}_R)^i \wedge \eta \wedge (\mathrm{id}_R)^{p-i} \wedge \mathrm{id}_N$ if $0 \le i \le p$. If M is an (R', R)-bimodule, then $B_*(M, R, N)$ is a simplicial R'-module.

We will discuss the geometric realization of simplicial spectra in X§§1-2, and we agree to write

$$B(M, R, N) = |B_*(M, R, N)|.$$

By X.1.5, geometric realization carries simplicial R-modules to R-modules. By XII.1.2 and X.1.2, there is a natural map

$$\psi : B(R, R, N) \longrightarrow N \tag{7.3}$$

of R-modules that is a homotopy equivalence of S-modules. More generally, by use of the product on R and its action on the given modules, we obtain a natural map of S-modules

$$\psi : B(M, R, N) \longrightarrow M \wedge_R N. \tag{7.4}$$

Proposition 7.5. *For a cell R-module M and any R-module N,*

$$\psi : B(M, R, N) \longrightarrow M \wedge_R N$$

is a weak equivalence of S-modules.

PROOF. If $\underline{M}$ is the constant simplicial R-module at M, then, by X.1.3 and the isomorphism $M \wedge_R R \cong M$, we have

$$M \wedge_R B(R,R,N) \cong |\underline{M} \wedge_R B_*(R,R,N)| \cong B(M,R,N).$$

Moreover, under this isomorphism, $\mathrm{id} \wedge_R \psi$ agrees with ψ. Since ψ of (7.3) is a weak equivalence of R-modules, the conclusion follows from III.3.8. □

For the bar construction to be useful calculationally, the simplicial spectrum $B_*(M,R,N)$ must be proper, in the sense of X.2.1 and X.2.2. By the following result, which is part of IX.2.8, we lose no generality by assuming this.

PROPOSITION 7.6. *If R is a q-cofibrant S-algebra, then $B_*(M,R,N)$ is a proper simplicial S-module.*

By X.2.9, when $B_*(M,R,N)$ is proper, we can use the simplicial filtration of $B(M,R,N)$ to construct a well-behaved spectral sequence that converges to $E_*B(M,R,N)$ for any spectrum E. When E is a commutative ring spectrum, we can use flatness hypotheses to identify the E_2 term. Recall that, in algebra, if A is an algebra over a commutative ring k, then there is a notion of a relatively flat A-module F, for which the functor $(-)\otimes_A F$ is exact when applied to k-split exact sequences. The obvious examples are the relatively free A-modules $A \otimes_k L$ for k-modules L. There is a concomitant relative torsion product $\mathrm{Tor}_*^{(A,k)}(M,N)$, and similarly for graded algebras over commutative graded rings. When k is a field, these reduce to ordinary absolute torsion products.

THEOREM 7.7. *Let E be a commutative ring spectrum. Let R be an S-algebra such that (R,S) is of the homotopy type of a relative CW S-module. Let M be a right and N a left cell R-module such that $B_*(M,R,N)$ is proper. If $E_*(R)$ and either $E_*(M)$ or $E_*(N)$ is E_*-flat, then the bar construction spectral sequence converging to*

$$E_*B(M,R,N) \cong E_*(M \wedge_R N)$$

satisfies

$$E^2_{p,q} = \mathrm{Tor}_{p,q}^{(E_*(R),E_*)}(E_*(M),E_*(N))$$

PROOF. Our hypotheses ensure that we can use smash products over S and internal smash products interchangeably when computing homology and homotopy groups. Our flatness hypotheses ensure that

$$E_*(M \wedge_S R^p \wedge_S N) \cong E_*(M) \otimes_{E_*} E_*(R)^{\otimes p} \otimes_{E_*} E_*(N),$$

where the p-fold tensor power is taken over E_*. This determines the E-homology of the spectrum of p-simplices of $B_*(M,R,N)$. Since $B_*(M,R,N)$ is proper, it follows that the complex that computes E^2 (see X.2.9) is the standard bar complex for the computation of the relative torsion product. □

Intuitively, interchanging the roles of M and N in the proof of Proposition 7.5, we see that the filtration quotients

$$F_pB(M,R,R)/F_{p-1}B(M,R,R)$$

play a role similar to that played by the terms $\mathbb{F}K_p$ in the construction of the spectral sequence of Theorem 6.2.

As promised, we have the following result, which shows that the spectral sequence of (7.1) is a special case.

THEOREM 7.8. *Let G be a topological monoid, X a right G-space, and Y a left G-space. Then $\Sigma^\infty G_+$ is an S-algebra, $\Sigma^\infty X_+$ is a right $\Sigma^\infty G_+$-module, and $\Sigma^\infty Y_+$ is a left $\Sigma^\infty G_+$-module. Moreover, there is a natural isomorphism of S-modules*

$$\Sigma^\infty B(X,G,Y)_+ \cong B(\Sigma^\infty X_+, \Sigma^\infty G_+, \Sigma^\infty Y_+),$$

and $B_*(\Sigma^\infty X_+, \Sigma^\infty G_+, \Sigma^\infty Y_+)$ is proper if G is nondegenerately based.

PROOF. The first statement is immediate from I.8.2 and II.1.2, together with the obvious identification

$$X_+ \wedge Y_+ \cong (X \times Y)_+$$

for spaces X and Y. The product on $\Sigma^\infty G_+$ is induced from the product on G,

$$\Sigma^\infty G_+ \wedge_S \Sigma^\infty G_+ \cong \Sigma^\infty (G \times G)_+ \longrightarrow \Sigma^\infty G_+,$$

and similarly for the actions on $\Sigma^\infty X_+$ and $\Sigma^\infty Y_+$. The second statement follows from the fact that the functors Σ^∞ and geometric realization commute, by X.1.3, and that Σ^∞ preserves properness; see X.2.1. We obtain an identification of simplicial spectra

$$\Sigma^\infty B_*(X,G,Y)_+ \cong B_*(\Sigma^\infty X_+, \Sigma^\infty G_+, \Sigma^\infty Y_+)$$

by applying Σ^∞ to the spaces

$$(X \times G^p \times Y)_+ \cong X_+ \wedge (G_+)^p \wedge Y,$$

where $(G_+)^p$ is the p-fold smash power. If G is non-degenerately based, then $B_*(X,G,Y)_+$ is a proper simplicial based space. $\square$

CHAPTER V

R-ring spectra and the specialization to MU

In this chapter, we think of the derived category of R-modules as an analog of the stable homotopy category. From this point of view, we have the notion of an R-ring spectrum, which is just like the classical notion of a ring spectrum in the stable homotopy category. We shall study such homotopical structures in this chapter.

We first show how to construct quotients M/IM and localizations $M[X^{-1}]$ of modules over a commutative S-algebra R. We then study when these constructions inherit a structure of R-ring spectrum from an R-ring spectrum structure on M.

When specialized to MU, our results give more highly structured versions of spectra that in the past have been constructed by means of the Baas-Sullivan theory of manifolds with singularities or the Landweber exact functor theorem. At least at odd primes, we obtain an entirely satisfactory, and surprisingly simple, treatment of MU-ring structures on the resulting MU-modules.

1. Quotients by ideals and localizations

Let R be a commutative S-algebra. We assume that all given R-modules M are of the homotopy types of cell R-modules, but we must keep in mind that R itself will not be of the homotopy type of a cell R-module. By III.1.4, we have a canonical weak equivalence of R-modules $\zeta : S_R \longrightarrow R$, where the sphere R-module $S_R = \mathbb{F}_R S$ is the free R-module generated by S, and we implicitly replace R by S_R when performing constructions on R regarded as an R-module. Concomitantly, we must sometimes replace the unit isomorphism $R \wedge_R M \simeq M$ by its composite with the weak equivalence $\zeta \wedge \mathrm{id}$. This is consistent with the standard practice of replacing spectra by weakly equivalent CW spectra without change of notation.

We shall work throughout in the derived category $\mathscr{D}_R$ of R-modules. To deduce S-module or spectrum level conclusions from our R-module level arguments, we must apply the forgetful functors $\mathscr{D}_R \longrightarrow \mathscr{D}_S$ and $\mathscr{D}_S \longrightarrow \bar{h}\mathscr{S}$. The process is routine, but it does entail reapproximating cell R-modules by CW S-modules or CW spectra since, in general, cell R-modules need not be of the homotopy types of CW S-modules or of CW spectra.

We are interested in homotopy groups, and we make use of the isomorphisms

$$(1.1) \qquad \pi_n(M) = h\mathscr{S}(S^n, M) \cong h\mathscr{M}_S(S^n_S, M) \cong h\mathscr{M}_R(S^n_R, M)$$

to represent elements as maps of R-modules, where, as usual, $S^n_R = \mathbb{F}_R S^n$. We write $M_* = \pi_*(M)$, and we do not distinguish notationally between a representative map of spectra $S^n \longrightarrow M$ and a representative map of R-modules $S^n_R \longrightarrow M$.

By III.3.7 and standard properties of spectrum level spheres ([38, pp 386-389]), we have a system of equivalences of R-modules

$$(1.2) \qquad S^q_R \wedge_R S^r_R \simeq S^{q+r}_R$$

that is associative, commutative, and unital up to a coherent system of homotopy equivalences and is compatible with suspension as q and r vary. For a pairing of R-modules $L \wedge_R M \longrightarrow N$, we therefore obtain a pairing of homotopy groups

$$L_* \otimes_{R_*} M_* \longrightarrow N_*.$$

Of course, $L \wedge_R M$ is an R-module since R is commutative.

For $x \in R_n$, thought of as an R-map $S^n_R \longrightarrow R$, we have the R-map

$$(1.3) \qquad S^n_R \wedge_R M \xrightarrow{x \wedge \mathrm{id}} R \wedge_R M \cong M.$$

This map of R-modules realizes multiplication by x on M_*. We agree to write $\Sigma^n M$ for $S^n_R \wedge_R M$ and to write $x : \Sigma^n M \longrightarrow M$ for the map (1.3) throughout this chapter. By III.3.7, $S^n_R \wedge_R M$ is isomorphic as an R-module to $S^n_S \wedge_S M$ and, by I.6.7 and II.1.9, $S^n_S \wedge_S M$ is weakly equivalent as a spectrum to $S^n \wedge M$. Therefore, on passage to $h\mathscr{S}$, the R-module $\Sigma^n M$ is a model for the spectrum level suspension of M.

DEFINITION 1.4. Define M/xM to be the cofiber of the map (1.3) and let $\rho : M \longrightarrow M/xM$ be the canonical map. Inductively, for a finite sequence $\{x_1, \dots, x_n\}$ of elements of R_*, define

$$M/(x_1, \dots, x_n)M = N/x_nN, \quad \text{where} \quad N = M/(x_1, \dots, x_{n-1})M.$$

For a (countably) infinite sequence $X = \{x_i\}$, define M/XM to be the telescope of the $M/(x_1, \dots, x_n)M$, where the telescope is taken with respect to the successive canonical maps ρ.

We have the following analogue of the universal property of quotients by principal ideals in algebra.

LEMMA 1.5. *Let N be an R-module such that $x : \Sigma^n N \longrightarrow N$ is zero and let $\alpha : M \longrightarrow N$ be a map of R-modules. Then there is a map of R-modules $\tilde{\alpha} : M/xM \longrightarrow N$ such that $\tilde{\alpha} \circ \rho = \alpha$, and $\tilde{\alpha}$ is unique if $[\Sigma^{n+1} M, N]_R = 0$.*

PROOF. This is obvious from the diagram

$$\begin{array}{ccccccc}
\Sigma^n M & \xrightarrow{x} & M & \xrightarrow{\rho} & M/xM & \longrightarrow & \Sigma^{n+1} M \\
\downarrow{\scriptstyle \alpha} & & \downarrow{\scriptstyle \alpha} & & & & \\
\Sigma^n N & \xrightarrow[x]{} & N, & & & &
\end{array}$$

in which the row is the cofiber sequence that defines M/xM. □

Clearly we have a long exact sequence

$$(1.6)\quad \cdots \longrightarrow \pi_{q-n}(M) \xrightarrow{x} \pi_q(M) \xrightarrow{\rho_*} \pi_q(M/xM) \longrightarrow \pi_{q-n-1}(M) \longrightarrow \cdots .$$

If x is not a zero divisor for $\pi_*(M)$, then ρ_* induces an isomorphism of R_*-modules

$$(1.7)\qquad \pi_*(M)/x\pi_*(M) \cong \pi_*(M/xM).$$

If $\{x_1, \dots, x_n\}$ is a regular sequence for $\pi_*(M)$, in the sense that x_i is not a zero divisor for $\pi_*(M)/(x_1, \dots, x_{i-1})\pi_*(M)$ for $1 \le i \le n$, then

$$(1.8)\qquad \pi_*(M)/(x_1, \dots, x_n)\pi_*(M) \cong \pi_*(M/(x_1, \dots, x_n)M),$$

and similarly for a possibly infinite regular sequence $X = \{x_i\}$. We shall see in a moment that M/XM is independent of the ordering of the elements of the set X. If I denotes the ideal generated by a regular sequence X, then, by Corollary 2.10 below, M/XM is independent of the choice of regular sequence (under reasonable hypotheses) and it is reasonable to define

$$(1.9)\qquad M/IM = M/XM.$$

However, this notation must be used with caution since, if we fail to restrict attention to regular sequences X, the homotopy type of M/XM will depend on the set X and not just on the ideal it generates. For example, quite different modules are obtained if we repeat a generator x_i of I in our construction.

As in algebra, we can describe the construction on general R-modules M as the smash product of M with the construction on R regarded as an R-module. We write R/X or R/I instead of R/XR or R/IR.

LEMMA 1.10. *For a sequence X of elements of R_*, there is a natural isomorphism in $\mathscr{D}_R$*

$$(R/X) \wedge_R M \longrightarrow M/XM.$$

In particular, for a finite sequence $X = \{x_1, \dots, x_n\}$,

$$R/(x_1, \dots, x_n) \simeq (R/x_1) \wedge_R \cdots \wedge_R (R/x_n),$$

and R/X is therefore independent of the ordering of the elements of X.

PROOF. Working on the point-set level, we have an isomorphism of cofiber sequences :

$$\begin{array}{ccccc} S_R^n \wedge_R R \wedge_R M & \xrightarrow{x\wedge \mathrm{id}} & R \wedge_R M & \xrightarrow{\rho \wedge \mathrm{id}} & (R/xR) \wedge_R M \\ \downarrow{\scriptstyle \mathrm{id}\wedge\lambda} & & \downarrow{\scriptstyle \lambda} & & \downarrow \\ S_R^n \wedge_R M & \xrightarrow{x} & M & \xrightarrow{\rho} & M/xM. \end{array}$$

We only claim an isomorphism in $\mathscr{D}_R$ since, working homotopically, we should replace R by S_R and use the weak equivalence $\zeta \wedge \mathrm{id} : S_R \wedge_R M \longrightarrow R \wedge_R M$ to obtain a composite weak equivalence $(S_R/xS_R) \wedge_R M \longrightarrow (R/xR) \wedge_R M \longrightarrow M/xM$. The rest follows by iteration and use of the commutativity of $\wedge_R$. $\square$

We turn next to localizations of R-modules at subsets $X = \{x_i\}$ of R_*. We restrict attention to countable sets for notational convenience, but this restriction can easily be removed. Let $\{y_i\}$ be any cofinal sequence of products of the x_i, such as that specified inductively by $y_1 = x_1$ and $y_i = x_1 \cdots x_i y_{i-1}$. If $y_i \in R_{n_i}$, we may represent y_i by an R-map $S_R^0 \longrightarrow S_R^{-n_i}$, which we also denote by y_i. Let $q_0 = 0$ and, inductively, $q_i = q_{i-1} + n_i$. The map of R-modules

$$S_R^0 \wedge_R M \xrightarrow{y_i \wedge \mathrm{id}} S_R^{-n_i} \wedge_R M$$

represents y_i. Smashing over R with $S_R^{-q_{i-1}}$ and using equivalences (1.2), we obtain a sequence of maps of R-modules

$$S_R^{-q_{i-1}} \wedge_R M \longrightarrow S_R^{-q_i} \wedge_R M. \tag{1.11}$$

DEFINITION 1.12. Define the localization of M at X, denoted $M[X^{-1}]$, to be the telescope of the sequence of maps (1.11). Since $M \simeq S_R^0 \wedge_R M$ in $\mathscr{D}_R$, we may regard the inclusion of the initial stage $S_R^0 \wedge_R M$ of the telescope as a natural map $\lambda : M \longrightarrow M[X^{-1}]$.

Again, we have an analogue of the standard universal property of localization in algebra.

LEMMA 1.13. *Let N be an R-module such that $x_i : \Sigma^{k_i} N \longrightarrow N$, $\deg x_i = k_i$, is an equivalence for each i and let $\alpha : M \longrightarrow N$ be a map of R-modules. Then there is a unique map of R-modules $\tilde{\alpha} : M[X^{-1}] \longrightarrow N$ such that $\tilde{\alpha} \circ \lambda = \alpha$.*

PROOF. Passage to telescopes gives $\tilde{\alpha} : M[X^{-1}] \longrightarrow N[X^{-1}] \simeq N$. The $\lim^1$ term is zero in the exact sequence

$$0 \longrightarrow \lim{}^1 [\Sigma^{1-q_i} M, N]_R \longrightarrow [M[X^{-1}], N]_R \longrightarrow \lim [\Sigma^{-q_i} M, N]_R \longrightarrow 0$$

since the maps of the inverse system are isomorphisms. Therefore $\tilde{\alpha}$ is unique. $\square$

Since homotopy groups commute with localization, by III.1.7, we see immediately that λ induces an isomorphism of R_*-modules

$$\pi_*(M[X^{-1}]) \cong \pi_*(M)[X^{-1}]. \tag{1.14}$$

Arguing as in Lemma 1.10, we see that the localization of M is the smash product of M with the localization of R.

LEMMA 1.15. *For a set X of elements of R_*, there is a natural weak equivalence*

$$R[X^{-1}] \wedge_R M \longrightarrow M[X^{-1}].$$

Moreover, $R[X^{-1}]$ is independent of the ordering of the elements of X. For sets X and Y, $R[(X \cup Y)^{-1}]$ is equivalent to the composite localization $R[X^{-1}][Y^{-1}]$.

PROOF. The independence of ordering is shown by use of the union of any two given cofinal sequences. The last statement is shown by use of the usual Fubini type theorem for iterated homotopy colimits. □

2. Localizations and quotients of R-ring spectra

Again, fix a commutative S-algebra R. Since $\mathscr{D}_R$ is a symmetric monoidal category under $\wedge_R$ with unit R, we have the notion of a monoid or a commutative monoid in $\mathscr{D}_R$. These are the analogs of associative or of associative and commutative ring spectra in classical stable homotopy theory. As there, we must allow weaker structures.

DEFINITION 2.1. An R-ring spectrum A is an R-module A with unit $\eta : R \longrightarrow A$ and product $\phi : A \wedge_R A \longrightarrow A$ in $\mathscr{D}_R$ such that the following left and right unit diagram commutes in $\mathscr{D}_R$:

$$\begin{array}{ccccc} R \wedge_R A & \xrightarrow{\eta \wedge \mathrm{id}} & A \wedge_R A & \xleftarrow{\mathrm{id} \wedge \eta} & A \wedge_R R \\ & {\scriptstyle\lambda} \searrow & \downarrow {\scriptstyle\phi} & \swarrow {\scriptstyle\lambda\tau} & \\ & & A. & & \end{array}$$

Of course, by neglect of structure, an R-ring spectrum A is a ring spectrum in the sense of classical stable homotopy theory; its unit is the composite of the unit of R and the unit of A and its product is the composite of the product of A and the canonical map

$$A \wedge A \simeq A \wedge_S A \longrightarrow A \wedge_R A.$$

Similarly, for an R-ring spectrum A, we have the evident homotopical notion of an A-module spectrum M. Here, in conformity with the definition just given, we only require that the action $\mu : A \wedge_R M \longrightarrow M$ satisfy the unit condition $\mu \circ (\eta \wedge \mathrm{id}) = \mathrm{id}$ in $\mathscr{D}_R$. When A is associative, it is conventional to insist that M also satisfy the evident associativity condition. These structures play a role in the study of our new derived categories of R-modules that is analogous to the role played by ring spectra and their module spectra in classical stable homotopy theory. When $R = S$, I.6.7 and II.1.9 imply that S-ring spectra and their module spectra are equivalent to classical ring spectra and their module spectra.

LEMMA 2.2. *If A and B are R-ring spectra, then so is $A \wedge_R B$. If A and B are associative or commutative, then so is $A \wedge_R B$.*

We ask the behavior of quotients and localizations with respect to R-ring structures. For localizations, the answer is immediate. We shall give a sharper point-set level analogue of the following result in VIII§2.

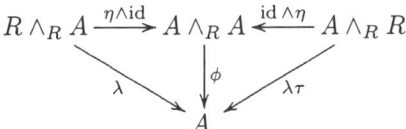

PROPOSITION 2.3. *Let X be a set of elements of R_*. If A is an R-ring spectrum, then $A[X^{-1}]$ is an R-ring spectrum such that $\lambda : A \longrightarrow A[X^{-1}]$ is a map of R-ring spectra. If A is associative or commutative, then so is $A[X^{-1}]$.*

PROOF. Since $A[X^{-1}] \simeq R[X^{-1}] \wedge_R A$, it suffices to prove that $R[X^{-1}]$ is an associative and commutative R-ring spectrum with unit λ. Lemma 1.15 gives an equivalence

$$R[X^{-1}] \wedge_R R[X^{-1}] \simeq R[X^{-1}][X^{-1}] \simeq R[X^{-1}]$$

under R, and this equivalence is the required product. □

This doesn't work for quotients since $(R/X)/X$ is not equivalent to R/X. However, we can analyze the problem by analyzing the deviation, and, by Lemma 1.10, we may as well work one element at a time. We have a necessary condition for R/x to be an R-ring spectrum that will be familiar from classical stable homotopy theory. We generally write η and ϕ for the units and products of R-ring spectra; as stated before, we write Σ^n for the module theoretic suspension functor $S^n_R \wedge_R (-)$.

LEMMA 2.4. *Let A be an R-ring spectrum. If A/xA admits an R-ring spectrum structure such that $\rho : A \longrightarrow A/xA$ is a map of R-ring spectra, then $x : A/xA \longrightarrow A/xA$ is null homotopic as a map of R-modules.*

PROOF. We have the following commutative diagram (where we omit suspension coordinates from the labels of maps):

$$\begin{array}{ccccc}
\Sigma^n R \wedge_R (A/xA) & \xrightarrow{\eta\wedge\mathrm{id}} & \Sigma^n(A/xA) \wedge_R (A/xA) & \xrightarrow{x\wedge\mathrm{id}} & (A/xA) \wedge_R (A/xA) \\
& \searrow^{\lambda} & \downarrow{\scriptstyle\phi} & & \downarrow{\scriptstyle\phi} \\
& & \Sigma^n(A/xA) & \xrightarrow[x]{\qquad} & A/xA.
\end{array}$$

In view of the following commutative diagram, its top composite is null homotopic:

$$\begin{array}{ccccc}
& & \Sigma^n A & \xrightarrow{x} & A \\
& \nearrow^{\eta} & \downarrow{\scriptstyle\rho} & & \downarrow{\scriptstyle\rho} \\
\Sigma^n R & \xrightarrow[\eta]{} & \Sigma^n(A/xA) & \xrightarrow[x]{} & A/xA.
\end{array}$$

□

Thus, for example, the Moore spectrum $S/2$ is not an S-algebra since the map $2 : S/2 \longrightarrow S/2$ is not null homotopic.

We need a lemma in order to obtain an R-ring spectrum structure on R/x in appropriate generality.

LEMMA 2.5. *Let $\rho : R \longrightarrow M$ be any map of R-modules. Then*

$$(\rho \wedge \mathrm{id}) \circ \rho = (\mathrm{id} \wedge \rho) \circ \rho : R \longrightarrow M \wedge_R M.$$

PROOF. Since $\lambda = \lambda \circ \tau : R \wedge_R R \longrightarrow R$, the following diagram commutes:

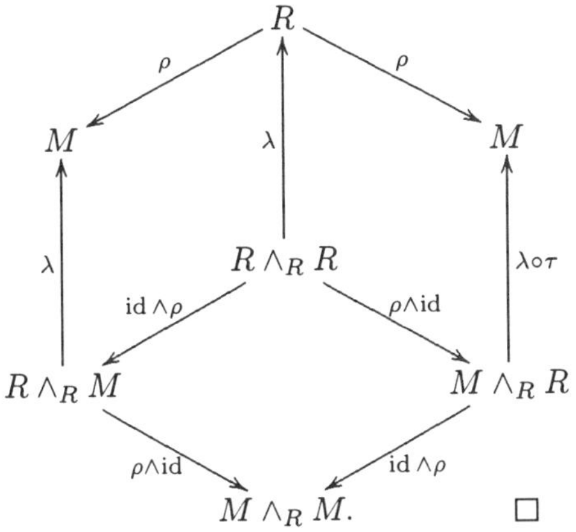

□

THEOREM 2.6. *Let $x \in R_n$, where $\pi_{n+1}(R/x) = 0$ and $\pi_{2n+1}(R/x) = 0$. Then R/x admits a structure of R-ring spectrum with unit $\rho : R \longrightarrow R/x$. Therefore A/XA admits a structure of R-ring spectrum such that $\rho : A \longrightarrow A/XA$ is a map of R-ring spectra for every R-ring spectrum A and every sequence X of elements of R_* such that $\pi_{n+1}(R/x) = 0$ and $\pi_{2n+1}(R/x) = 0$ for all $x \in X$, where* $\deg x = n$.

PROOF. Consider the following diagram in the derived category $\mathscr{D}_R$:

(2.7)

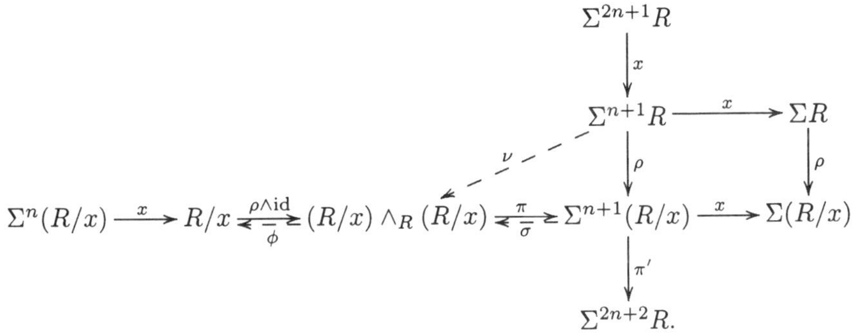

The map x is that specified by (1.3). The bottom row is the cofiber sequence that results from the equivalence

$$(R/x) \wedge_R (R/x) \simeq (R/x)/x$$

of Lemma 1.10, and the column is also a cofiber sequence. The composite $x \circ \rho$ is null homotopic since $\rho \circ x$ is null homotopic and the square commutes. Therefore there is a map ν such that $\pi \circ \nu = \rho$, and ν is unique since $\pi_{n+1}(R/x) = 0$. Since $\pi \circ \nu \circ x = \rho \circ x = 0$, $\nu \circ x$ factors through a map $\Sigma^{2n+1}R \longrightarrow R/x$. Since $\pi_{2n+1}(R/x) = 0$, such maps are null homotopic. Thus $\nu \circ x$ is null homotopic. Therefore there is a map σ such that $\sigma \circ \rho = \nu$. Now $\pi \circ \sigma \circ \rho = \pi \circ \nu = \rho$, hence $(\pi \circ \sigma - \mathrm{id})\rho = 0$. Therefore $\pi \circ \sigma - \mathrm{id}$ factors through a map $\Sigma^{2n+2}R \longrightarrow$

$\Sigma^{n+1}(R/x)$. Again, such maps are null homotopic. Therefore $\pi \circ \sigma = \mathrm{id}$. Thus the bottom cofiber sequence splits (proving in passing that $x : \Sigma^n(R/x) \longrightarrow R/x$ is null homotopic, as it must be). A choice ϕ of a splitting gives a product on R/x. The unit condition $\phi\circ(\rho\wedge\mathrm{id}) = \mathrm{id}$ is automatic. To see that $\phi\circ(\mathrm{id}\wedge\rho) = \mathrm{id}$, we observe that, by the lemma,

$$(\phi \circ (\mathrm{id}\wedge\rho) - \mathrm{id}) \circ \rho = \phi \circ (\mathrm{id}\wedge\rho - \rho\wedge\mathrm{id}) \circ \rho = 0.$$

Therefore $\phi \circ (\mathrm{id}\wedge\rho) - \mathrm{id}$ factors through a map $\Sigma^{n+1}R \longrightarrow R/x$. Again, such maps are null homotopic, hence $\phi \circ (\mathrm{id}\wedge\rho) = \mathrm{id}$. This completes the proof that R/x is an R-ring spectrum with unit ρ. The rest follows from Lemmas 1.10 and 2.2. □

The product on R/x can be described a little more concretely. The wedge sum

$$(2.8) \qquad (\rho\wedge\mathrm{id})\vee\sigma : (R/x)\vee\Sigma^{n+1}(R/x) \longrightarrow (R/x)\wedge_R(R/x)$$

is an equivalence. The product ϕ restricts to the identity on the first wedge summand and to the trivial map on the second wedge summand. Thus the product is determined by the choice of σ, and two choices of σ differ by a composite

$$(2.9) \qquad \Sigma^{n+1}(R/x) \xrightarrow{\pi'} \Sigma^{2n+2}R \longrightarrow (R/x)\wedge_R(R/x).$$

By the splitting (2.8) and the assumption that $\pi_{n+1}(R/x) = 0$, we can view the second map as an element of $\pi_{2n+2}(R/x)$. If x is not a zero divisor, then $\pi'_* = 0$ on homotopy groups and any two products have the same effect on homotopy groups.

Before continuing our discussion of these R-ring spectra, we insert the following easy consequence of the mere existence of the R-ring structure.

COROLLARY 2.10. *Assume that $R_i = 0$ if i is odd. Let X and Y be regular sequences in R_* that generate the same ideal. Then there is an equivalence of R-modules $\xi : R/X \longrightarrow R/Y$ under R.*

PROOF. It suffices to construct a map ξ under R since it will automatically induce an isomorphism on homotopy groups. Each x_i is an R_*-linear combination of the y_j and each $y_j : R/Y \longrightarrow R/Y$ is zero. By Lemma 1.5, we obtain a unique map $\xi_i : R/x_i \longrightarrow R/Y$ under R. Since R/Y is an R-ring spectrum, we may first take the smash product of these maps and then use the product (associated conveniently) on R/Y and passage to telescopes (if X is infinite) to obtain ξ. □

3. The associativity and commutativity of R-ring spectra

We assume given an R-ring spectrum A. For $x \in R_n$ as in Theorem 2.6, we give $A/xA \simeq (R/x)\wedge_R A$ the product induced by one of our constructed products on R/x and the given product on A. We refer to any such product as a "canonical" product on A/xA. Of course, we do not claim that every product on A/xA is canonical. Observe that, by first using the product on A, the product on A/xA can be factored through

$$\phi\wedge_R\mathrm{id} : (R/x)\wedge_R(R/x)\wedge_R A \longrightarrow (R/x)\wedge_R A.$$

This allows us to smash any diagram giving information about the product on R/x with A and so obtain information about the product on A/xA. Obviously any diagram so constructed is a diagram of right A-module spectra via the product action of A on itself. This smashing with A can kill obstructions. Clearly, a map of A-modules $\Sigma^q A \longrightarrow M$ is determined by its restriction $S^q \longrightarrow M$ along the unit of A regarded as a map of spectra (or S-modules), which is just an element of $\pi_q(M)$. These considerations lead to the following result.

THEOREM 3.1. *Let $x \in R_n$, where $\pi_{n+1}(R/x) = 0$ and $\pi_{2n+1}(R/x) = 0$. Let A be an R-ring spectrum and assume that $\pi_{2n+2}(A/xA) = 0$. Then there is a unique canonical product on A/xA. If A is commutative, then A/xA is commutative. If A is associative and $\pi_{3n+3}(A/xA) = 0$, then A/xA is associative.*

PROOF. The second arrow of (2.9) becomes zero after smashing with A since it is then given by an element of $\pi_{2n+2}(A/xA) = 0$. This proves the uniqueness statement. The commutativity statement follows since if ϕ is a canonical product on A/xA, then so is $\phi\tau$. However, it may be worth displaying the obstruction that lies in $\pi_{2n+2}(A/xA)$. Looking at the splitting (2.8), we see that ϕ is commutative on the wedge summand R/x by the unit property. For the summand $\Sigma^{n+1}R$, consider the diagram

$$\begin{array}{ccccc} \Sigma^{n+1}R & \xrightarrow{\rho} & \Sigma^{n+1}(R/x) & \xrightarrow{(\phi-\phi\circ\tau)\circ\sigma} & R/x. \\ & & \Big\downarrow{\scriptstyle \pi'} & \nearrow{\scriptstyle \gamma} & \\ & & \Sigma^{2n+2}R & & \end{array}$$

The horizontal composite is null homotopic since $\pi_{n+1}(R/x) = 0$. Thus there exists γ such that the triangle commutes. It is the obstruction to the commutativity of R/x, and smashing with A gives the obstruction to the commutativity of A/xA.

For the associativity, consider the splitting displayed in the following diagram:

$$\begin{array}{c} (R/x) \vee \Sigma^{n+1}(R/x) \vee \Sigma^{n+1}(R/x) \vee \Sigma^{2n+2}(R/x) \\ \Big\downarrow{\scriptstyle \simeq} \\ [(R/x) \vee \Sigma^{n+1}(R/x)] \vee \Sigma^{n+1}[(R/x) \vee \Sigma^{n+1}(R/x)] \\ \Big\downarrow{\scriptstyle [(\rho\wedge \mathrm{id})\vee\sigma]\vee\Sigma^{n+1}[(\rho\wedge\mathrm{id})\vee\sigma]} \\ [(R/x) \wedge_R (R/x)] \vee \Sigma^{n+1}[(R/x) \wedge_R (R/x)] \\ \Big\downarrow{\scriptstyle \simeq} \\ (R/x) \wedge_R [(R/x) \vee \Sigma^{n+1}(R/x)] \\ \Big\downarrow{\scriptstyle \mathrm{id}\wedge[(\rho\wedge\mathrm{id})\vee\sigma]} \\ (R/x) \wedge_R (R/x) \wedge_R (R/x). \end{array}$$

The question of associativity can be considered separately on the restrictions of the iterated product to the four wedge summands. Via easy diagram chases, we see that, under the splitting (2.8) and unit isomorphisms, the natural maps

$$\rho \wedge \mathrm{id} \wedge \mathrm{id} : R \wedge_R (R/x) \wedge_R (R/x) \longrightarrow (R/x) \wedge_R (R/x) \wedge_R (R/x)$$

and

$$\mathrm{id} \wedge \rho \wedge \mathrm{id} : (R/x) \wedge_R R \wedge_R (R/x) \longrightarrow (R/x) \wedge_R (R/x) \wedge_R (R/x)$$

correspond to the inclusions of the first and third and first and second wedge summands, respectively. Therefore the unital property of ϕ and the unital and associativity properties of $\wedge_R$ imply that the restriction of ϕ to the first three wedge summands is associative. Let ω denote the displayed inclusion of the fourth wedge summand and consider the diagram

$$\begin{array}{ccccc} \Sigma^{2n+2}R & \xrightarrow{\ \rho\ } & \Sigma^{2n+2}(R/x) & \xrightarrow{[\phi\circ(\phi\wedge\mathrm{id})-\phi\circ(\mathrm{id}\wedge\phi)]\circ\omega} & R/x \\ & & \Big\downarrow{\scriptstyle \pi'} & & \\ & & \Sigma^{3n+3}R & & \end{array}$$

Call the horizontal composite α. If α is nullhomotopic then the deviation from associativity $[\phi\circ(\phi\wedge\mathrm{id})-\phi\circ(\mathrm{id}\wedge\phi)]\circ\omega$ factors through a map $\Sigma^{3n+3}R \longrightarrow R/x$. Thus if $\pi_{3n+3}(R/x) = 0$, then the element $\alpha \in \pi_{2n+2}(R/x)$ is the obstruction to the associativity of R/x. If both relevant homotopy groups become zero after smashing with A, we can conclude that A/xA is associative if A is associative. □

We can iterate the argument to arrive at the following fundamental conclusion.

THEOREM 3.2. *Assume that $R_i = 0$ if i is odd and let X be a sequence of non zero divisors in R_* such that $\pi_*(R/X)$ is concentrated in degrees congruent to zero mod 4. Then R/X has a unique canonical structure of R-ring spectrum, and it is commutative and associative.*

PROOF. We first observe that for an element $x \in \pi_q(R)$, an R-module M, and an R-module N such that $x : \Sigma^q N \longrightarrow N$ is null homotopic, the map $\rho : M \longrightarrow M/xM$ induces an epimorphism

$$\rho^* : [M/xM, N]_R \longrightarrow [M, N]_R$$

since the action $x^* : [M, N]_R \longrightarrow [\Sigma^q M, N]_R$ can be computed from the action on N and is therefore zero. Let X_n be the subsequence consisting of the first n elements of the sequence X. Then R/X is defined to be the telescope of the R/X_n, and Lemma 2.4 implies that multiplication by x_n is null homotopic on R/X for each n. Since $R/X \wedge_R R/X$ is equivalent to the telescope of the $R/X_n \wedge_R R/X_n$, we obtain a product on R/X from a canonical product on the R/X_n by passage to telescopes. Moreover, by the Mittag-Leffler criterion, our first observation implies that all relevant $\lim^1$ terms are zero. Thus it suffices to show that any two products on X_n become equal and the commutativity and associativity diagrams for R/X_n become commutative upon composition with the map $R/X_n \longrightarrow R/X$, and we may proceed by induction on n. The conclusion follows since the obstructions to uniqueness, commutativity, and associativity of each R/x_n become trivial when we map to R/X. □

4. The specialization to MU-modules and algebras

It was observed in [48] that MU can be constructed as an E_∞ ring spectrum, and we apply $S \wedge_{\mathscr{L}} (-)$ to convert it to a commutative S-algebra. Of course, its homotopy groups are concentrated in even degrees, and every non-zero element is a non zero divisor. Thus Proposition 2.3, Theorem 2.6, and Theorem 3.2 combine to give the following result.

THEOREM 4.1. *Let X be a regular sequence in MU_*, let I be the ideal generated by X, and let Y be any sequence in MU_*. Then there is an MU-ring spectrum $(MU/X)[Y^{-1}]$ and a natural map of MU-ring spectra (the unit map)*

$$\eta : MU \longrightarrow (MU/X)[Y^{-1}]$$

such that

$$\eta_* : MU_* \longrightarrow \pi_*((MU/X)[Y^{-1}])$$

realizes the natural homomorphism of MU_*-algebras

$$MU_* \longrightarrow (MU_*/I)[Y^{-1}].$$

If MU_*/I is concentrated in degrees congruent to zero mod 4, then there is a unique canonical product on $(MU/X)[Y^{-1}]$, and this product is commutative and associative.

In comparison with constructions of this sort based on the Baas-Sullivan theory of manifolds with singularities or on Landweber's exact functor theorem (where it applies), we have obtained a simpler proof of a substantially stronger result. We emphasize that an MU-ring spectrum is a much richer structure than just a ring spectrum and that commutativity and associativity in the MU-ring spectrum sense are much more stringent conditions than mere commutativity and associativity of the underlying ring spectrum. Observe that the assumption that X is regular is used only to obtain the calculational description of η_*.

We illustrate by explaining how BP appears in this context. Fix a prime p and write $(-)_p$ for localization at p. Let BP be the Brown-Peterson spectrum at p. We are thinking of Quillen's idempotent construction, and we have the splitting maps $i : BP \longrightarrow MU_p$ and $e : MU_p \longrightarrow BP$. These are maps of commutative and associative ring spectra such that $e \circ i = \mathrm{id}$. Let I be the kernel of the composite

$$MU_* \longrightarrow MU_{p*} \longrightarrow BP_*.$$

Then I is generated by a regular sequence X, and our MU/X is a canonical integral version of BP. For the moment, let $BP' = (MU/X)_p$. Let $\zeta : BP \longrightarrow BP'$ be the composite

$$BP \xrightarrow{\ i\ } MU_p \xrightarrow{\ \zeta_p\ } BP'.$$

It is immediate that ξ is an equivalence. In effect, since we have arranged that ζ_p has the same effect on homotopy groups as e, ξ induces the identity map of $(MU_*/I)_p$ on homotopy groups. By the splitting of MU_p and the fact that self-maps of MU_p are determined by their effect on homotopy groups [2, II.9.3], maps $MU_p \longrightarrow BP$ are determined by their effect on homotopy groups. This implies that $\xi \circ e = \zeta_p : MU_p \longrightarrow BP'$. The product on BP is the composite

$$BP \wedge BP \xrightarrow{i \wedge i} MU_p \wedge MU_p \xrightarrow{\phi} MU_p \xrightarrow{e} BP.$$

Since ζ_p is a map of MU-ring spectra and thus of ring spectra, a trivial diagram chase now shows that the equivalence $\xi : BP \longrightarrow BP'$ is a map of ring spectra.

We conclude that our BP' is a model for BP that is an MU-ring spectrum, commutative and associative if $p > 2$. The situation for $p = 2$ is interesting. We conclude from the equivalence that BP' is commutative and associative as a ring spectrum, although we do not know that it is commutative or associative as an MU-algebra.

Recall that $\pi_*(BP) = \mathbb{Z}_{(p)}[v_i | deg(v_i) = 2(p^i - 1)]$, where the generators v_i come from $\pi_*(MU)$ (provided that we use the Hazewinkel generators). We list a few of the spectra derived from BP, with their coefficient rings. Let $\mathbb{F}_p$ denote the field with p elements.

$BP\langle n \rangle$	$\mathbb{Z}_{(p)}[v_1, \ldots, v_n]$	$E(n)$	$\mathbb{Z}_{(p)}[v_1, \ldots, v_n, v_n^{-1}]$
$P(n)$	$\mathbb{F}_p[v_n, v_{n+1}, \ldots]$	$B(n)$	$\mathbb{F}_p[v_n^{-1}, v_n, v_{n+1}, \ldots]$
$k(n)$	$\mathbb{F}_p[v_n]$	$K(n)$	$\mathbb{F}_p[v_n, v_n^{-1}]$

By the method just illustrated, we can construct canonical integral versions of the $BP\langle n \rangle$ and $E(n)$. All of these spectra fit into the context of Theorem 4.1. If $p > 2$, they all have unique canonical commutative and associative MU-ring spectra structures. Further study is needed when $p = 2$, but we leave that to the interested reader. In any case, our theory makes it unnecessary to appeal to Baas-Sullivan theory or to Landweber's exact functor theorem for the construction and analysis of spectra such as these.

We have started with MU because it appears in nature with a canonical structure as a commutative S-algebra. However, it is also possible to start with BP, using the second author's result that BP admits a commutative S-algebra structure; in fact, it admits uncountably many different ones [34].

CHAPTER VI

Algebraic *K*-theory of S-algebras

In this chapter we apply the basic constructions of algebraic K-theory to the new categories of modules over S-algebras. We show how to construct a K-theory spectrum KR for each S-algebra R in such a way that K becomes a functor from the category of S-algebras to the stable category. When R is a connective commutative S-algebra, so is KR. We prove that weakly equivalent S-algebras have equivalent K-theories, and we prove a Morita invariance result. When R is connective we are able to give an alternate description of this K-theory in terms of Quillen's plus construction, a "plus equals $S_\bullet$" theorem. When $R = Hk$ is an Eilenberg-Mac Lane S-algebra, this K-theory is essentially Quillen's algebraic K-theory of the ring k. When $R = \Sigma^\infty|GSX|_+$ is the suspension spectrum of the space obtained by adjoining a disjoint basepoint to the geometric realization of the loop group of the singular complex of a topological space X, this K-theory is Waldhausen's algebraic K-theory of the space X.

1. Waldhausen categories and algebraic *K*-theory

We first review the basic definitions of Waldhausen [71] that we shall use.

DEFINITION 1.1. A category with w-cofibrations $\mathscr{C}$ is a (small) category with preferred zero object "$*$", together with a chosen subcategory $\mathrm{co}(\mathscr{C})$ that satisfies the following three axioms:

(i) Any isomorphism in $\mathscr{C}$ is a morphism in $\mathrm{co}(\mathscr{C})$; in particular, $\mathrm{co}(\mathscr{C})$ contains all the objects of $\mathscr{C}$.
(ii) For every object A in $\mathscr{C}$, the unique map $* \longrightarrow A$ is in $\mathrm{co}(\mathscr{C})$.
(iii) If $A \longrightarrow B$ is a map in $\mathrm{co}(\mathscr{C})$, and $A \longrightarrow C$ is a map in $\mathscr{C}$, then the pushout $B \amalg_A C$ exists in $\mathscr{C}$ and the canonical map $C \longrightarrow B \amalg_A C$ is in $\mathrm{co}(\mathscr{C})$; in particular, $\mathscr{C}$ has finite coproducts.

We call the morphisms in $\mathrm{co}(\mathscr{C})$ w-cofibrations, and often use the feathered arrow "$\rightarrowtail$" to denote them in diagrams. Although Waldhausen called these arrows "cofibrations" we will consistently use the term "w-cofibration" so that there will be no confusion with our standard use of the word "cofibration" to mean those maps that satisfy the homotopy extension property (HEP).

DEFINITION 1.2. A Waldhausen category (in [71], "a category with cofibrations and weak equivalences") is a category with w-cofibrations $\mathscr{C}$ and a chosen subcategory $\mathrm{w}(\mathscr{C})$ of $\mathscr{C}$ that satisfies the following axioms:

(i) Any isomorphism in $\mathscr{C}$ is a morphism in $\mathrm{w}(\mathscr{C})$; in particular, $\mathrm{w}(\mathscr{C})$ contains all the objects of $\mathscr{C}$.

(ii) Given any commutative diagram in $\mathscr{C}$

$$\begin{array}{ccccc} C & \longleftarrow & A & \rightarrowtail & B \\ \downarrow & & \downarrow & & \downarrow \\ C' & \longleftarrow & A' & \rightarrowtail & B' \end{array}$$

in which the vertical maps are in $\mathrm{w}(\mathscr{C})$ and the feathered arrows are in $\mathrm{co}(\mathscr{C})$, the induced map $B \amalg_A C \longrightarrow B' \amalg_{A'} C'$ is in $\mathrm{w}(\mathscr{C})$.

We call the morphisms in $\mathrm{w}(\mathscr{C})$ weak equivalences, and often use the arrows "$\xrightarrow{\sim}$" to denote them. We say that the weak equivalences are saturated or that $\mathscr{C}$ is a saturated Waldhausen category if whenever f and g are composable arrows in $\mathscr{C}$ and any two of f, g, and gf are weak equivalences then so is the third.

DEFINITION 1.3. A functor between Waldhausen categories is exact if it preserves all of the above structure; i.e. it must send w-cofibrations to w-cofibrations, weak equivalences to weak equivalences, the preferred zero object to the preferred zero object, and it must preserve the pushouts along a w-cofibration.

We now have all the necessary ingredients to describe Waldhausen's $S_\bullet$ construction [71]. Let $\mathscr{C}$ be a Waldhausen category. For each $n \geq 0$, define a category $S_n\mathscr{C}$ as follows. An object of $S_n\mathscr{C}$ consists of n composible arrows in $\mathrm{co}(\mathscr{C})$ starting from the preferred zero object $*$,

$$* = A_0 \overset{\alpha_0}{\rightarrowtail} A_1 \overset{\alpha_1}{\rightarrowtail} \cdots \overset{\alpha_{n-1}}{\rightarrowtail} A_n,$$

together with objects $A_{i,j}$ for $0 \leq i \leq j \leq n$ and maps $a_{i,j}\colon A_j \longrightarrow A_{i,j}$ such that $A_{i,i} = *$, $A_j = A_{0,j}$ with $a_{0,j}$ the identity map, and the diagrams

$$\begin{array}{ccc} A_i & \overset{\alpha_{j-1}\circ\cdots\circ\alpha_i}{\rightarrowtail} & A_j \\ \downarrow & & \downarrow{\scriptstyle a_{i,j}} \\ * & \rightarrowtail & A_{i,j} \end{array}$$

are pushouts for $0 \leq i < j \leq n$. A morphism of $S_n\mathscr{C}$ from $\{A_j, A_{i,j}, \alpha_j, a_{i,j}\}$ to $\{A'_j, A'_{i,j}, \alpha'_j, a'_{i,j}\}$ is a sequence of maps $f_j\colon A_j \longrightarrow A'_j$ such that the diagram

$$\begin{array}{ccccccc} A_0 & \overset{\alpha_0}{\rightarrowtail} & A_1 & \overset{\alpha_1}{\rightarrowtail} & \cdots & \overset{\alpha_{n-1}}{\rightarrowtail} & A_n \\ {\scriptstyle f_0}\downarrow & & \downarrow{\scriptstyle f_1} & & & & \downarrow{\scriptstyle f_n} \\ A'_0 & \underset{\alpha'_0}{\rightarrowtail} & A'_1 & \underset{\alpha'_1}{\rightarrowtail} & \cdots & \underset{\alpha'_{n-1}}{\rightarrowtail} & A'_n \end{array}$$

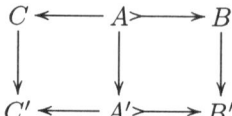

commutes. Observe that by the universal property of pushouts, we have induced maps $A_{i,j} \longrightarrow A'_{i,j}$ making all the appropriate diagrams commute. We give $S_n\mathscr{C}$ the structure of a Waldhausen category by defining a map $\{f_0, \dots, f_n\}$ to be a w-cofibration (resp. weak equivalence) if each f_j is a w-cofibration (resp. weak equivalence) of $\mathscr{C}$. Observe that when $\{f_0, \dots, f_n\}$ is a w-cofibration (resp. weak equivalence) all the induced maps $A_{i,j} \longrightarrow A'_{i,j}$ are w-cofibrations (resp. weak equivalences). Notice that $S_0\mathscr{C}$ is the trivial category and that $S_1\mathscr{C}$ is isomorphic to $\mathscr{C}$.

For $0 \leq i \leq n$, define $d_k\colon S_n\mathscr{C} \longrightarrow S_{n-1}\mathscr{C}$ to be the functor that drops the k-th row and k-th column from the matrix $\{A_{i,j}\}$. More precisely, d_0 sends the object $\{A_j, A_{i,j}, \alpha_j, a_{i,j}\}$ of $S_n\mathscr{C}$ to the object $\{B_j, B_{i,j}, \beta_j, b_{i,j}\}$ of $S_{n-1}\mathscr{C}$ where $B_j = A_{1,j+1}$, $B_{i,j} = A_{i+1,j+1}$, and the maps β_j and $b_{i,j}$ are the maps induced from α_{j+1} and $a_{i+1,j+1}$ by the universal property of the pushout. For $k > 0$, the functor d_k is defined similarly. For $0 \leq k \leq n$, define $s_k\colon S_n\mathscr{C} \longrightarrow S_{n+1}\mathscr{C}$ to be the functor that repeats the k-th row and k-th column in the matrix $\{A_{i,j}\}$. More precisely, s_k sends the object $\{A_j, A_{i,j}, \alpha_j, a_{i,j}\}$ of $S_n\mathscr{C}$ to the object $\{B_j, B_{i,j}, \beta_j, b_{i,j}\}$ of $S_{n+1}\mathscr{C}$ where

$$
\begin{aligned}
B_j &= \begin{cases} A_j & \text{if } j \leq k \\ A_{j-1} & \text{if } j > k \end{cases} \\
B_{i,j} &= \begin{cases} A_{i,j} & \text{if } j \leq k \\ A_{i,j-1} & \text{if } j > k \text{ and } i \leq k \\ A_{i-1,j-1} & \text{if } i > k \end{cases} \\
\beta_j &= \begin{cases} \alpha_j & \text{if } j < k \\ \mathrm{id} & \text{if } j = k \\ \alpha_{j-1} & \text{if } j > k \end{cases} \\
b_{i,j} &= \begin{cases} a_{i,j} & \text{if } j \leq k \\ a_{i,j-1} & \text{if } j > k \text{ and } i \leq k \\ a_{i-1,j-1} & \text{if } i > k. \end{cases}
\end{aligned}
$$

Observe that the functors d_k and s_k satisfy the simplicial identities and the collection $\{S_n\mathscr{C}\}$ assembles into a simplicial category, which we denote $S_\bullet\mathscr{C}$. Furthermore, the functors d_k and s_k are exact and $S_\bullet$ has the structure of a simplicial Waldhausen category. In particular, we can iterate this construction to form the bisimplicial Waldhausen category $S_\bullet^{(2)}\mathscr{C} = S_\bullet S_\bullet\mathscr{C}$, and the polysimplicial Waldhausen categories $S_\bullet^{(n)}\mathscr{C} = S_\bullet \cdots S_\bullet\mathscr{C}$. We abbreviate the notation for the category of weak equivalences in $S_\bullet^{(n)}\mathscr{C}$ to $wS_\bullet^{(n)}\mathscr{C}$. We are interested in the classifying spaces of these categories, the spaces $|wS_\bullet^{(n)}\mathscr{C}|$. Since $S_0\mathscr{C}$ is the trivial category, we see that $|wS_\bullet\mathscr{C}|$ is connected, and it is not much harder to see that in general $|wS_\bullet^{(n)}\mathscr{C}|$ is $(n-1)$-connected. Observe that the identifications of $\mathscr{C}$ with $S_1\mathscr{C}$ and more generally $S_\bullet^{(n)}\mathscr{C}$ with $S_1S_\bullet^{(n)}\mathscr{C}$ induce maps $\Sigma|w\mathscr{C}| \longrightarrow |wS_\bullet\mathscr{C}|$ and $\Sigma|wS_\bullet^{(n)}\mathscr{C}| \longrightarrow |wS_\bullet^{(n+1)}\mathscr{C}|$ which are inclusions of subcomplexes. It is a fundamental observation of [71, §1.3, 1.5.3] that in the sequence

$$|w\mathscr{C}| \longrightarrow \Omega|wS_\bullet\mathscr{C}| \longrightarrow \Omega^2|wS_\bullet^{(2)}\mathscr{C}| \longrightarrow \cdots$$

all maps beyond the first are homotopy equivalences. This motivates the following definition.

DEFINITION 1.4. The algebraic K-theory of the Waldhausen category $\mathscr{C}$ is the spectrification of the Σ-cofibrant prespectrum $\{|w\mathscr{C}|, |wS_\bullet\mathscr{C}|, |wS_\bullet^{(2)}\mathscr{C}|, \dots\}$. We denote this spectrum by the symbol $K\mathscr{C}$. The algebraic K-groups of $\mathscr{C}$ are the homotopy groups of this spectrum, $K_n\mathscr{C} = \pi_n K\mathscr{C} = \pi_{n+1}|wS_\bullet\mathscr{C}|$. In particular, $K_n\mathscr{C} = 0$ for $n < 0$.

Waldhausen observes that in the special case when $\mathscr{C}$ is an exact category (where the w-cofibrations are the admissible monos, and the weak equivalences are the isomorphisms) the algebraic K-groups defined above agree with those defined by Quillen [58]. In fact, the basic properties of the Q-construction are all easily provable in terms of the $S_\bullet$ construction [67] (see also [53]).

Observe that an exact functor $\mathscr{C} \longrightarrow \mathscr{D}$ induces an exact functor $S_\bullet\mathscr{C} \longrightarrow S_\bullet\mathscr{D}$ and hence exact functors $S_\bullet^{(n)}\mathscr{C} \longrightarrow S_\bullet^{(n)}\mathscr{D}$. This induces a map of prespectra $\{wS_\bullet^{(n)}\mathscr{C}\} \longrightarrow \{wS_\bullet^{(n)}\mathscr{D}\}$ and hence a map of spectra $K\mathscr{C} \longrightarrow K\mathscr{D}$. If the map $|wS_\bullet\mathscr{C}| \longrightarrow |wS_\bullet\mathscr{D}|$ is a weak equivalence, then the maps $|wS_\bullet^{(n)}\mathscr{C}| \longrightarrow |wS_\bullet^{(n)}\mathscr{D}|$ are weak equivalences and therefore homotopy equivalences. Since these prespectra are Σ-cofibrant, maps between them that are spacewise homotopy equivalences induce homotopy equivalences of their spectrifications. In other words, an exact functor that induces a weak equivalence on $|wS_\bullet -|$ induces a homotopy equivalence of K-theory spectra. For this reason, although Waldhausen defines the algebraic K-theory of a Waldhausen category $\mathscr{C}$ to be the space $\Omega|wS_\bullet\mathscr{C}|$, all the results we use from [71] apply to the K-theory spectra, even when they are stated only for the K-theory spaces, and we will use them this way without further comment; moreover, whenever we assert a result about K-theory, we mean the result about the K-theory spectra unless otherwise noted.

2. Cylinders, homotopies, and approximation theorems

Let $\mathrm{MOR}\,\mathscr{C}$ be the category whose objects are the morphisms of $\mathscr{C}$ and whose morphisms are the commutative diagrams. For A, A', B, B' objects of $\mathscr{C}$ and $a\colon A \longrightarrow A'$ and $b\colon B \longrightarrow B'$ maps of $\mathscr{C}$, a and b are objects of $\mathrm{MOR}\,\mathscr{C}$. If $f\colon A \longrightarrow B$ and $f'\colon A' \longrightarrow B'$ are maps in $\mathscr{C}$ that make the diagram

$$\begin{array}{ccc} A & \xrightarrow{f} & B \\ {\scriptstyle a}\downarrow & & \downarrow{\scriptstyle b} \\ A' & \xrightarrow[f']{} & B' \end{array}$$

commute, then (f, f') is a map in $\mathrm{MOR}\,\mathscr{C}$. When $\mathscr{C}$ is a Waldhausen category, we can give $\mathrm{MOR}\,\mathscr{C}$ the structure of a Waldhausen category by saying that a map (f, f') is a w-cofibration (resp. weak equivalence) of $\mathrm{MOR}\,\mathscr{C}$ if both f and f' are w-cofibrations (resp weak equivalences) of $\mathscr{C}$. We shall also need to define a Waldhausen category $C\,\mathrm{MOR}\,\mathscr{C}$, the underlying category of which is the full subcategory of $\mathrm{MOR}\,\mathscr{C}$ of objects $a\colon A \longrightarrow A'$ that are morphisms in $\mathrm{co}(\mathscr{C})$. The weak equivalences in $C\,\mathrm{MOR}\,\mathscr{C}$ are the weak equivalences of $\mathrm{MOR}\,\mathscr{C}$ that lie in

$C\,\text{Mor}\,\mathscr{C}$. A map (f, f') from $a\colon A \rightarrowtail A'$ to $b\colon B \rightarrowtail B'$ is a w-cofibration of $C\,\text{Mor}\,\mathscr{C}$ if f is a w-cofibration of $\mathscr{C}$ and the induced map $A' \amalg_A B \longrightarrow B'$ is a w-cofibration in $\mathscr{C}$. Verification that $C\,\text{Mor}\,\mathscr{C}$ is a Waldhausen category can be found in [71, 1.1.1].

Definition 2.1. (cf. [71, 1.6]) Let $\mathscr{C}$ be a Waldhausen category. A cylinder functor is a functor $T\colon \text{Mor}\,\mathscr{C} \longrightarrow \mathscr{C}$ together with natural transformations i_1, i_2, and p that make the following diagram commute for a morphism $f\colon A \longrightarrow B$ in $\mathscr{C}$:

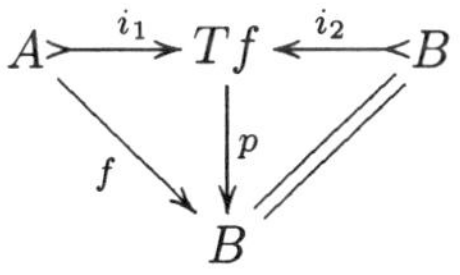

and that satisfies the following properties:

(i) $i_1 \amalg i_2\colon A \amalg B \rightarrowtail Tf$ is in $\text{co}(\mathscr{C})$.
(ii) The functor $(A \longrightarrow B) \mapsto (A \amalg B \xrightarrow{i_1 \amalg i_2} Tf)$ is an exact functor $\text{Mor}\,\mathscr{C} \longrightarrow C\,\text{Mor}\,\mathscr{C}$.
(iii) $T(* \longrightarrow B) = B$, with p and i_2 the identity map.

We say that the cylinder functor satisfies the cylinder axiom if in addition $p\colon Tf \longrightarrow B$ is in $\text{w}(\mathscr{C})$ for all morphisms f. We will often refer to i_1 and i_2 as face maps and to p as the collapse map.

The importance of cylinder functors is shown by Waldhausen's approximation theorem [71, 1.6.7].

Theorem 2.2 (Approximation Theorem). *Let $\mathscr{A}$ and $\mathscr{B}$ be saturated Waldhausen categories, where $\mathscr{A}$ has a cylinder functor that satisfies the cylinder axiom. Let $F\colon \mathscr{A} \longrightarrow \mathscr{B}$ be an exact functor such that*

(i) *If f is a morphism in $\mathscr{A}$ such that $F(f)$ is in $\text{w}(\mathscr{B})$, then f is in $\text{w}(\mathscr{A})$.*
(ii) *For any object $A \in \mathscr{A}$ and any map $f\colon FA \longrightarrow B$, there exists a map $g\colon A \longrightarrow A'$ in $\mathscr{A}$ and a weak equivalence $h\colon FA' \longrightarrow B$ in $\text{w}(\mathscr{B})$ such that $f = h \circ F(g)$:*

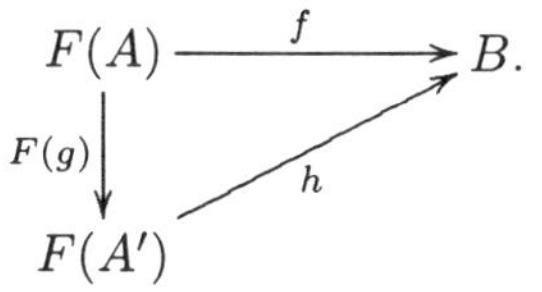

Then F induces a homotopy equivalence $K\mathscr{A} \longrightarrow K\mathscr{B}$.

Remark 2.3. In [71] it is required that g be a w-cofibration, but [68] points out that this requirement is unnecesary since we can use cylinders to replace an arbitrary map with a w-cofibration.

Often, we have the situation where it is easy to make the diagram in 2.2(ii) commute up to some kind of homotopy. By integrating the idea of homotopy into Waldhausen's language of K-theory, we can prove two easy but extremely useful corollaries of Waldhausen's approximation theorem. To this end, we offer the following definitions.

DEFINITIONS 2.4. Let $\mathscr{C}$ be a Waldhausen category with cylinder functor T. Observe that T gives an exact functor $I\colon \mathscr{C} \longrightarrow \mathscr{C}$ by restriction along the exact functor $\mathbf{1}\colon \mathscr{C} \longrightarrow \mathrm{MOR}\,\mathscr{C}$ that sends an object to its identity morphism. We call (W, j_1, j_2, q) a cylinder object of the object X if $W = IX$, $q = p$ (the collapse map) and either $j_1 = i_1$ and $j_2 = i_2$, or $j_1 = i_2$ and $j_2 = i_1$. We say that (W, j_1, j_2, q) is a generalized cylinder object of the object X, if W is the pushout over alternate face maps of a sequence of cylinder objects, $j_1, j_2\colon X \rightarrowtail W$ are the two unused face maps, and $q\colon W \longrightarrow X$ is the gluing of the collapse maps; in particular observe that $q \circ j_i = 1_X$ for $i = 1, 2$. We call two maps $f_1, f_2\colon X \longrightarrow Y$ homotopic if there exists a generalized cylinder object W of X and a map $\phi\colon W \longrightarrow Y$ such that $\phi \circ j_i = f_i$ for $i = 1, 2$. It is easy to see that this specifies an equivalence relation.

Let us say that an exact functor $F\colon \mathscr{C} \longrightarrow \mathscr{D}$ between Waldhausen categories with cylinder functors preserves cylinder objects if there is a natural isomorphism $\alpha\colon FI_{\mathscr{C}} \cong I_{\mathscr{D}}F$ such that $\alpha \circ F(i_k) = i_k$ and $p \circ \alpha = F(p)$:

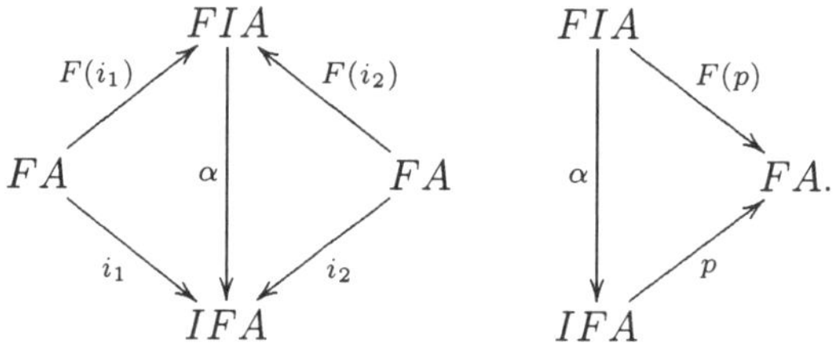

Observe that a functor that preserves cylinder objects also preserves generalized cylinder objects, in the sense that α gives an isomorphism of FW to a generalized cylinder object W' with $\alpha \circ F(j_k) = j'_k$ and $q' \circ \alpha = F(q)$. It is easy to see that when F preserves cylinder objects, F also preserves the relation of homotopy of morphisms.

THEOREM 2.5 (HOMOTOPY APPROXIMATION THEOREM). *Let $\mathscr{A}$ and $\mathscr{B}$ be small saturated Waldhausen categories with cylinder functors satisfying the cylinder axiom. Let $F\colon \mathscr{A} \longrightarrow \mathscr{B}$ be an exact functor that preserves cylinder objects and such that*

(i) *If f is a morphism in $\mathscr{A}$ such that $F(f)$ is in $\mathrm{w}(\mathscr{B})$, then f is in $\mathrm{w}(\mathscr{A})$.*
(ii) *For any object $A \in \mathscr{A}$ and any map $f\colon FA \longrightarrow B$, there exists a map $a\colon A \longrightarrow X$ in $\mathscr{A}$ and a weak equivalence $e\colon FX \longrightarrow B$ in $\mathrm{w}(\mathscr{B})$ such that f is homotopic to $e \circ F(a)$.*

Then F induces a homotopy equivalence $K\mathscr{A} \longrightarrow K\mathscr{B}$.

PROOF. We produce an object A' and maps g, h that satisfy condition (ii) of Waldhausen's approximation theorem.

We have assumed that f is homotopic to $e \circ F(a)$, so there exists a generalized cylinder object (W', j'_1, j'_2, q') of FA and a map $\psi\colon W' \longrightarrow B$ with $\psi \circ j'_1 = f$ and $\psi \circ j'_2 = e \circ Fa$. We can construct in $\mathscr{A}$ the generalized cylinder object W

with the same gluings of faces; then we have $\alpha\colon FW \cong W'$ with $\alpha \circ F(j_i) = j'_i$, since F preserves cylinder objects.

Let $A' = W \amalg_a X$, and let g be the evident map $A \longrightarrow A'$ induced by $j_1\colon A \longrightarrow W$. Then α induces an isomorphism $FA' \longrightarrow W' \amalg_{F(a)} FX$, which we will denote by $\tilde{\alpha}$. Consider the map $\tilde{h} = \psi \amalg_{F(a)} e\colon W' \amalg_{F(a)} FX \longrightarrow B$. The inclusion $FX \longrightarrow W' \amalg_{F(a)} FX$ is a weak equivalence by property 1.2.(ii), since it is the pushout of the following weak equivalence of diagrams

$$\begin{array}{ccccc} FX & \xleftarrow{F(a)} & FA & \xrightarrow{\mathrm{id}} & FA \\ \sim\downarrow \mathrm{id} & & \sim\downarrow \mathrm{id} & & \sim\downarrow j'_1 \\ FX & \xleftarrow[F(a)]{} & FA & \xrightarrow[j'_1]{} & W'. \end{array}$$

The composite of this inclusion with $\tilde{h}$ is the weak equivalence e, so we conclude that $\tilde{h}$ is a weak equivalence since $\mathscr{B}$ is saturated by assumption. Choosing h to be the weak equivalence $\tilde{h} \circ \tilde{\alpha}$ makes diagram 2.2(ii) commute. □

The next corollary of Waldhausen's approximation theorem requires some preliminary definitions.

DEFINITIONS 2.6. We say that a map $f\colon X \longrightarrow Y$ is a homotopy equivalence if there exists a morphism $g\colon Y \longrightarrow X$ so that $f \circ g$ and $g \circ f$ are homotopic to the respective identity morphisms. In this case, we call g a homotopy inverse to f. We say that $\mathscr{C}$ is a category with w-cofibrations and homotopy equivalences or a Waldhausen homotopy category (WH category for short) if $\mathscr{C}$ is a saturated Waldhausen category with cylinder functor satisfying the cylinder axiom such that the weak equivalences are the homotopy equivalences.

We need to make an observation about the "derived" category of a WH category, the category formed by inverting the homotopy equivalences. First note that if f is a homotopy equivalence and g a homotopy inverse to f, then $g \circ f$ and $f \circ g$ are both identity morphisms in the derived category. From this, it is easy to see that every map in the derived category of $\mathscr{C}$ is represented by a map in $\mathscr{C}$.

One can ask when two maps give the same map in the derived category. First observe that for homotopic maps f_1, f_2 and homotopic maps g_1, g_2, the compositions $f_1 \circ g_1$ and $f_2 \circ g_2$ are homotopic; so we can form the homotopy category, $h\mathscr{C}$, whose objects are the objects of $\mathscr{C}$ and whose maps are homotopy classes of maps. It is straightforward to verify that homotopic maps represent the same map in the derived category, and that if two maps represent the same map in the derived category then they are homotopic. We conclude that the natural map from $\mathscr{C}$ to its derived category factors through $h\mathscr{C}$, and that the map from $h\mathscr{C}$ to the derived category is actually an isomorphism (not merely equivalence) of categories.

The next result is now an immediate corollary to Theorem 2.5.

COROLLARY 2.7. *Let $\mathscr{A}$ and $\mathscr{B}$ be WH categories. Suppose $F\colon \mathscr{A} \longrightarrow \mathscr{B}$ is an exact functor that preserves cylinder objects and that passes to an equivalence on the derived categories. Then F induces a homotopy equivalence $K\mathscr{A} \longrightarrow K\mathscr{B}$.*

PROOF. We reduce to Theorem 2.5: Condition (i) is clear. For condition (ii), choose $X \in \mathscr{A}$ so that FX is isomorphic to B in the derived category of $\mathscr{B}$. Since every map in the derived category of $\mathscr{B}$ is represented by an actual map in $\mathscr{B}$, we can choose $e\colon FX \longrightarrow B$ that represents this isomorphism. Then e is a homotopy equivalence; let e' be a homotopy inverse. Now there exists a map $a\colon A \longrightarrow X$ so that $Fa\colon FA \longrightarrow FX$ represents the same map as $e' \circ f$ in the derived category of $\mathscr{B}$. We conclude that f is homotopic to $e \circ Fa$. □

3. Application to categories of R-modules

For an S-algebra R, let $\mathscr{C}_R$ be the full subcategory $\mathscr{M}_R$ consisting of the cell R-modules, and $\mathscr{CW}_R$ the category of CW R-modules and cellular maps. We denote by $f\mathscr{C}_R$ the full subcategory of $\mathscr{C}_R$ of finite cell R-modules and $f\mathscr{CW}_R$ the full subcategory of $\mathscr{CW}_R$ of finite CW R-modules; more precisely, we must choose small full subcategories containing at least one object of each isomorphism class, but the fact that the category of spectra has canonical colimits allows a strict interpretation of the definition of cell and CW R-modules under which the categories $f\mathscr{C}_R$ and $f\mathscr{CW}_R$ are already small. When $\mathscr{C}$ is one of the categories $f\mathscr{C}_R$ or $f\mathscr{CW}_R$, we can give $\mathscr{C}$ the structure of a WH category as follows. We define the category of w-cofibrations, $\mathrm{co}(\mathscr{C})$, to consist of those maps which are isomorphic (in $\mathrm{MOR}\,\mathscr{C}$) to the inclusion of a subcomplex, and the category of weak equivalences $\mathrm{w}(\mathscr{C})$ to consist of those maps in $\mathscr{C}$ which are homotopy equivalences. We take as our cylinder functor the ordinary mapping cylinder.

PROPOSITION 3.1. *These definitions specify structures of WH-categories on $f\mathscr{C}_R$ and $f\mathscr{CW}_R$ and the inclusion $f\mathscr{CW}_R \longrightarrow f\mathscr{C}_R$ is an exact functor which preserves cylinder objects. Furthermore, when R is connective, this inclusion induces a homotopy equivalence of K-theory spectra.*

PROOF. We check the definitions directly. Let $\mathscr{C}$ be either $f\mathscr{C}_R$ or $f\mathscr{CW}_R$ with $\mathrm{co}(\mathscr{C})$ and $\mathrm{w}(\mathscr{C})$ as above.

First we check that $\mathrm{co}(\mathscr{C})$ is a category. Let $f\colon A \longrightarrow B$ and $g\colon B \longrightarrow C$ be arrows in $\mathscr{C}$. These are isomorphic to inclusions of subcomplexes, and without loss of generality we can assume that f is the inclusion of a subcomplex and that g is isomorphic to an inclusion $g'\colon B' \longrightarrow C$ with the isomorphism on the codomain C the identity:

$$\begin{array}{ccc} B & \xrightarrow{\;b\;} & B' \\ {\scriptstyle g}\downarrow & & \downarrow{\scriptstyle g'} \\ C & = & C. \end{array}$$

By choosing different sequential filtrations if necessary we can assume that the map $B \longrightarrow C$ is sequentially cellular and therefore so is the map $B \longrightarrow B'$ (we adjust the sequential filtration on A as well if necessary so that it remains a subcomplex of B). Since C is built from B' by attaching cells, we can form an isomorphic complex D by attaching the same cells to B via b^{-1}. In the CW case, D is CW and the isomorphisms to and from C are cellular because we have assumed that the isomorphisms b and b^{-1} are cellular. Now A is a subcomplex of B, which is a subcomplex of D and the map $A \longrightarrow D$ is isomorphic to the map $A \longrightarrow C$.

Properties 1.1(i) and (ii) and 1.2(i) are clear as will be 1.2(ii) once we show 1.1(iii). Given a diagram in $\mathscr{C}$

$$\begin{array}{ccc} A & \overset{f}{\rightarrowtail} & B \\ {\scriptstyle g}\downarrow & & \\ C & & \end{array}$$

with f an arrow in co($\mathscr{C}$), we show that we can find a pushout in $\mathscr{C}$ so that the map from C is the inclusion of a subcomplex. We can assume without loss of generality that f is the inclusion of a subcomplex and that the map g is sequentially cellular. Since B is built from A by attaching cells, we can form a cell complex D by attaching these cells to C via $g\colon A \longrightarrow C$. In the CW case, D will be a CW complex since we have assumed that g is cellular. For categorical reasons, D must be a pushout of the above diagram, and by construction the map from C is the inclusion of a subcomplex.

If R is connective, we can approximate any finite cell R-module by a finite CW R-module, and the last statement follows by Theorem 2.5. □

DEFINITION 3.2. We define the algebraic K-theory of the S-algebra R to be the spectrum $Kf\mathscr{C}_R$, and we denote it by KR. We define the algebraic K-groups of R to be the homotopy groups of this spectrum, $K_nR = \pi_n KR = K_n f\mathscr{C}_R$.

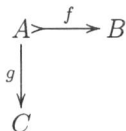

Although the categories $f\mathscr{C}_R$ and $f\mathscr{C}\mathscr{W}_R$ seem the most natural choices for K-theory, there are many other possibilities. Indeed, since pushouts along cofibrations in $\mathscr{M}_R$ preserve weak equivalences, it is easy to see that any subcategory of $\mathscr{M}_R$ that is a category with w-cofibrations such that all of the w-cofibrations are cofibrations becomes a Waldhausen category by taking the weak equivalences to be the ordinary weak equivalences. In particular, when $\mathscr{X}$ is a small full subcategory of $\mathscr{M}_R$ that contains the trivial R-module and is closed under pushouts along cofibrations, then $\mathscr{X}$ is a category with w-cofibrations the set of all cofibrations in $\mathscr{X}$, and in this way $\mathscr{X}$ becomes a Waldhausen category. We shall call the resulting Waldhausen category structure on $\mathscr{X}$ the standard Waldhausen structure. If $\mathscr{C}$ is a full subcategory of $\mathscr{C}_R$ that is small, contains the trivial R-module, and is closed under pushouts along maps isomorphic to inclusions of subcomplexes, then $\mathscr{C}$ is a category with w-cofibrations the set of maps in $\mathscr{C}$ isomorphic to the inclusion of a subcomplex and in this way becomes a Waldhausen category. We shall call the resulting Waldhausen category structure on $\mathscr{C}$ the standard cellular Waldhausen structure. If $\mathscr{X}$ or $\mathscr{C}$ is closed under smashing with I_+, then the mapping cylinder gives a cylinder functor satisfying the cylinder axiom, which we shall call the standard cylinder functor. Since a standard cellular Waldhausen category with the the standard cylinder functor is a WH category, we shall call such a category a standard WH category. It might at first appear that the standard Waldhausen structures are somewhat rare, but the following remark demonstrates that they are actually quite common.

REMARK 3.3 (SMALLEST STANDARD WALDHAUSEN CATEGORIES). If $\mathscr{A}$ is a set of objects of $\mathscr{M}_R$ that is not necessarily closed under pushouts along cofibrations, we can form a small category $\mathscr{B}$ containing $\mathscr{A}$ that is. We let $\mathscr{B}$ be the union of an expanding sequence of small categories $\mathscr{A}_0 \longrightarrow \mathscr{A}_1 \longrightarrow \mathscr{A}_2 \longrightarrow \cdots$, where $\mathrm{Ob}(\mathscr{A}_0) = \mathscr{A}$ and $\mathscr{A}_{n+1}$ is the full subcategory of $\mathscr{M}_R$ of objects that are pushouts of diagrams (with one leg a cofibration) in $\mathscr{A}_n$ (one choice of object for each such diagram). Since the set of all maps in $\mathscr{A}_n$ is small (by induction), $\mathscr{A}_{n+1}$ is a small category. It is easy to see that $\mathscr{B}$ has a kind of universal property: whenever a standard Waldhausen category contains a full subcategory equivalent to $\mathscr{A}$ (regarded as a full subcategory of $\mathscr{M}_R$), it must contain a full subcategory compatibly equivalent to $\mathscr{B}$. For this reason, we will refer to $\mathscr{B}$ as the smallest standard Waldhausen category containing $\mathscr{A}$. Observe that when all the objects of $\mathscr{A}$ have the weak homotopy types of finite cell complexes then so do all the objects of $\mathscr{B}$ (by Corollary I.6.5).

Often we will want our standard Waldhausen categories to have the standard cylinder functor. In forming $\mathscr{A}_{n+1}$ from $\mathscr{A}_n$ above we could also include $X \wedge I_+$ for each $X \in \mathscr{A}_n$. In this case, $\mathscr{A}_{n+1}$ will still be small, but now $\mathscr{B}$ will be closed under smashing with I_+, and hence have the standard cylinder functor. It is easy to see that $\mathscr{B}$ will now have a similar universal property with respect to standard Waldhausen categories with the standard cylinder functor. For this reason we will refer to the category constructed in this way as the smallest standard Waldhausen category with standard cylinder functor containing $\mathscr{A}$. Again, when all the objects of $\mathscr{A}$ have the weak homotopy types of finite cell complexes then so do all the objects of $\mathscr{B}$.

If $\mathscr{A} \subset \mathscr{C}_R$, we can do a similar construction but using maps isomorphic to inclusions of subcomplexes in place of cofibrations. Then the resulting category $\mathscr{B} \subset \mathscr{C}_R$ is a standard cellular Waldhausen category (a WH category if we include smashes with I_+) and has a similar universal property with respect to standard cellular Waldhausen categories. We shall not actually use this construction, but we could call $\mathscr{B}$ in this case the smallest standard cellular Waldhausen category containing $\mathscr{A}$ (or if we include smashes with I_+, smallest standard WH category containing $\mathscr{A}$). Furthermore, observe that if all the objects of $\mathscr{A}$ have the homotopy types of finite cell complexes, then so do all the objects of $\mathscr{B}$.

One advantage of the standard Waldhausen structures is that inclusions of subcategories are exact functors.

PROPOSITION 3.4. *Suppose $\mathscr{X}$ is a subcategory of $\mathscr{Y}$. If $\mathscr{X}$ and $\mathscr{Y}$ are both standard Waldhausen categories or both standard cellular Waldhausen categories, then the inclusion $\mathscr{X} \longrightarrow \mathscr{Y}$ is an exact functor. If $\mathscr{X}$ is a standard cellular Waldhausen category and $\mathscr{Y}$ is a standard Waldhausen category, then the inclusion $\mathscr{X} \longrightarrow \mathscr{Y}$ is an exact functor.*

Many standard Waldhausen categories have K-theory equivalent to $f\mathscr{C}_R$. The following proposition follows directly from Theorem 2.5 (and the Whitehead

Theorem) and will often apply to the smallest standard categories constructed above.

PROPOSITION 3.5. *Let $\mathscr{X}$ be a standard Waldhausen category with standard cylinder functor or a standard WH category. If $\mathscr{X}$ contains $f\mathscr{C}_R$, and if furthermore each object of $\mathscr{X}$ is weakly equivalent to a finite cell complex, then the induced map of K-theory spectra is a homotopy equivalence.*

The K-theory we consider, the K-theory of $f\mathscr{C}_R$, is best thought of as analogous to the K-theory of finitely generated free modules: indeed, since all objects are constructed from the sphere R-modules by a finite number of extensions by spheres, it follows immediately that the obvious homomorphism $K_0^f \pi_0 R \longrightarrow K_0 R$ (induced by $[n] \mapsto \vee_{i=1}^n S_R$) is surjective, where "$K_0^f \pi_0 R$" denotes K_0 of the finitely generated free modules of the ring $\pi_0 R$. When R is connective this homomorphism is an isomorphism whose inverse is given by the Euler characteristic of a CW object X, the alternating sum of the classes of $C_n(X)$, where C_* is the chain functor of IV§3. For this reason, categories that could be reasonable alternatives to the categories $f\mathscr{C}_R$ and $f\mathscr{C}\mathscr{W}_R$ would be those small categories of semi-finite cell R-modules that are standard WH categories. When such a category contains $f\mathscr{C}_R$, it follows from [68, 1.10.1], [71, 1.5.9] and the argument of [24, §1] (as observed in [68, 1.10.2]) that the inclusion will induce an isomorphism of homotopy groups of K-theory spectra except in dimension zero. Intuitively, whereas the K-theory of $f\mathscr{C}_R$ or $f\mathscr{C}\mathscr{W}_R$ is like the K-theory of the finitely generated free modules, we might think of the K-theory of semi-finite objects as analogous to the K-theory of the finitely generated projective modules.

We conclude this section by remarking that when R is an A_∞ ring spectrum but not an S-algebra, we can make analogous observations about the K-theory of categories of its modules. However the functor $S \wedge_{\mathscr{L}} (-)$ is an exact functor that converts such a category to the corresponding category of $S \wedge_{\mathscr{L}} R$-modules and induces a homotopy equivalence of K-theory spectra by Theorem 2.5. Thus, results about the K-theory of A_∞ ring spectra follow from results about the K-theory of S-algebras.

4. Homotopy invariance and Quillen's algebraic K-theory of rings

In this section we prove some properties of the K-theory of the category $f\mathscr{C}_R$ and compare with the K-theory of (discrete) rings.

We observe that K-theory as defined above gives a functor from the category of S-algebras to the stable category which has nice homotopical properties.

PROPOSITION 4.1. *If $\phi\colon A \longrightarrow B$ is a map of S-algebras, then the functor $B \wedge_A (-)\colon f\mathscr{C}_A \longrightarrow f\mathscr{C}_B$ (or $f\mathscr{C}\mathscr{W}_A \longrightarrow f\mathscr{C}\mathscr{W}_B$) is exact and preserves cylinder objects. This association makes K into a functor from the category of S-algebras to the stable category.*

PROOF. The first statement follows from III.4.1, the second from the isomorphisms $C \wedge_B (B \wedge_A (-)) \cong C \wedge_A (-)$ and $A \wedge_A (-) \cong \mathrm{id}$. □

PROPOSITION 4.2. *If $\phi\colon A \longrightarrow B$ is a map of S-algebras that is a weak equivalence, then $K\phi$ is a homotopy equivalence.*

PROOF. From III.4.2, $B \wedge_A (-)$ induces an equivalence of derived categories $\mathscr{D}_A \longrightarrow \mathscr{D}_B$, which restricts to an equivalence of the derived categories of finite cell complexes by an easy application of the Whitehead theorem. The result follows from Corollary 2.7. $\square$

We compare this K-theory with Quillen's algebraic K-theory. Let k be a ring, and let Hk denote the Eilenberg–Mac Lane S-algebra of k. We shall use the symbol $K^f k$ for the algebraic K-theory of the finitely generated free modules of k, a covering spectrum of Kk.

THEOREM 4.3. *KHk is homotopy equivalent to $K^f k$, naturally up to homotopy in k.*

PROOF. We can identify $K^f k$ with the K-theory of the WH category of finite free k-chain complexes with w-cofibrations the split monics, weak equivalences the quasi-isomorphisms, and the cylinder functor given by the usual mapping cylinder. (see, for example, [68, 1.11.7].) We will denote this WH category as $f\mathscr{C}\mathscr{W}_k$.

The functor $C_*\colon f\mathscr{C}\mathscr{W}_{Hk} \longrightarrow f\mathscr{C}\mathscr{W}_k$ of IV§2 is exact and preserves cylinder objects. By the Hurewicz theorem IV.3.6, a map between finite CW modules is a weak equivalence if and only if its image under C_* is a quasi-isomorphism, hence the theorem will follow from Theorem 2.2 if we can show that condition (ii) holds.

Given a finite free chain complex M_*, we can actually construct a CW Hk-module X whose cellular chain complex $C_*(X)$ is isomorphic to M_*. We proceed by induction. Since M_* is finite, M_i is zero below some m, and we take the i-skeleton of X, X^i to be the trivial Hk-module for $i < m$. Now assume that we have constructed X^n and an isomorphism $C_*(X^n) \longrightarrow M_*^{\leq n}$, where $M_*^{\leq n}$ denotes the brutal n-truncation of M_*, i.e. $M_i^{\leq n} = M_i$ for $i \leq n$ and $M_i^{\leq n} = 0$ for $i > n$. By IV.2.3, $\pi_n(X^n) \cong H_n(M_*^{\leq n})$, which is the kernel of the differential d_{n-1}, i.e. the cycles of M_n. Via this isomorphism and a choice of basis for M_{n+1}, the differential d_n specifies a homotopy class of maps from a wedge of S^n_{Hk} to X^n. Choose a representative of this homotopy class, and let X^{n+1} be the CW complex formed by attaching $(n+1)$-cells along this map. By construction $C_{n+1}(X^{n+1}) \cong M_{n+1}$, compatibly with the differentials.

Given an Hk-module A and a map $f\colon C_*(A) \longrightarrow M_*$, we show that we can find a map $a\colon A \longrightarrow X$ such that $C_*(a)$ agrees with f via the isomorphism contructed above. Assume we have constructed this map as far as the n-skeleton of A, i.e. we have $a_n : A^n \longrightarrow X$ such that $C_*(a_n) = f^{\leq n}$. Now A^{n+1} is formed from A^n by attaching a finite wedge $\vee CS^n_{Hk}$ along a map $\alpha\colon \vee S^n_{Hk} \longrightarrow A^n$. The map $f_{n+1}\colon C_{n+1}(A) \longrightarrow M_{n+1}$ specifies a homotopy class of maps

$$(\vee CS^n_{Hk}, \vee S^n_{Hk}) \longrightarrow (X^{n+1}, X^n) \longrightarrow (X, X^n),$$

whose class on $\vee S^n_{Hk}$ agrees with $[a_n \circ \alpha]$, since $\pi_n(X_n)$ coincides with the cycles of M_n. We choose a representative in the homotopy class whose restriction to

S^n_{Hk} is $a_n \circ \alpha$. This extends to a map

$$a_{n+1}\colon A^{n+1} \longrightarrow X^{n+1} \longrightarrow X,$$

and by construction $C_*(a_{n+1})$ agrees with $f^{\leq n+1}$. □

REMARK 4.4. For a connective S-algebra R with $k = \pi_0(R)$, the functor C_* of IV§3 is an exact functor $f\mathscr{C}\mathscr{W}_R \longrightarrow f\mathscr{C}\mathscr{W}_k$. The induced map of K-theory can be thought of as "discretization" and factors as $KR \longrightarrow KHk \longrightarrow K^f k$.

REMARK 4.5. Another question one may ask is how the K-theory of k compares with the K-theory of k regarded as an A_∞ ring, i.e. the K-theory of the category of finite cell A_∞ k-modules (as constructed in [35]). In fact, Propositions 4.1 and 4.2 have exact analogs in the theory of discrete A_∞ rings (with close analogs of the proofs). In particular, it follows from Corollary 2.7 that the natural quasi-isomorphism of the ring k with its A_∞ enveloping algebra induces a homotopy equivalence from the free K-theory of k to its A_∞-K-theory.

5. Morita equivalence

Next, we discuss Morita equivalence, the relationship of the category of R-modules to the category of modules over the analogue of a matrix ring of R. We introduce the shorthand notation $\vee_n X$ for $\vee_{i=1}^n X$, and we define

$$M_n R = F_R(\vee_n R, \vee_n R),$$

$$M_{n1} = F_R(R, \vee_n R) \cong \vee_n R,$$

and

$$M_{1n} = F_R(\vee_n R, R) \cong \prod_n R.$$

By III.6.12, we see that $M_n R$ is an S-algebra, M_{n1} an $(M_n R, R)$-bimodule, and M_{1n} an $(R, M_n R)$-bimodule. Classical Morita equivalence is the theorem that for a (discrete) ring R, tensoring with these two bimodules gives an equivalence between the category of R-modules and the category of $M_n R$-modules. The observation that this restricts to an equivalence between the categories of finitely generated projective modules proves that Quillen's algebraic K-theory is Morita invariant.

In the case we consider, it is unreasonable to hope for an equivalence between $\mathscr{M}_R$ and $\mathscr{M}_{M_n R}$ since products and coproducts are not isomorphic, but we can ask for an equivalence of $\mathscr{D}_R$ and $\mathscr{D}_{M_n R}$. Furthermore, because our K-theory is really the K-theory of free modules, we cannot expect the induced map of K-theory to give an isomorphism on K_0 in general (since, for a discrete ring, the image of the free module of rank one is a projective but not free module for $n > 1$), but we can ask for an isomorphism of the higher K-groups. In this section we find affirmative answers to each of these questions in the following theorems.

THEOREM 5.1 (MORITA EQUIVALENCE). *The derived functors of $M_{n1} \wedge_R (-)$ and $M_{1n} \wedge_{M_n R} (-)$ give an equivalence of categories $\mathscr{D}_R \simeq \mathscr{D}_{M_n R}$, which restricts to an equivalence of the derived categories of semi-finite objects.*

THEOREM 5.2 (MORITA INVARIANCE OF K-THEORY). *The point-set functor $M_{1n} \wedge_{M_nR} (-)$ induces a map of K-theory $KM_nR \longrightarrow KR$, which on homotopy groups (K-groups) sends a generator in dimension zero to n times a generator, and gives an isomorphism on the higher groups.*

We prove Theorem 5.1 by imitating as much as possible the proof of classical Morita equivalence. The following lemma gives a good start in this direction.

LEMMA 5.3. *The (R, R)-bimodules R and $M_{1n} \wedge_{M_nR} M_{n1}$ are isomorphic.*

PROOF. By a comparison of colimits, using the map of S-algebras $R \longrightarrow M_nR$, it is not hard to see that the following diagram is a coequalizer (cf. VII.1.9):

$$M_{1n} \wedge_R M_nR \wedge_R M_{n1} \rightrightarrows M_{1n} \wedge_R M_{n1} \longrightarrow M_{1n} \wedge_{M_nR} M_{n1}.$$

The evaluation map $M_{1n} \wedge_R M_{n1} \longrightarrow R$ coequalizes this diagram, so induces a map $M_{1n} \wedge_{M_nR} M_{n1} \longrightarrow R$, which is evidently an (R, R)-bimodule map. We show that this is an isomorphism by observing that

$$M_{1n} \wedge_R M_nR \wedge_R M_{n1} \rightrightarrows M_{1n} \wedge_R M_{n1} \longrightarrow R$$

is a split coequalizer of S-modules. The splitting is given by maps analogous to those in the discrete case: The map $R \cong R \wedge_R R \longrightarrow M_{1n} \wedge_R M_{n1}$ is the smash product of the map $R \longrightarrow \vee_n R$ that includes it as the first wedge summand with the map $R \longrightarrow F_R(\vee_n R, R)$ that is induced by the map $\vee_n R \longrightarrow R$ that collapses onto the first summand. The map

$$M_{1n} \wedge_R M_{n1} \cong M_{1n} \wedge_R M_{n1} \wedge_R R \longrightarrow M_{1n} \wedge_R M_nR \wedge_R M_{n1}$$

is the smash product of the identity on M_{1n}, the map $M_{n1} \longrightarrow M_nR$ induced by the map $\vee_n R \longrightarrow R$ collapsing onto the first summand in the first variable, and the inclusion of R as the first summand in M_{n1}. It is straightforward to verify that the composites are as required to split the diagram. □

PROOF OF THEOREM 5.1. We verify that the composite $\mathscr{D}_R \longrightarrow \mathscr{D}_{M_nR} \longrightarrow \mathscr{D}_R$ is naturally isomorphic to the identity. Let X be a cell R-module, and let Y be a cell M_nR-module approximation to $M_{n1} \wedge_R X$; we must show that the map

$$M_{1n} \wedge_{M_nR} Y \longrightarrow M_{1n} \wedge_{M_nR} M_{n1} \wedge_R X \cong X$$

is a weak equivalence. Observe that the obvious map $\vee_n M_{n1} \longrightarrow M_nR$ is a weak equivalence and a map of (M_nR, R)-bimodules. Since X is a cell R-module, the map $\vee_n M_{n1} \wedge_R X \longrightarrow M_nR \wedge_R X$ is a weak equivalence and the composite map $\vee_n Y \longrightarrow M_nR \wedge_R X$ is a homotopy equivalence. Now we conclude that $M_{1n} \wedge_{M_nR} \vee_n Y \longrightarrow \vee_n X$ must be a weak equivalence, since the map $\vee_n X \longrightarrow (\prod_n R) \wedge_R X \cong M_{1n} \wedge_R X$ is, but the induced map on homotopy groups is just the direct sum of n copies of the map we are interested in, so this map must also be a weak equivalence.

The reverse composite $\mathscr{D}_{M_nR} \longrightarrow \mathscr{D}_R \longrightarrow \mathscr{D}_{M_nR}$ is simpler. Let X be a cell M_nR-module. Since $M_{n1} \wedge_R (-)$ preserves weak equivalences, the composite functor can be represented by $X \mapsto M_{n1} \wedge_R M_{1n} \wedge_{M_nR} X$. Observe that the evaluation map $M_{n1} \wedge_R M_{1n} \longrightarrow M_nR$ is a weak equivalence of (M_nR, M_nR)-bimodules. On the underlying R-modules it is a map $\vee_n \prod_n R \longrightarrow \prod_n \vee_n R$ inducing an isomorphsim $\oplus \prod R_* \longrightarrow \prod \oplus R_*$. This induces the natural isomorphism (in $\mathscr{D}_{M_nR}$) to the identity.

Since the derived categories of semi-finite objects are full subcategories of the derived categories $\mathscr{D}_R$ and $\mathscr{D}_{M_nR}$, we see that this equivalence restricts, since both functors send wedge summands of finite objects to wedge summands of finite objects. □

Let $\mathscr{C}$ be the smallest standard Waldhausen category with standard cylinder functor containing $f\mathscr{C}_R$ and the image of $f\mathscr{C}_{M_nR}$. By Proposition 3.5, the K-theory of $\mathscr{C}$ is homotopy equivalent to KR. Let $\mathscr{I}$ be the full subcategory of $\mathscr{C}$ of objects weakly equivalent to objects in the image of $f\mathscr{C}_{M_nR}$. Since pushouts along cofibrations are homotopy equivalent to homotopy pushouts, which $M_{1n} \wedge_{M_nR} (-)$ preserves, it is easy to check that $\mathscr{I}$ is closed under pushouts along cofibrations and is therefore a standard Waldhausen category with standard cylinder functor; moreover, the functor $M_{1n} \wedge_{M_nR} (-)\colon f\mathscr{C}_{M_nR} \longrightarrow \mathscr{I}$ is exact.

LEMMA 5.4. *The exact functor* $M_{1n} \wedge_{M_nR} (-)\colon f\mathscr{C}_{M_nR} \longrightarrow \mathscr{I}$ *induces a homotopy equivalence of K-theory.*

PROOF. We apply Theorem 2.5: given $A \in f\mathscr{C}_{M_nR}$, $B \in \mathscr{I}$, and a map $f\colon M_{1n} \wedge_{M_nR} A \longrightarrow B$, we find $X \in f\mathscr{C}_{M_nR}$, a weak equivalence $e\colon M_{1n} \wedge_{M_nR} X \longrightarrow B$ and a map $a\colon A \longrightarrow X$, such that $e \circ M_{1n} \wedge_{M_nR} a$ is homotopic to f. By assumption, B is weakly equivalent to the image of some $X \in f\mathscr{C}_{M_nR}$, so $M_{n1} \wedge_R B$ is an M_nR-module weakly equivalent to $M_{n1} \wedge_R M_{1n} \wedge_{M_nR} X$, which in turn is weakly equivalent to X. Thus, by the Whitehead Theorem, there exists a weak equivalence $\epsilon\colon X \longrightarrow M_{n1} \wedge_R B$. Since the natural map $i\colon M_{n1} \wedge_R M_{1n} \wedge_{M_nR} A \longrightarrow M_nR \wedge_{M_nR} A \cong A$ is a weak equivalence, it has a homotopy retraction r, and there exists a map $a\colon A \longrightarrow X$ such that $\epsilon \circ a$ is homotopic to $(M_{n1} \wedge_R f) \circ r$, again by the Whitehead theorem. Thus the solid line part of the following diagram commutes up to homotopy.

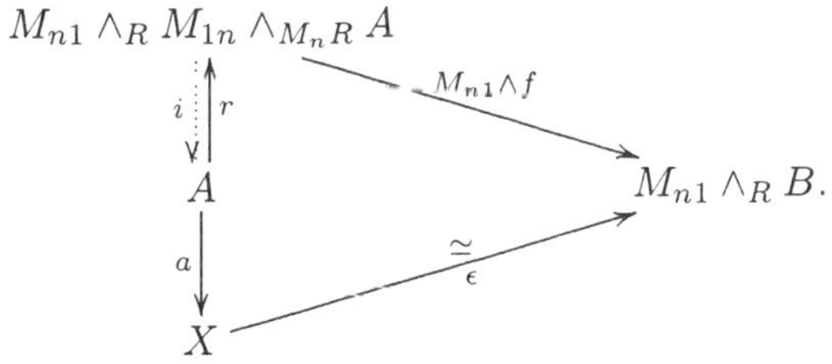

We apply the functor $M_{1n} \wedge_{M_nR} (-)$. The isomorphism constructed in Lemma 5.3 induces a natural transformation $\mu\colon \mathrm{id} \longrightarrow M_{1n} \wedge_{M_nR} M_{n1} \wedge_R (-)$, from which we get the diagram

$$\begin{array}{ccc} M_{1n}\wedge_{M_nR}A & & \\ {\scriptstyle (M_{1n}\wedge i)\circ\mu}\Big\downarrow\Big\uparrow{\scriptstyle M_{1n}\wedge r} & \searrow^{f} & \\ M_{1n}\wedge_{M_nR}A & & B. \\ {\scriptstyle M_{1n}\wedge a}\Big\downarrow & \nearrow^{\simeq}_{\mu^{-1}\circ(M_{1n}\wedge\epsilon)} & \\ M_{1n}\wedge_{M_nR}X & & \end{array}$$

By the associativity of the multiplication pairing, the diagram

$$\begin{CD} M_{1n}\wedge_{M_nR}M_{n1}\wedge_R M_{1n} @>{\cong}>> R\wedge_R M_{1n} \\ @V{M_{1n}\wedge i}VV @VV{\cong}V \\ M_{1n}\wedge_{M_nR}M_nR @>>{\cong}> M_{1n} \end{CD}$$

must commute, and we conclude that $(M_{1n}\wedge_{M_nR}i)\circ\mu$ is the identity. Now, letting $e=\mu^{-1}\circ M_{1n}\wedge_{M_nR}\epsilon$, we see that $e\circ M_{1n}\wedge_{M_nR}a$ is homotopic to f as required. □

LEMMA 5.5. *$\mathscr{I}$ is closed under extensions in $\mathscr{C}$.*

PROOF. We need to show that for a cofibration sequence $A\rightarrowtail B\twoheadrightarrow C$ in $\mathscr{C}$, if A and C are in $\mathscr{I}$, then B is also in $\mathscr{I}$. It suffices to consider the case when A, B, and C are celluar R-modules, since any cofibration sequence can be replaced by a weakly equivalent one of this form. Using the proof of the last lemma, the map $*\longrightarrow C$ allows us to find X in $f\mathscr{C}_{M_nR}$ together with a weak equivalence $e\colon M_{1n}\wedge_{M_nR}X\longrightarrow C$. Composing with the map $c\colon C\longrightarrow\Sigma A$ implied by the cofibration sequence, and applying once again the proof of the last lemma, we find an object Y in $f\mathscr{C}_{M_nR}$, a map $a\colon X\longrightarrow Y$, and a weak equivalence $f\colon M_{1n}\wedge_{M_nR}Y\longrightarrow\Sigma A$ that make the following square homotopy commute:

$$\begin{CD} M_{1n}\wedge_{M_nR}X @>{M_{1n}\wedge a}>> M_{1n}\wedge_{M_nR}Y \\ @V{e}V{\simeq}V @V{\simeq}V{f}V \\ C @>>{c}> \Sigma A. \end{CD}$$

We conclude that the induced map on cofibers $C(M_{1n}\wedge_{M_nR}a)\longrightarrow C(c)$ is a weak equivalence. The functor $M_{1n}\wedge_{M_nR}(-)$ commutes with smashing over S on the right with S_S^{-1} and smashing on the right with the space S^1, both composites of which are homotopic to the identity in $f\mathscr{C}_{M_nR}$. We conclude that the map $C(M_{1n}\wedge_{M_nR}a)\wedge_S S_S^{-1}\longrightarrow C(c)\wedge_S S_S^{-1}$ is a weak equivalence. But $C(c)\wedge_S S_S^{-1}$ is homotopy equivalent to B and $C(M_{1n}\wedge_{M_nR}a)\wedge_S S_S^{-1}\cong M_{1n}\wedge_{M_nR}(C(a)\wedge_S S_S^{-1})$ is in the image of $f\mathscr{C}_{M_nR}$, hence B is in $\mathscr{I}$. □

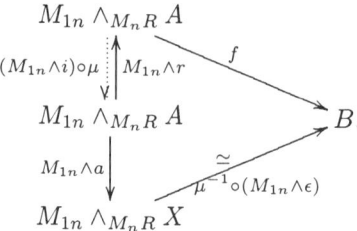

PROOF OF THEOREM 5.2. The inclusion $\mathscr{I}\longrightarrow\mathscr{C}$ is an exact functor. It is easy to see that on K_0 it sends a generator (S_{M_nR}) to n times a generator $(M_{1n}\wedge_{M_nR}M_nR\wedge_S S_S\cong M_{1n}\wedge_S S_S\simeq\vee_n S_R)$. We want to see that it induces an isomorphism of the higher K-groups. Let $\mathscr{J}$ be the full subcategory of $\mathscr{C}$ of all objects whose class in $K_0\mathscr{C}$ is in the image of $K_0\mathscr{I}$ (so in particular $\mathscr{I} \subset \mathscr{J}$). It follows from the relations that define K_0 that $\mathscr{J}$ inherits the structure of a standard Waldhausen category with standard cylinder functor. By Proposition 3.4, the inclusions $\mathscr{I} \longrightarrow \mathscr{J}$ and $\mathscr{J} \longrightarrow \mathscr{C}$ are exact functors.

We use the argument of [24, §1] to show that $\mathscr{I}$ is strictly cofinal in $\mathscr{J}$ (in the sense of [71, 1.5.9]). We define an equivalence relation on the objects of $\mathscr{C}$ by letting A and A' be equivalent if there exists some $X \in \mathscr{I}$ such that $A \vee X$ is weakly equivalent to $A' \vee X$. Let G be the set of equivalence classes under this relation. Then G is a group under the operation "$\vee$" with the inverse of A represented by $\vee_{n-1}A$. We have an obvious homomorphism $G \longrightarrow K_0\mathscr{C}/K_0\mathscr{I}$; we construct an inverse to this homomorphism. If $A \rightarrowtail B \twoheadrightarrow C$ is a cofibration sequence in $\mathscr{C}$, then $\vee_n A \rightarrowtail B \vee (\vee_{n-1}A) \vee (\vee_{n-1}C) \twoheadrightarrow \vee_n C$ is a cofibration sequence. But $\vee_n A$ and $\vee_n C$ are in $\mathscr{I}$, so $B \vee (\vee_{n-1}A) \vee (\vee_{n-1}C)$ is in $\mathscr{I}$ since $\mathscr{I}$ is closed under extensions in $\mathscr{C}$; therefore, $B \vee (\vee_{n-1}A) \vee (\vee_{n-1}C)$ represents the identity in G and hence B represents the same element as $A \vee C$ in G. If A is weakly equivalent to A' then they represent the same element in G. We see that the association of an object to its class in G satisfies the universal relations that define $K_0\mathscr{C}$ and so specifies a map $K_0\mathscr{C} \longrightarrow G$. This map clearly factors through a map $K_0\mathscr{C}/K_0\mathscr{I} \longrightarrow G$ that is evidently inverse to the map above. Thus, we see that $\mathscr{J}$ consists of the objects whose class in G is the identity, so we conclude that for any $X \in \mathscr{J}$, there exists $Y \in \mathscr{I}$ such that $X \vee Y$ is weakly equivalent to an object of $\mathscr{I}$ and hence $X \vee Y$ is an object of $\mathscr{I}$.

Now by [71, 1.5.9], $\mathscr{I} \longrightarrow \mathscr{J}$ induces a homotopy equivalence of K-theory, but by [68, 1.10.1], $K_i\mathscr{J} \longrightarrow K_i\mathscr{C}$ is an isomorphism for $i > 0$. □

6. Multiplicative structure in the commutative case

In this section, we prove the following theorem (cf. [65]).

THEOREM 6.1. *If R is a connective commutative S-algebra then KR is homotopy equivalent to an E_∞ ring spectrum and therefore weakly equivalent to a commutative S-algebra.*

This result and the results of the next section depend on the following technical lemma, which the reader may recognize as a simple application of the theory of [71, §1.6–1.8] to our new categories. Although we believe that Theorem 6.1 may generalize to non-connective commutative S-algebras, this lemma is peculiar to the connective case and relies on the existence of the ordinary homology theories of IV§3. For this lemma, we need R to be a connective but not necessarily commutative S-algebra. Let $\mathscr{C}$ be a standard Waldhausen category of R-modules that contains $f\mathscr{C}_R$ and that only contains objects of the weak homotopy type of finite cell R-modules. We see by Proposition 3.5 that $K\mathscr{C}$ is homotopy equivalent to KR. We denote by $\mathscr{C}^m$ the full subcategory of objects weakly equivalent to a finite wedge of S^m_R. Observe that for each m, the category of weak equivalences of $\mathscr{C}^m$ is a symmetric monoidal category under the operation of wedge, and denote the associated spectrum as $k\mathscr{C}^m$. Suspension induces a system of maps of spectra $k\mathscr{C}^m \longrightarrow k\mathscr{C}^{m+1}$.

LEMMA 6.2. *The homotopy colimit of the system $\{k\mathscr{C}^m\}$ is homotopy equivalent to $K\mathscr{C}$.*

PROOF. The Hurewicz theorem IV.3.6 allows us to identify $\mathscr{C}^m$ with the full subcategory of $\mathscr{C}$ of objects whose ordinary homology H_*^R is zero in all dimensions except m, and in dimension m is a finitely generated free module. Let $\tilde{\mathscr{C}}^m$ be the full subcategory of $\mathscr{C}$ of objects whose ordinary homology H_*^R is zero in all dimensions except m and in dimension m is a finitely generated stably free module, i.e. is isomorphic to the kernel of a surjective map of finitely generated free modules. Let $\mathscr{C}^{\geq n}$ be the full subcategory of $\mathscr{C}$ of those objects which are $(n-1)$-connected. By the Hurewicz theorem IV.3.6 these are exactly the objects whose homology is zero in dimensions less than n. We give the categories $\mathscr{C}^m$ and $\tilde{\mathscr{C}}^m$ Waldhausen structures by defining the w-cofibrations to be the w-cofibrations of $\mathscr{C}$ whose quotients lie in the subcategory in question. The categories $\mathscr{C}^{\geq n}$ have the structure of standard Waldhausen categories with the standard cylinder functor.

Suspension gives exact functors $\tilde{\mathscr{C}}^m \longrightarrow \tilde{\mathscr{C}}^{m+1}$ and $\mathscr{C}^{\geq n} \longrightarrow \mathscr{C}^{\geq n}$. For $m \geq n$, the inclusion of $\tilde{\mathscr{C}}^m$ in $\mathscr{C}^{\geq n}$ is an exact functor; for fixed n, these inclusions induce a map

$$\operatorname*{hocolim}_{m\to\infty} |wS_\bullet\tilde{\mathscr{C}}^m| \longrightarrow \operatorname*{hocolim}_{\Sigma} |wS_\bullet\mathscr{C}^{\geq n}|,$$

where the colimit on the right is taken over repeated application of suspension. Next observe that ordinary homology H_*^R restricted to $\mathscr{C}^{\geq n}$ is a homology theory in the sense of [71, §1.7] (at least after shifting the indexing), and that the categories $\tilde{\mathscr{C}}^m$ form categories of "spherical objects" for $\mathscr{C}^{\geq n}$ for the class of finitely generated stably free modules. Since this theory satisfies the "Hypothesis" of [71, 1.7.1], we conclude that the map above is a weak equivalence. On the other hand the inclusions $\mathscr{C}^{\geq n} \longrightarrow \mathscr{C}^{\geq n+1}$ are exact functors which induce cofibrations $|wS_\bullet\mathscr{C}^{\geq n}| \longrightarrow |wS_\bullet\mathscr{C}^{\geq n+1}|$, whose colimit is $|wS_\bullet\mathscr{C}|$. Taking the colimit (over n) of the homotopy equivalence above, we get a homotopy equivalence

$$\operatorname*{hocolim}_{m\to\infty} |wS_\bullet\tilde{\mathscr{C}}^m| \longrightarrow \operatorname*{hocolim}_{\Sigma} |wS_\bullet\mathscr{C}|.$$

The maps on the right are all homotopy equivalences (by [71, 1.6.2]), so we conclude that there exists a homotopy equivalence $\operatorname{hocolim} K\tilde{\mathscr{C}}^m \longrightarrow K\mathscr{C}$.

We apply the Strict Cofinality Theorem [71, 1.5.9] to conclude that $K\tilde{\mathscr{C}}^m$ is homotopy equivalent to $K\mathscr{C}^m$. Now we are reduced to comparing $K\mathscr{C}^m$ with $k\mathscr{C}^m$. According to [71, 1.8.1], it suffices to observe that cofibrations in $\mathscr{C}^m$ are "splittable up to weak equivalence". Given a cofibration $A \rightarrowtail B$, we can find a basis of the free module $H_m(B)$ that represents the union of bases for H_mA and $H_m(B/A)$. The Hurewicz theorem IV.3.6 now specifies homotopy classes of weak equivalences from the wedge of A and a wedge of spheres to B and to $A \vee B/A$, relative to the maps from A. $\square$

PROOF OF THEOREM 6.1 Let $\mathscr{C}$ be the smallest standard Waldhausen category with standard cylinder functor containing R, S_R^n for all n, and all finite smash products over R of these. It is easy to check that the bifunctor $(-)\wedge_R(-)$

restricts to $\mathscr{C}$ (up to equivalence), so $\mathscr{C}$ is a symmetric bimonoidal category under coproduct and smash product over R. Let $\mathscr{C}^0$ be as in the lemma above. Then $\mathscr{C}^0$ is the full subcategory of $\mathscr{C}$ of objects weakly equivalent to a finite wedge of S_R. Since smash product over R with R and with S_R^n preserve weak equivalences, so do smash products over R with any object of $\mathscr{C}$, and the smash product over R of objects in $\mathscr{C}^0$ is weakly equivalent to a finite wedge of S_R and therefore is an object of $\mathscr{C}^0$. Thus the smash product over R restricts to a bifunctor on $\mathscr{C}^0$ that makes $\mathscr{C}^0$ a symmetric bimonoidal category. By the work of [50], we can construct $k\mathscr{C}^0$ functorially as an E_∞ ring spectrum.

Next observe that suspension and $S_R^{-1}\wedge_R(-)$ give functors $\mathscr{C}^m \longrightarrow \mathscr{C}^{m+1}$ and $\mathscr{C}^{m+1} \longrightarrow \mathscr{C}^m$ for which both composites are weakly equivalent to the identity. We conclude that suspension gives a homotopy equivalence $k\mathscr{C}^m \longrightarrow k\mathscr{C}^{m+1}$, and that $k\mathscr{C}^0$ is homotopy equivalent to $K\mathscr{C}$ by the previous lemma. □

7. The plus construction description of KR

We have observed that the category $f\mathscr{C}_R$ gives a K-theory $Kf\mathscr{C}_R$ that has some right to be called the algebraic K-theory of R. This section is devoted to a comparison with another possible definition, based on Quillen's plus construction. In what follows, R is a fixed connective S-algebra, and $k = \pi_0 R$. We shall make use of classifying spaces of the topological monoids $\mathscr{M}_R(X,X)$. Unfortunately even when $X = S_R$, we cannot guarantee that the inclusion of the identity element is a cofibration. There are well-known ways of overcoming this difficulty, e.g. whiskering the monoids [45] or using thickened realizations [66]. In this and the next section, we shall take advantage of such techniques implicitly wherever necessary without further comment.

Let $\tilde{M}_n R$ be the topological space $f\mathscr{C}_R(\bigvee_n S_R, \bigvee_n S_R)$; then $\pi_0 \tilde{M}_n R \cong M_n(k)$, the ordinary matrix ring of the ring k. Let $\widetilde{GL_n R}$ be the space consisting of those connected components of $\tilde{M}_n R$ whose image in $M_n(k)$ is invertible. Then $\widetilde{GL_n R}$ is a topological monoid; indeed, it is the monoid of homotopy equivalences in $\tilde{M}_n R$. We can consider its classifying space $B\widetilde{GL_n R}$. We have the inclusion $i_n\colon \widetilde{GL_n R} \longrightarrow \widetilde{GL_{n+1} R}$ obtained by sending the last wedge summand to the last wedge summand via the identity map, and it induces $Bi_n\colon B\widetilde{GL_n R} \longrightarrow B\widetilde{GL_{n+1} R}$. Let $B\widetilde{GLR}$ be the telescope of these maps.

Now $\pi_1 B\widetilde{GLR} \cong GL(k)$ has a perfect normal subgroup, so we can form $B\widetilde{GLR}^+$ (Quillen's plus construction). We shall see shortly that $K_0^f k \times B\widetilde{GLR}^+$ is an infinite loop space. Define K^+R to be the connective spectrum obtained by delooping $K_0^f k \times B\widetilde{GLR}^+$. We prove the following "plus equals $S_\bullet$" theorem.

THEOREM 7.1. *K^+R is weakly equivalent to KR.*

First we need to specify the infinite loop space structure on $K_0^f k \times B\widetilde{GLR}^+$. For this, we observe that $K_0^f k \times B\widetilde{GLR}^+$ is the group completion of the classifying space of the topological category $\mathscr{W}$ whose (discrete) set of objects is the finite wedges of S_R and whose space of morphisms is the set of homotopy equivalences topologized as a subspace of the space of morphisms of $\mathscr{M}_R$. Call this group completion B. In the case when $K_0^f k$ is the integers, the classifying space of $\mathscr{W}$

is the disjoint union of the $\widetilde{BGL_n R}$ and we may apply the remarks of [66, §4] to conclude that we have a homology isomorphism to B from the telescope of maps $\coprod \widetilde{BGL_n R}$ to itself induced by the maps $Bi_n\colon \widetilde{BGL_n R} \longrightarrow \widetilde{BGL_{n+1} R}$. This telescope is easily seen to be $K_0^f k \times \operatorname{hocolim}_n \widetilde{BGL_n R}$. We conclude that $B \simeq K_0^f k \times (\operatorname{hocolim}_n \widetilde{BGL_n R})^+$.

In the pathological case when $K_0^f k$ is not the integers, i.e. when there exists a homotopy equivalence $\vee_j S_R \simeq \vee_k S_R$ for $j \neq k$, we still have a homology isomorphism to the group completion B from the telescope T of maps from $B\mathscr{W}$ to itself induced by addition of an identity map on the wedges of sphere R-modules. Proposition 7.2 below allows us to see that $B\mathscr{W}$ is homotopy equivalent to a disjoint union of of some of the $\widetilde{BGL_n R}$, one choice for each isomorphism class of finitely generated free $\pi_0 R$-modules. Now we see that the telescope T is homotopy equivalent to $K_0^f k \times \operatorname{hocolim}_n \widetilde{BGL_n R}$, and we conclude that $B \simeq K_0^f k \times (\operatorname{hocolim}_n \widetilde{BGL_n R})^+$.

To identify the homotopy type of $B\mathscr{W}$ in the pathological case above, we need the following proposition. We will need a similar result again later, and we have written this proposition in the minimal possible generality necessary to handle both cases. The proposition says essentially that if the morphisms in a category are all homotopy equivalences (in a certain sense), then the classifying space of the monoid of endomorphisms of any object is homotopy equivalent to its connected component in the classifying space of the category. Because this proposition has obvious generalizations with more general scope than its use in this section, we break our rule of not mentioning the necessary cofibration assumptions. As always the reader has the choice of deleting the cofibration assumption by using a whiskering technique or employing the thickened realization.

PROPOSITION 7.2. *(cf. [71, 2.2.7]) Let $\mathscr{C}$ be a topological category with discrete set of objects such that the identity morphism (from objects to morphisms) is a cofibration. Let X be an object of $\mathscr{C}$ and denote by $\mathscr{C}_X$ the full subcategory of $\mathscr{C}$ containing X. Suppose that for each morphism $f\colon Y \longrightarrow Z$ in $\mathscr{C}$, there is some $f'\colon Z \longrightarrow Y$ so that $f' \circ f$ and $f \circ f'$ each lie in the same path component of $\mathscr{C}(Y,Y)$ and $\mathscr{C}(Z,Z)$ as the respective identity elements. Then the inclusion $\mathscr{C}_X \longrightarrow \mathscr{C}$ induces a homotopy equivalence of the classifying space of $\mathscr{C}_X$ with the connected component of its image in the classifying space of the category $\mathscr{C}$.*

PROOF. First observe that Quillen's "Theorem A" [58] holds with essentially the same proof for continuous functors between topological categories with discrete object sets whose identity map (objects to morphisms) is a cofibration.

Since the connected component of the image of $\mathscr{C}_X$ in the classifying space of $\mathscr{C}$ is the classifying space of the connected component (as a graph) of $\mathscr{C}$ that contains X, we can reduce to this smaller category and assume without loss of generality that $\mathscr{C}$ is connected (as a graph). Applying Quillen's Theorem A (dual formulation), we are reduced to showing that for every Y in $\mathscr{C}$, the topological category $\mathscr{C}_X/Y$ is contractible. But if $f\colon Y \longrightarrow Z$ is morphism in $\mathscr{C}$, then we have $f'\colon Z \longrightarrow Y$ and paths $\gamma\colon f' \circ f \rightsquigarrow 1_Y$ and $\gamma'\colon f \circ f' \rightsquigarrow 1_Z$. We can interpret the morphisms f and f' as continuous functors $\mathscr{C}_X/Y \longrightarrow \mathscr{C}_X/Z$,

$\mathscr{C}_X/Z \longrightarrow \mathscr{C}_X/Y$, and the paths γ and γ' as continuous functors $\mathscr{C}_X/Y \times I \longrightarrow \mathscr{C}_X/Y$, $\mathscr{C}_X/Z \times I \longrightarrow \mathscr{C}_X/Z$. Passing to the classifying spaces we see that the paths represent homotopies $B(\mathscr{C}_X/Y) \times I \longrightarrow B(\mathscr{C}_X/Y)$ and $B(\mathscr{C}_X/Z) \times I \longrightarrow B(\mathscr{C}_X/Z)$ from the composites $Bf' \circ Bf$ and $Bf \circ Bf'$ to the repective identities. In short, $\mathscr{C}_X/Y$ and $\mathscr{C}_X/Z$ are homotopy equivalent. Since we have reduced to the case when $\mathscr{C}$ is connected (as a graph), we see that $\mathscr{C}_X/Y$ is homotopy equivalent to $\mathscr{C}_X/X$. The lemma is established by the observation that $\mathscr{C}_X/X$ has a final object and therefore is contractible. □

We begin to compare K^+R to $Kf\mathscr{C}_R$. One obvious obstacle is that we have defined K^+R in terms of a topological category and $Kf\mathscr{C}_R$ in terms of a discrete one. Let $w\mathscr{C}^0$ denote the (discrete) full subcategory of $\mathrm{w}(f\mathscr{C}_R)$ whose objects are homotopy equivalent to wedges of S_R; the set of morphisms is the set of homotopy equivalences. Using arguments similar to [71, 2.2], we relate $w\mathscr{C}^0$ to both $\mathscr{W}$ and $wS_\bullet f\mathscr{C}_R$.

LEMMA 7.3. *(cf.* [71, 2.2.5]*) There is a chain of weak equivalences relating the classifying spaces of the categories $\mathscr{W}$ and $w\mathscr{C}^0$. Each map in the chain is a map of E_∞ spaces.*

PROOF. For each k, let $\mathscr{W}\Delta_k$ be the (discrete) category whose objects are the objects of $w\mathscr{C}^0$ and whose morphisms $\mathscr{W}\Delta_k(X,Y)$ consist of the set of continuous maps $\Delta[k] \longrightarrow f\mathscr{C}_R(X,Y)$ whose image lands in the component of a weak equivalence, where $\Delta[k]$ denotes the standard topological k-simplex. In light of the adjunction $\mathscr{T}(\Delta[k]_+, f\mathscr{C}_R(X,Y)) \cong f\mathscr{C}_R(X \wedge \Delta[k]_+, Y)$, we see that this is the same as the set of weak equivalences $X \wedge \Delta[k]_+ \longrightarrow Y$. This is a simplicial category. Let $N_{j,k}$ be the nerve of this category.

If we realize $N_{j,k}$ in the k direction, we obtain a simplicial space that is the nerve of a topological category with a discrete set of objects. We denote this category as $|\mathscr{W}\Delta|$. In particular, the objects of $|\mathscr{W}\Delta|$ are the objects of $w\mathscr{C}^0$ and the morphism space $|\mathscr{W}\Delta|(X,Y)$ is the geometric realization of the total singular complex of the subspace of $f\mathscr{C}_R(X,Y)$ consisting of those components which contain homotopy equivalences. For each $X \in \mathscr{W}$, let $|\mathscr{W}\Delta|_X$ be the full subcategory of $|\mathscr{W}\Delta|$ consisting of the single object X. By the previous proposition, the inclusion $|\mathscr{W}\Delta|_X \longrightarrow |\mathscr{W}\Delta|$ induces a homotopy equivalence from the classifying space of $|\mathscr{W}\Delta|_X$ to its connected component in the classifying space of $|\mathscr{W}\Delta|$. On the other hand we have a natural weak equivalence of monoids $|\mathscr{W}\Delta|(X,X) \longrightarrow \mathscr{W}(X,X)$, giving a weak equivalence of their classifying spaces. Let $|\mathscr{W}\Delta S_R|$ be the full subcategory of $|\mathscr{W}\Delta|$ consisting of the finite wedges of S_R. Then we have weak equivalences $\|N_{j,k}\| \overset{\sim}{\longleftarrow} \|\mathscr{W}\Delta S_R\| \overset{\sim}{\longrightarrow} |\mathscr{W}|$.

Next we produce a weak equivalence between $w\mathscr{C}^0$ and $\mathscr{W}\Delta_*$. The map $\Delta[k]_+ \longrightarrow S^0$ induces a functor $F\colon w\mathscr{C}^0 \longrightarrow \mathscr{W}\Delta_k$ that is the identity on objects. Let $G\colon \mathscr{W}\Delta_k \longrightarrow w\mathscr{C}^0$ be the functor induced by the map $S^0 \longrightarrow \Delta[k]_+$ that sends the non-basepoint to the zeroth vertex of $\Delta[k]$. Then GF is the identity functor on $w\mathscr{C}^0$. We show that FG is homotopic to the identity. Let $H\colon \mathscr{W}\Delta_k \longrightarrow \mathscr{W}\Delta_k$ be the functor that takes X to $X \wedge I_+$ and that on morphisms is induced by a map $I \times \Delta[k] \longrightarrow \Delta[k]$ that is the identity on the bottom face $\{0\} \times \Delta[k]$ and sends the whole top face $\{0\} \times \Delta[k]$ to the zeroth vertex of

$\Delta[k]$. There are obvious natural transformations id $\longrightarrow H$ and $FG \longrightarrow H$ given by the inclusion of bottom face and the inclusion of top face, from which we conclude that FG is homotopic to the identity. We may regard $w\mathscr{C}^0$ as a simplicial category constant in the k direction. The functors F are compatible with the faces and degeneracies (in k), and therefore assemble to a simplicial functor $w\mathscr{C}^0 \longrightarrow \mathscr{W}\Delta_*$ that induces a homotopy equivalence upon passage to classifying spaces.

It is easy to see that the simplicial maps above realize to maps of E_∞ spaces as they are induced by functors that preserve wedges. □

PROOF OF THEOREM 7.1. If we let $\mathscr{C}$ be the category $f\mathscr{C}_R$, then $w\mathscr{C}^0$ is exactly the subcategory of weak equivalences of the category $\mathscr{C}^0$ defined above Lemma 6.2, the associated spectrum of which we denoted $k\mathscr{C}^0$. Again suspension and $S_R^{-1} \wedge_R (-)$ give functors $\mathscr{C}^m \longrightarrow \mathscr{C}^{m+1}$ and $\mathscr{C}^{m+1} \longrightarrow \mathscr{C}^m$ whose composites are weakly equivalent to the identity. We conclude that the maps in the homotopy colimit are homotopy equivalences and that $k\mathscr{C}^0$ is homotopy equivalent to KR. On the other hand, the previous proposition shows that K^+R is weakly equivalent to $k\mathscr{C}^0$. □

REMARK 7.4. Note that we only needed the connectivity hypothesis to show the relationship between $k\mathscr{C}^0$ and $Kf\mathscr{C}_R$. More generally we do have a homotopy equivalence $k\mathscr{C}^0 \simeq K^+R$ (the spectrum whose zeroth space is $K_0^f k \times \widetilde{BGLR^+}$), but there is no reason to expect that the map $k\mathscr{C}^0 \longrightarrow Kf\mathscr{C}_R$ will be a homotopy equivalence. In particular $k\mathscr{C}^0$ cannot see any relationships between spheres of different dimensions. For example, if $K_0^f k = \mathbb{Z}$, but $S_R \simeq S_R^1$, then $|wS_\bullet f\mathscr{C}_R|$ is contractible but $|wN_\bullet\mathscr{C}^0|$ is not.

REMARK 7.5. We should also observe that this allows another interpretation of the discretization map: π_0 applied to the simplicial space $N\mathscr{W}$ gives an E_∞ map $KR(0) \simeq K_0^f k \times \widetilde{BGLR^+} \longrightarrow K_0^f k \times BGLk^+ \simeq K^f k(0)$, which evidently coincides with the discretization map and is a weak equivalence in the case when $R = Hk$.

REMARK 7.6 (MONOMIAL MATRICES). Let $\mathscr{V}$ be the subcategory of $\mathscr{W}$ of those maps $\vee_n S_R \longrightarrow \vee_n S_R$ that are wedges of n maps $S_R \longrightarrow S_R$ in any order. Thinking of $\mathscr{W}(\vee_n S_R, \vee_n S_R)$ as analogous to $GL_n R$, then $\mathscr{V}(\vee_n S_R, \vee_n S_R)$ is analogous to the subgroup of monomial matrices, those matrices with a single non-zero entry in each row and column. Let $R^\times$ denote the monoid $\mathscr{V}(S_R, S_R) = \mathscr{W}(S_R, S_R)$. Then $\mathscr{V}(\vee_n S_R, \vee_n S_R)$ is isomorphic to the monoid $\Sigma_n \int R^\times$ and the classifying space of $\mathscr{V}$ is isomorphic to the disjoint union of the classifying spaces of these monoids; moreover, under this isomorphism the E_∞ space structure induced by wedge sums becomes the E_∞ space structure induced by block sums. We conclude that the group completion of the classifying space of $\mathscr{V}$ is homotopic to $QBR^\times{}_+$, and that $\mathscr{V} \longrightarrow \mathscr{W}$ induces a map of spectra $\Sigma^\infty BR^\times{}_+ \longrightarrow KR$.

REMARK 7.7. (Naturality) Let $A \longrightarrow B$ be a map of S-algebras. We saw in Propostion 4.1 that the functor $B \wedge_A (-)$ induces a map of K-theory spectra $KA \longrightarrow KB$. This also restricts to a continuous functor of topological categories $\mathscr{W}_A \longrightarrow \mathscr{W}_B$ that induces a map of the plus contruction spectra above. We conclude that these two maps represent the same map in the stable category, since this functor commutes up to natural isomorphism with the functors used in comparing K^+ with K.

8. Comparison with Waldhausen's K-theory of spaces

Now we compare the new algebraic K-theory with Waldhausen's algebraic K-theory of spaces. For this, let X be a connected pointed topological space, and let $G = |GSX|$, the geometric realization of the Kan loop group of the based singular complex of X. This is a topological group with non-degenerate identity. We let $R = \Sigma^\infty(G_+)$ (where the plus subscript is union with a disjoint basepoint) and we note that R is an S-algebra (IV.7.8) with $k = \pi_0 R = \mathbb{Z}[\pi_0 G]$.

DEFINITION 8.1. Let $\mathcal{H}_n^m$ denote the topological monoid of pointed G-equivariant homotopy equivalences of $\bigvee_n \Sigma^m G_+$ with itself, and let $B\mathcal{H}_n^m$ denote its classifying space. We have monoid maps $\mathcal{H}_n^m \longrightarrow \mathcal{H}_n^{m+1}$, and $\mathcal{H}_n^m \longrightarrow \mathcal{H}_{n+1}^m$ which are induced by suspension and by addition of an identity map on the last wedge summand and which are cofibrations. The algebraic K-theory of the space X is defined to be the space $A(X) = K_0^f\mathbb{Z}[\pi_0 G] \times (\operatorname{colim}_{m,n} B\mathcal{H}_n^m)^+$. This is obviously equivalent to Waldhausen's definition [71, 2.2.1]. We shall also use the symbol $A(X)$ to denote the spectrum associated to its delooping, and under this interpretation we will prove the following result.

THEOREM 8.2. *The spectra $K\Sigma^\infty G_+$ and $A(X)$ are homotopy equivalent, naturally in X.*

Observe that the functors Σ_m^∞ give maps of topological monoids

$$\mathcal{H}_n^m \longrightarrow \mathscr{S}(\textstyle\bigvee_n \Sigma^\infty G_+, \bigvee_n \Sigma^\infty G_+)$$

which are easily seen to be compatible with suspension and addition of an identity map. Composing with the functors $\mathbb{L}$ and $S \wedge_{\mathscr{L}} (-)$, we obtain maps of topological monoids

$$\mathcal{H}_n^m \longrightarrow \mathscr{M}_S(\textstyle\bigvee_n S \wedge_{\mathscr{L}} \mathbb{L}\Sigma^\infty G_+, \bigvee_n S \wedge_{\mathscr{L}} \mathbb{L}\Sigma^\infty G_+).$$

We denote this composite functor by L_n^m. The observation that the functor $G_+ \wedge (-)$ is naturally isomorphic to the functor $R \wedge_S (-)$ immediately implies that L_n^m sends G-equivariant maps to R-module maps; therefore, we can interpret L_n^m as a map of topological monoids $\mathcal{H}_n^m \longrightarrow \widetilde{M}_n R$. Since for $m \geq 2$, $\mathcal{H}_n^m$ consists of the subspace of those connected components of $\operatorname{Map}_G(\bigvee_n \Sigma^m G_+, \bigvee_n \Sigma^m G_+)$ which π_0 maps to $GL_n(R)$, we see that L_n^m restricts to a map of monoids $\mathcal{H}_n^m \longrightarrow \widetilde{GL_n R}$. We will show that in the colimit this map is a homotopy equivalence.

PROPOSITION 8.3. *The map of topological monoids*

$$L_n\colon \operatorname{colim}_m \mathcal{H}_n^m \longrightarrow \widetilde{GL_n R}$$

is a homotopy equivalence.

PROOF. We have defined L_n^m via a composition of functors so that it would be easy to see that it is a map of monoids; we rewrite this composition to make it easier to analyze homotopically.

Consider the map of spaces

$$f_m\colon \mathscr{T}(\bigvee_n S^m, \bigvee_n \Sigma^m G_+) \longrightarrow \mathscr{M}_S(\bigvee_n S_S, \bigvee_n S_R)$$

(for fixed n) induced by the composite of the functors Σ_m^∞, $\mathbb{L}$, and $S \wedge_S (-)$. The colimit of the f_m is the composite of the maps

$$\begin{aligned} \operatorname{colim}_m \mathscr{T}(\bigvee_n S^m, \bigvee_n \Sigma^m G_+) &\longrightarrow \mathscr{S}(\bigvee_n S, \bigvee_n \Sigma^\infty G_+) \\ &\longrightarrow \mathscr{S}[\mathbb{L}](\bigvee_n \mathbb{L}S, \bigvee_n \mathbb{L}\Sigma^\infty G_+) \\ &\longrightarrow \mathscr{M}_S(\bigvee_n S_S, \bigvee_n S_R), \end{aligned}$$

each of which is a homotopy equivalence. Via the obvious isomorphisms, the map L_n agrees with the restriction of this map to the connected components that consist of weak equivalences, and so it is also a homotopy equivalence. □

Since the inclusion of the identity in G is a cofibration, we see that induced map $\operatorname{colim}_m B\mathcal{H}_n^m \longrightarrow B\widetilde{GL}_n R$ is a homotopy equivalence, and hence the induced map on the plus constructions of the telescopes is a homotopy equivalence.

PROOF OF THEOREM 8.1. We need to show that we have a map of spectra. But the infinite loop space structure on $A(X)$ comes from the operation wedge on the colimit of the topological categories whose objects are finite wedges of $\Sigma^m G_+$ (for each m) and whose maps are the $\mathcal{H}_n^m$. The functors L_n^m assemble to a continuous functor from this colimit to the category $\mathscr{W}$ which commutes with wedges and which coincides with the above homotopy equivalence on the plus constructions. We conclude that the map constructed above

$$K_0^f \mathbb{Z}[\pi_0 G] \times (\operatorname{colim}_{m,n} B\mathcal{H}_n^m)^+ \longrightarrow K_0^f \mathbb{Z}[\pi_0 G] \times B\widetilde{GL}R^+$$

is a map of E_∞ spaces.

Since G is a CW space, $\Sigma^\infty G_+$ is a CW spectrum, so $\tilde{M}_n R$, $\widetilde{GL}_n R$, $B\widetilde{GL}_n R$, $B\widetilde{GL}R$, and $B\widetilde{GL}R^+$ have the homotopy type of CW spaces; therefore, the plus construction of the previous section produces a spectrum homotopy equivalent to KR. We conclude that the spectrum $A(X)$ is homotopy equivalent to KR. □

REMARK 8.4 (LINEARIZATION). The map $R \longrightarrow HZ \wedge_S R$ is a map of S-algebras and a rational equivalence. The map

$$HZ \wedge_S (-)\colon \mathscr{W}_R(\vee_n S_R, \vee_m S_R) \longrightarrow \mathscr{W}_{HZ\wedge_S R}(\vee_n S_{HZ\wedge_S R}, \vee_m S_{HZ\wedge_S R})$$

induces an equivalence on rational homology. We conclude that the induced map $KR \longrightarrow K(HZ \wedge_S R)$ is a rational equivalence. A comparison of the categories of modules for the S-algebra $HZ \wedge_S R$ and the simplicial ring $\mathbb{Z}[GSX]$ would then give a linearization result. We save this and other observations along these lines for a future paper.

CHAPTER VII

R-algebras and topological model categories

In Chapter II, we set up the ground category of S-modules, and we developed the theory of S-algebras and their modules by exploiting the good formal properties of that category. In Chapter III, we set up a ground category of modules over a commutative S-algebra R that enjoys the same formal properties as the category of S-modules, and the previous three chapters gave applications of that theory. As we discuss in Section 1, we can go on to define R-algebras and their modules simply by changing ground categories from $\mathscr{M}_S$ to $\mathscr{M}_R$.

At this point, we face a homotopical problem. We want to use point-set level constructions, such as bar constructions and constructions of topological Hochschild homology, that involve taking smash powers of a commutative R-algebra A. To make homotopical use of these constructions, we need to know that the underlying R-modules of these smash powers represent their smash powers in the derived category of R-modules. However, A need not have the homotopy type of a cell R-module, so we must approximate it by a weakly equivalent R-algebra with better properties. We first attacked this problem by use of the bar construction of Chapter XII, but we shall here deal with it by use of Quillen model categories.

Thus we shall prove that all of our various categories of A_∞ and E_∞ ring spectra, R-algebras, commutative R-algebras, and modules over any of these are complete and cocomplete, tensored and cotensored, topologically enriched categories that admit canonical (closed) model structures in the sense of Quillen [57]. Since cofibrations and fibrations in the classical sense are important in our theory, we shall use the terms q-cofibration and q-fibration for the model category concepts.

The proofs that our categories are so richly structured are almost entirely formal, and these formal structures do not solve or even address the motivating homotopical problem since forgetful functors need not preserve q-cofibrant homotopy types. However, we shall see that the problem can be solved by combining the formal theory with the homotopical analysis of the linear isometries operad.

Much of the formal theory in this chapter is based on ideas and results originally due to Hopkins and McClure (in part in [32], but we have also benefited from many profitable conversations) or to McClure, Schwänzl and Vogt [54].

1. R-algebras and their modules

We fix a commutative S-algebra R and work in the symmetric monoidal category $\mathscr{M}_R$ of R-modules.

DEFINITION 1.1. An R-algebra is a monoid in $\mathscr{M}_R$. A commutative R-algebra is a commutative monoid in $\mathscr{M}_R$.

As in algebra, we obtain free R-algebras by "extension of scalars" from S to R. To show this, we use an alternative description of R-algebras and commutative R-algebras, which again is the same as in algebra. Say that a map $\eta : R \longrightarrow A$ of R-algebras is central if the following diagram commutes:

$$\begin{array}{ccccc} R \wedge_S A & & \xrightarrow{\tau} & & A \wedge_S R \\ \downarrow{\scriptstyle \eta\wedge\mathrm{id}} & & & & \downarrow{\scriptstyle \mathrm{id}\wedge\eta} \\ A \wedge_S A & \xrightarrow[\phi]{} & A & \xleftarrow[\phi]{} & A \wedge_S A \end{array}$$

We learned the following interpretation of this definition from McClure.

REMARK 1.2. The center of an associative k-algebra A with product ϕ can be written as the equalizer displayed in the diagram

$$C(A) \longrightarrow A \underset{\widetilde{\phi\tau}}{\overset{\tilde{\phi}}{\rightrightarrows}} \mathrm{Hom}_k(A, A);$$

here $\tilde{\phi}(a)(b) = ab$ and $\widetilde{\phi\tau}(a)(b) = ba$. This suggests that the center $C(A)$ of an S-algebra A should be defined as the equalizer displayed in the diagram

$$C(A) \longrightarrow A \underset{\widetilde{\phi\tau}}{\overset{\tilde{\phi}}{\rightrightarrows}} F_S(A, A).$$

The definition of a central map $\eta : R \longrightarrow A$ then says precisely that η factors through $C(A)$.

LEMMA 1.3. *An R-algebra A is an S-algebra with a central map $R \longrightarrow A$ of S-algebras. A commutative R-algebra A is a commutative S-algebra with a map $R \longrightarrow A$ of S-algebras.*

PROOF. Trivially, if A is an R-algebra, then its unit $\eta : R \longrightarrow A$ is a central map of R-algebras. Conversely, if A is an S-algebra and $\eta : R \longrightarrow A$ is a map of S-algebras, then A is a left R-module via the composite

$$R \wedge_S A \xrightarrow{\eta\wedge\mathrm{id}} A \wedge_S A \xrightarrow{\phi} A.$$

There is a symmetrically defined right action of R on A that makes A an (R, R)-bimodule. Centrality ensures that the left and right actions agree under the commutativity isomorphism of their domains. The product of A therefore factors through $A \wedge_R A$ to give the required R-algebra structure. □

We leave the proofs of the next few results as exercises; as in the proofs above, one first writes down the proof of the algebraic analogue and then replaces tensor products with smash products.

PROPOSITION 1.4. *If Q is an S-algebra, then $R \wedge_S Q$ is the free R-algebra generated by Q, hence $R \wedge_S \mathbb{T}M$ is the free R-algebra generated by an S-module M. If Q is a commutative S-algebra, then $R \wedge_S Q$ is the free commutative R-algebra generated by Q, hence $R \wedge_S \mathbb{P}M$ is the free commutative R-algebra generated by M.*

REMARK 1.5. We may think of $R \wedge_S (S \wedge_{\mathscr{L}} \mathbb{B}X)$ as the "free" R-algebra generated by a spectrum X and $R \wedge_S (S \wedge_{\mathscr{L}} \mathbb{C}X)$ as the "free" commutative R-algebra generated by a spectrum X. However, in view of II.1.3 (see also III§1), this is a misnomer since the right adjoints of these functors from the category of spectra to the category of R-algebras or commutative R-algebras are weakly equivalent rather than equal to the obvious forgetful functors.

PROPOSITION 1.6. *Let $f : R \longrightarrow R'$ and $g : R \longrightarrow R''$ be maps of commutative S-algebras. Then $R' \wedge_R R''$ is both the coproduct of R' and R'' in the category of commutative R-algebras and the pushout of f and g in the category of commutative S-algebras. More generally, let $f : A \longrightarrow A'$ and $g : A \longrightarrow A''$ be maps of commutative R-algebras. Then $A' \wedge_A A''$ is the pushout of f and g in the category of commutative R-algebras.*

As in algebra, we can define the notion of a module over an R-algebra A, but it turns out to be equivalent to the notion of a module over A regarded just as an S-algebra. Recall III.3.1.

DEFINITION 1.7. Let A be an R-algebra. A left or right A-module is a left or right A-object in $\mathscr{M}_R$.

The free A-module generated by an S-module M is

$$A \wedge_R (R \wedge_S M) \cong A \wedge_S M.$$

This gives an isomorphism of monads that implies the following result.

LEMMA 1.8. *Let A be an R-algebra. A module over A regarded as an S-algebra is the same thing as a module over A regarded as an R-algebra. That is, an action $A \wedge_S M \longrightarrow M$ necessarily factors through an action $A \wedge_R M \longrightarrow M$.*

Similarly, if M and N are A-modules, then $M \wedge_A N$ is the same whether defined using a coequalizer diagram in the category of R-modules or in the category of S-modules.

LEMMA 1.9. *Let A be an R-algebra, and let M be a right and N a left A-module. Then $M \wedge_A N$ can be identified with the coequalizer $M \wedge_{(A,R)} N$ displayed in the diagram*

$$M \wedge_R A \wedge_R N \underset{\mathrm{id} \wedge_R \nu}{\overset{\mu \wedge_R \mathrm{id}}{\rightrightarrows}} M \wedge_R N \longrightarrow M \wedge_{(A,R)} N,$$

The analogous result holds for function A-modules.

PROOF. The proof is a formal categorical chase of the following schematic diagram:

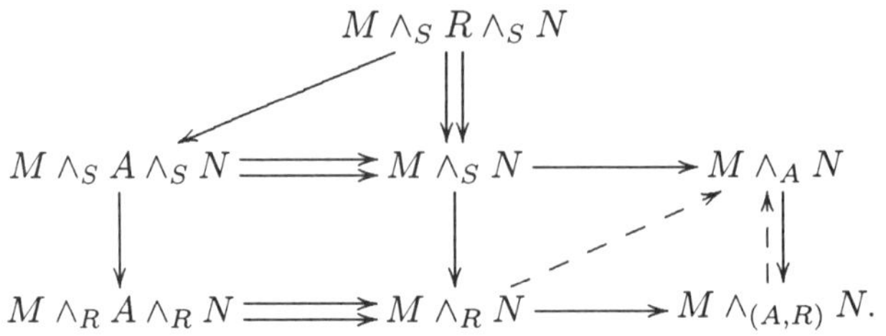

Here the left vertical arrow is an epimorphism, and this implies that the diagonal dotted arrow factors through the dotted right vertical arrow. □

Although we have only one notion of an A-module, it is helpful to think of its study as divided into an "absolute theory", in which we take the ground ring to be S, and a "relative theory", in which we take the ground ring to be R. The absolute theory is a special case of the study of modules over algebras that we developed in Chapter III. In particular, III.1.4 shows that $\mathbb{F}_A X$ is weakly equivalent to $A \wedge X$ for a CW spectrum X. Here the free functor $\mathbb{F}_A$ is isomorphic to the composite functor $A \wedge_R (R \wedge_S \mathbb{F}_S)$ from spectra to A-modules. Again, the term free is a misnomer since the right adjoint of $\mathbb{F}_S$ is only weakly equivalent to the forgetful functor. The theory of cell and CW A-modules and the definition of the derived category of A-modules are part of the absolute theory.

The previous lemma shows that the absolute smash product $\wedge_A$ and function module functors F_A are isomorphic to the relative functors, so that $M \wedge_A N$ and $F_A(M,N)$ are R-modules. Of course, if A is a commutative R-algebra, then these are A-modules and duality theory applies. In the relative theory, if we replace (R,S) by (A,R), with concomitant changes of notations for various functors, then all of the statements in Chapter III which make sense remain true. Note, for example, that we have relative versions of III.3.10 and of the pairings discussed in III§6. The results on pairings give the following generalization of III.6.12.

PROPOSITION 1.10. *Let R be a commutative S-algebra, A be an R-algebra, and M and N be A-modules. Then $F_A(M,M)$ is an R-algebra and $F_A(M,N)$ is an $(F_A(N,N), F_A(M,M))$-bimodule.*

Of course, the case $A = R$ is of particular interest.

2. Tensored and cotensored categories of structured spectra

As in I§1, consider the categories $\mathscr{P}$ and $\mathscr{S}$ of prespectra and spectra indexed on a universe U. It was proven in [38, p.17-18] that these categories are topologically enriched, in the sense that their Hom sets are based topological spaces such that composition is continuous. For prespectra D and D', $\mathscr{P}(D,D')$ is topologized as a subspace of the product over indexing spaces V of the function spaces $F(DV, D'V)$. Since maps between spectra are just maps between their underlying prespectra, this fixes the topology on $\mathscr{S}(E,E')$. It was also observed in [38, p.18] that all of the functors introduced in that volume are continuous and all of the adjunctions proven in it are given by homeomorphisms of Hom sets.

For example, by [38, I.3.3], there are natural homeomorphisms

$$\mathscr{S}(E \wedge X, E') \cong \mathscr{T}(X, \mathscr{S}(E, E')) \cong \mathscr{S}(E, F(X, E')) \tag{2.1}$$

for spaces X and spectra E and E', where $\mathscr{T}$ denotes the category of based spaces. In categorical language [33, §3.7], (2.1) states that $\mathscr{S}$ is tensored with tensors $E \wedge X$ and cotensored with cotensors $F(X, E)$. Adjoining disjoint basepoints to unbased spaces X, we obtain similar homeomorphisms involving the category $\mathscr{U}$ of unbased spaces. We give a formal definition in the unbased context.

DEFINITION 2.2. Let $\mathscr{E}$ be a category enriched over the category $\mathscr{U}$ of unbased spaces. Then $\mathscr{E}$ is tensored if there is a functor $\otimes_{\mathscr{E}} : \mathscr{E} \times \mathscr{U} \longrightarrow \mathscr{E}$, continuous in both variables, together with a natural homeomorphism

$$\mathscr{E}(E \otimes_{\mathscr{E}} X, E') \cong \mathscr{U}(X, \mathscr{E}(E, E'))$$

for spaces X and objects E and E' of $\mathscr{E}$. We write $\otimes$ for $\otimes_{\mathscr{E}}$ when $\mathscr{E}$ is clear from the context. Dually, $\mathscr{E}$ is cotensored if there is a functor $F_{\mathscr{E}} : \mathscr{U}^{op} \times \mathscr{E} \longrightarrow \mathscr{E}$, continuous in both variables, together with a natural homeomorphism

$$\mathscr{U}(X, \mathscr{E}(E, E')) \cong \mathscr{E}(E, F_{\mathscr{E}}(X, E')).$$

As in the motivating example (2.1), $F_{\mathscr{E}}$ will always admit an explicit description. The tensors are more interesting and less familiar. We will give a way of describing them for many spaces X in the next section.

Again, by the argument illustrated in [38, p.18-19], colimits and limits of spectra are continuous (a better term would be topological). This means that the isomorphisms

$$\mathscr{S}(\operatorname{colim} E_i, F) \cong \lim \mathscr{S}(E_i, F) \tag{2.3}$$

and

$$\mathscr{S}(F, \lim E_i) \cong \lim \mathscr{S}(F, E_i) \tag{2.4}$$

are homeomorphisms.

The continuity can also be deduced categorically. There are valuable general notions of indexed colimits and limits in enriched categories, which are defined and discussed in Kelly [33, §3.1]. Indexed colimits include tensors with spaces and continuous colimits as special cases, and dually for limits. We shall not repeat the general definition, since we shall not have occasion to use it, and we shall rely on the following result of Kelly [33, 3.69-3.73] to deduce the existence of indexed colimits and limits.

DEFINITION 2.5. A category $\mathscr{E}$ enriched over $\mathscr{U}$ is topologically cocomplete if it has all indexed colimits and topologically complete if it has all indexed limits.

THEOREM 2.6 (KELLY). *Let $\mathscr{E}$ be a category enriched over the category of based or unbased spaces. Then $\mathscr{E}$ is topologically cocomplete if it is cocomplete and admits tensor products and is topologically complete if it is complete and admits cotensor products. In particular, the given colimits and limits are continuous.*

Our various categories of structured ring, module, and algebra spectra inherit subspace topologies on their Hom sets. Thus they are all topologically enriched. All of the functors and adjunctions that we have constructed in this paper are continuous, by the cited arguments of [38, p.18-19]. We claim that our various categories of rings, modules, and algebras are topologically cocomplete and complete.

For modules, this is immediate from II.1.4, III.1.1, and inspection. If R is an S-algebra, M is an R-module, and X is a based space, then

$$(2.7)\quad \mathscr{M}_R(M \wedge X, M') \cong \mathscr{T}(X, \mathscr{M}_R(M, M')) \cong \mathscr{M}_R(M, S \wedge_{\mathscr{L}} F(X, M')).$$

We deduce the first isomorphism from the first isomorphism of (2.1) by first writing $\mathscr{M}_S(M, M')$ as the equalizer of a pair of maps $\mathscr{S}(M, M') \longrightarrow \mathscr{S}(\mathbb{L}M, M')$ and then writing $\mathscr{M}_R(M, M')$ as the equalizer of a pair of maps $\mathscr{M}_S(M, M') \longrightarrow \mathscr{M}_S(R \wedge_S M, M')$. We deduce the second isomorphism from the first by use of the isomorphisms

$$M \wedge X \cong M \wedge_S \Sigma^\infty X \quad \text{and} \quad S \wedge_{\mathscr{L}} F(X, M) \cong F_S(\Sigma^\infty X, M).$$

PROPOSITION 2.8. *For any S-algebra R, $\mathscr{M}_R$ is topologically cocomplete and complete. Its tensors $M \wedge X$ and all other indexed colimits are created in $\mathscr{M}_S$ or, equivalently, in $\mathscr{S}$. Its cotensors $F_S(\Sigma^\infty X, M)$ and all other indexed limits are created in $\mathscr{M}_S$ or, equivalently, by applying the functor $S \wedge_{\mathscr{L}} (-)$ to indexed limits created in $\mathscr{S}$.*

Now consider the categories of R-algebras and of commutative R-algebras. We agree to denote these categories by $\mathscr{A}_R$ and $\mathscr{C}\mathscr{A}_R$, respectively. We must enrich these categories over $\mathscr{U}$, since there are no "trivial maps" to take as basepoints of Hom sets. We have already observed in II§7 that the categories $\mathscr{A}_R$ and $\mathscr{C}\mathscr{A}_R$ are complete and cocomplete. Continuing that discussion, we obtain the following result. The proof works equally well in the categories of A_∞ and E_∞ ring spectra, where the result is due to Hopkins and McClure [32] and, in the E_∞ case, is the main technical result of McClure, Schwänzl, and Vogt [54, Thm A].

THEOREM 2.9. *For any commutative S-algebra R, the categories $\mathscr{A}_R$ of R-algebras and $\mathscr{C}\mathscr{A}_R$ of commutative R-algebras are topologically cocomplete and complete. Their cotensors and all other indexed limits are created in $\mathscr{M}_S$ or, equivalently, by applying the functor $S \wedge_{\mathscr{L}} (-)$ to indexed limits created in $\mathscr{S}$.*

PROOF. By II.7.1 (compare II.4.5), we have monads $\mathbb{T}$ and $\mathbb{P}$ in the category of R-modules whose algebras are the R-algebras and commutative R-algebras, and these monads are continuous (e.g., by inspection when $R = S$ and use of Proposition 1.4). Now II.7.2 and II.7.4 apply to show that $\mathscr{A}_R$ and $\mathscr{C}\mathscr{A}_R$ are cocomplete. In the commutative case, the construction of colimits is quite simple since Proposition 1.6 gives coproducts and pushouts, and it is trivial to construct coequalizers from them. Moreover, by an easy bootstrap argument from the continuity of colimits in the ground category of spectra, coequalizers in $\mathscr{A}_R$ and $\mathscr{C}\mathscr{A}_R$ are continuous. Now II.7.2 and the following categorical result complete the proof. □

PROPOSITION 2.10. *Let $\mathbb{T} : \mathscr{C} \longrightarrow \mathscr{C}$ be a continuous monad defined on a topologically enriched category $\mathscr{C}$ and let $\mathscr{C}[\mathbb{T}]$ be the category of algebras over $\mathbb{T}$. Assume that $\mathscr{C}$ is topologically cocomplete and complete.*

(i) *The forgetful functor $\mathscr{C}[\mathbb{T}] \longrightarrow \mathscr{C}$ creates all indexed limits.*
(ii) *If $\mathbb{T}$ preserves reflexive coequalizers, then $\mathscr{C}[\mathbb{T}]$ has all indexed colimits.*

PROOF. Part (i) is the enriched version of [43, VI.2, Ex 2]. Our version of (ii) seems to be new, although part of the proof is due to Hopkins [32]. Reflexive coequalizers are defined in II.6.5. By II.7.4, we know that $\mathscr{C}[\mathbb{T}]$ is cocomplete in the ordinary sense. By Kelly's theorem (Theorem 2.6), we need only construct tensors in $\mathscr{C}[\mathbb{T}]$. Thus let (C, ξ) be a $\mathbb{T}$-algebra and X be a space. Let $C \otimes X$ denote their tensor in $\mathscr{C}$. Define $\nu : \mathbb{T}C \otimes X \longrightarrow \mathbb{T}(C \otimes X)$ to be the adjoint of the composite map of spaces

$$X \longrightarrow \mathscr{C}(C, C \otimes X) \xrightarrow{\mathbb{T}} \mathscr{C}(\mathbb{T}C, \mathbb{T}(C \otimes X)),$$

where the first arrow is adjoint to the identity map $C \otimes X \longrightarrow C \otimes X$. Define $C \otimes_{\mathscr{C}[\mathbb{T}]} X$ to be the coequalizer in $\mathscr{C}$ displayed in the following diagram:

$$\mathbb{T}(\mathbb{T}C \otimes X) \underset{\mu \circ \mathbb{T}\nu}{\overset{\mathbb{T}(\xi \otimes \mathrm{id})}{\rightrightarrows}} \mathbb{T}(C \otimes X) \longrightarrow C \otimes_{\mathscr{C}[\mathbb{T}]} X.$$

Clearly the parallel arrows are both maps of $\mathbb{T}$-algebras. We claim that this diagram is a reflexive coequalizer in $\mathscr{C}$. It will follow from II.6.6 that the diagram is a coequalizer in $\mathscr{C}[\mathbb{T}]$. We check the adjunction homeomorphism

$$\mathscr{C}[\mathbb{T}](C \otimes_{\mathscr{C}[\mathbb{T}]} X, C') \cong \mathscr{U}(X, \mathscr{C}[\mathbb{T}](C, C'))$$

required of a tensor by using the fact that $\mathscr{C}[\mathbb{T}](C, C')$ is the equalizer in $\mathscr{U}$ of

$$\mathscr{C}(C, C') \xrightarrow{\mathscr{C}(\xi, \mathrm{id})} \mathscr{C}(\mathbb{T}C, C')$$

and

$$\mathscr{C}(C, C') \xrightarrow{\mathbb{T}} \mathscr{C}(\mathbb{T}C, \mathbb{T}C') \xrightarrow{\mathscr{C}(\mathrm{id}, \xi')} \mathscr{C}(\mathbb{T}C, C').$$

To see that the displayed coequalizer is reflexive, as claimed, consider the map

$$\mathbb{T}(\eta \otimes \mathrm{id}) : \mathbb{T}(C \otimes X) \longrightarrow \mathbb{T}(\mathbb{T}C \otimes X).$$

Clearly $\mathbb{T}(\xi \otimes \mathrm{id}) \circ \mathbb{T}(\eta \otimes \mathrm{id}) = \mathrm{id}$. Less obviously, $\mu \circ \mathbb{T}\nu \circ \mathbb{T}(\eta \otimes \mathrm{id}) = \mathrm{id}$. This follows by adjunction from the commutative diagram

$$\begin{array}{ccc} X & \xrightarrow{\tilde{\eta}} & \mathscr{C}(C, \mathbb{T}(C \otimes X)) \\ \downarrow & \nearrow_{\mathscr{C}(\mathrm{id}, \eta)} & \uparrow{\scriptstyle \mathscr{C}(\eta, \mathrm{id})} \\ \mathscr{C}(C, C \otimes X) & \xrightarrow[\mathbb{T}]{} & \mathscr{C}(\mathbb{T}, \mathbb{T}(C \otimes X)). \end{array}$$

Here the commutativity of the upper left triangle defines the adjoint $\tilde{\eta}$, and the lower right triangle commutes by the naturality, $\eta \circ f = \mathbb{T}f \circ \eta$, of η on maps $f : C \longrightarrow C \otimes X$. Thus $\mathbb{T}(\eta \otimes \mathrm{id})$ gives the required reflection. □

In particular, $F_S(\Sigma^\infty X_+, A)$ is the cotensor of a space X and an R-algebra or commutative R-algebra A. The diagonal on X and the product on A induce the product on $F_S(\Sigma^\infty X_+, A)$. The following instance of a general categorical observation explains the relationship between the smash product $A \wedge X_+$ in the category of R-modules and the tensor $A \otimes X$ in the category of R-algebras or commutative R-algebras.

PROPOSITION 2.11. *For R-algebras A and spaces X there is a natural map of R-modules*

$$\omega : A \wedge X_+ \longrightarrow A \otimes X$$

such that ω is the canonical isomorphism if $X = \{*\}$ and the following transitivity diagrams commute:

$$\begin{array}{ccccc} (A \wedge X_+) \wedge Y_+ & \xrightarrow{\omega \wedge \mathrm{id}} & (A \otimes X) \wedge Y_+ & \xrightarrow{\omega} & (A \otimes X) \otimes Y \\ \cong \downarrow & & & & \downarrow \cong \\ A \wedge (X \times Y)_+ & & \xrightarrow[\omega]{} & & A \otimes (X \times Y). \end{array}$$

For $x \in X$, let $i_x : A \longrightarrow A \wedge X_+$ be the map induced by the inclusion $\{x\}_+ \longrightarrow X_+$. A map $f : A \wedge X_+ \longrightarrow B$ of spectra into an R-algebra B such that each composite $f \circ i_x : A \longrightarrow B$ is a map of R-algebras uniquely determines a map of R-algebras $\tilde{f} : A \otimes X \longrightarrow B$ such that $f = \tilde{f} \circ \omega$. The same statement holds for commutative R-algebras.

PROOF. We have a natural map

$$\mathscr{A}_R(A \otimes X, B) \cong \mathscr{U}(X, \mathscr{A}_R(A, B)) \longrightarrow \mathscr{U}(X, \mathscr{M}_R(A, B)) \cong \mathscr{M}_R(A \wedge X_+, B),$$

and ω is the image of the identity map of $A \otimes X$. The rest is easy diagram chasing, using the natural map $\mathscr{M}_R(A \wedge X_+, B) \longrightarrow \mathscr{S}(A \wedge X_+, B)$ for the last statement. □

REMARK 2.12. For R-algebras A and B, the previous result says that a map $A \otimes X \longrightarrow B$ of R-algebras determines and is determined by a map $A \wedge X_+ \longrightarrow B$ of spectra that is pointwise a map of R-algebras. A similar construction and result apply whenever one has a tensored category $\mathscr{E}$ with a continuous forgetful functor to spectra. For objects A and B in $\mathscr{E}$, we define a homotopy to be a map $h : A \otimes I \longrightarrow B$. Then h is induced by a homotopy $A \wedge I_+ \longrightarrow B$ through maps in $\mathscr{E}$.

3. Geometric realization and calculations of tensors

To prepare for our construction of model structures and our study of thh, we explain how to calculate tensors $E \otimes X$ for certain spaces X, and we use this calculation to study pushouts and cofibrations in the context of R-algebras. Our main tool is geometric realization, and the reader is urged to read the first two sections of Chapter X, which give a down to earth study of the geometric realization of simplicial spectra, before reading this section.

Fix a topologically complete and cocomplete category $\mathscr{E}$ with a continuous forgetful functor to spectra. We have the notion of a simplicial object E_* in $\mathscr{E}$. There are two notions of the geometric realization of such an object. We can first forget down to the category of simplicial spectra and take the geometric realization $|E_*|$ there, or we can rework the definition and carry out the construction entirely in $\mathscr{E}$, obtaining the internal geometric realization $|E_*|_{\mathscr{E}}$. Explicitly, $|E_*|_{\mathscr{E}}$ is the coend

$$|E_*|_{\mathscr{E}} = \int^{\Delta} E_q \otimes_{\mathscr{E}} \Delta_q. \tag{3.1}$$

The following relationships between these two kinds of geometric realization generalize and clarify observations of McClure, Schwänzl, and Vogt [54, 4.3, 4.4] about the category of E_∞ ring spectra. We defer the proofs to the end of the section.

PROPOSITION 3.2. *Let X_* be a simplicial space and let $A \in \mathscr{E}$. Then there is a natural isomorphism*

$$A \otimes_{\mathscr{E}} |X_*| \cong |A \otimes_{\mathscr{E}} X_*|_{\mathscr{E}},$$

of objects of $\mathscr{E}$.

The realization of underlying simplicial spectra is more amenable to homotopical analysis than the internal realization. In favorable cases, the realization $|E_*|$ will again be an object of $\mathscr{E}$, but this is not formal. We shall prove in X§1 that this holds for all of the categories of interest to us. In such cases, the two geometric realizations are isomorphic. In particular, the following result holds.

PROPOSITION 3.3. *Let R be any commutative S-algebra, such as $R = S$. For simplicial R-algebras A_*, there is a natural isomorphism of R-algebras*

$$|A_*| \cong |A_*|_{\mathscr{A}_R},$$

and similarly for simplicial commutative R-algebras.

COROLLARY 3.4. *For R-algebras A and simplicial spaces X_*, there is a natural isomorphism of R-algebras*

$$A \otimes_{\mathscr{A}_R} |X_*| \cong |A \otimes_{\mathscr{A}_R} X_*|,$$

and similarly for commutative R-algebras.

In the following discussion, we let $\mathscr{E}$ denote either $\mathscr{A}_R$ or $\mathscr{C}\mathscr{A}_R$ and write $\otimes$ for $\otimes_{\mathscr{E}}$. We use the term R-algebra in either case. The computation of $A \otimes |X_*|$ just given applies particularly effectively to simplicial sets X_*, regarded as discrete simplicial spaces. We have a categorical coproduct $\amalg$ in $\mathscr{E}$. This is $\wedge_R$ in the commutative case, but it is the "free product" in the non-commutative case. In the commutative case, the codiagonal map $\nabla : A \amalg A \longrightarrow A$ is the product on A. In both cases, the unit $\eta : R \longrightarrow A$ is the unique map from the initial object. Since a discrete set $\underline{n}$ with n points is the coproduct of its elements and the functor $A \otimes (-)$ preserves coproducts, $A \otimes \underline{n}$ is the coproduct of n copies of A. To calculate $A \otimes |X_*|$, we need only identify the induced face and degeneracy operators on coproducts of copies of A in terms of the structure maps ∇ and η.

In order to understand homotopy theory in $\mathscr{E}$, we need to understand $A \otimes I$. We shall describe it in terms of a bar construction that is defined on R-algebras. Recall that we defined the bar construction $B(M, R, N)$ for a commutative S-algebra R and R-modules M and N in IV.7.2. We shall later use the evident generalization in which we replace R and its modules by a commutative R-algebra A and its modules. We here introduce a variant that applies equally well to either commutative or non-commutative R-algebras. In the commutative case, it is just the specialization of the cited generalization in which the given A-modules are restricted to be commutative A-algebras.

DEFINITION 3.5. Let A be an R-algebra, and let $f : A \longrightarrow A'$ and $g : A \longrightarrow A''$ be maps of R-algebras. These maps and the identity maps of A' and A'' determine maps of R-algebras

$$\mu : A' \amalg A \longrightarrow A' \quad \text{and} \quad \nu : A \amalg A'' \longrightarrow A''$$

Define a simplicial R-algebra $\beta_*^R(A', A, A'')$ by replacing $\wedge_S$ and ϕ by $\amalg$ and ∇ in IV.7.2. Then define an R-algebra $\beta^R(A', A, A'')$ by

$$\beta^R(A', A, A'') = |\beta_*^R(A', A, A'')|.$$

There is an evident natural map of R-algebras

$$\psi : \beta^R(A', A, A'') \longrightarrow A' \amalg_A A''$$

from the bar construction to the displayed pushout.

Define the double mapping cylinder R-algebra $M(A', A, A'')$ by

$$M(A', A, A'') = A' \amalg_A (A \otimes I) \amalg_A A'' \tag{3.6}$$

and observe that the map $I \longrightarrow \{pt\}$ induces a collapse map

$$\psi : M(A', A, A'') \longrightarrow A' \amalg_A A''.$$

We have the following identification of these two constructions.

PROPOSITION 3.7. *Let A be an R-algebra with given maps to R-algebras A' and A''. Then there is a natural isomorphism*

$$\beta^R(A, A, A) \cong A \otimes I$$

of R-algebras over A and under $A \amalg A$, and there is a natural isomorphism

$$\beta^R(A', A, A'') \cong M(A', A, A'')$$

of R-algebras over $A' \amalg_A A''$ and under $A' \amalg A''$.

PROOF. Let I_* be the standard simplicial 1-simplex with realization I. It has $p+2$ p-simplices, and a simple comparison of its face and degeneracy operations (e.g., [44, p.14]) with those of the bar construction shows that we have a natural identification of simplicial R-algebras

$$\beta^R_*(A, A, A) \cong A \otimes I_*.$$

In fact, one can see this quite directly, since the only non-degenerate simplices of I_* are a 1-simplex Δ_1 and its faces, and similarly for $\beta^R_*(A, A, A)$. The rest follows. □

We use this to obtain a result about cofibrations that will be at the heart of our construction of model structures on $\mathscr{E}$ in Section 6. Let $\mathbb{T} : \mathscr{M}_R \longrightarrow \mathscr{E}$ be the free R-algebra functor; thus $\mathbb{T}$ must be interpreted as $\mathbb{P}$ in the commutative case. We shall prove in XII.2.3 that the functor $\mathbb{T}$ preserves cofibrations of R-modules. Since $\mathbb{T}$ preserves tensors and pushouts and since $R = \mathbb{T}(*)$, we have

$$\mathbb{T}CM \cong R \amalg_{\mathbb{T}M} (\mathbb{T}M \otimes I).$$

PROPOSITION 3.8. *For an R-module M and a map of R-algebras $\mathbb{T}M \longrightarrow A$, the natural map of R-algebras*

$$\psi : M(\mathbb{T}CM, \mathbb{T}M, A) \longrightarrow \mathbb{T}CM \amalg_{\mathbb{T}M} A$$

is homotopic rel A to an isomorphism.

PROOF. For a based space X, it is trivial to see that the map

$$CX \cup_X (X \wedge I_+) \longrightarrow CX$$

that retracts the cylinder onto the base of the cone is homotopic to a homeomorphism. Working in the category of R-modules, the same argument works with X replaced by M. Applying the functor $\mathbb{T}$, the cited map then becomes the map

$$\rho : R \amalg_{\mathbb{T}M} (\mathbb{T}M \otimes I) \amalg_{\mathbb{T}M} (\mathbb{T}M \otimes I) \longrightarrow R \amalg_{\mathbb{T}M} (\mathbb{T}M \otimes I)$$

that retracts the second copy of $\mathbb{T}M \otimes I$ onto the base of the first. We have

$$M(\mathbb{T}CM, \mathbb{T}M, A) \cong R \amalg_{\mathbb{T}M} (\mathbb{T}M \otimes I) \amalg_{\mathbb{T}M} (\mathbb{T}M \otimes I) \amalg_{\mathbb{T}M} A,$$

and ψ is obtained by applying the functor $(-) \amalg_{\mathbb{T}M} A$ to ρ. The conclusion follows. □

PROPOSITION 3.9. *For any pushout diagram of R-algebras*

$$\begin{array}{ccc} \mathbb{T}M & \longrightarrow & A \\ \big\downarrow & & \big\downarrow{\scriptstyle i} \\ \mathbb{T}CM & \longrightarrow & B, \end{array}$$

the map i is a cofibration of R-modules and therefore of spectra.

PROOF. The essential point is just that the unit map $\eta : R \longrightarrow \mathbb{T}CM$ is the inclusion of a wedge summand of R-modules and a retract of R-algebras. From this, we find that the induced map $A \longrightarrow \mathbb{T}CM \amalg A$ of R-algebras is also the inclusion of a wedge summand of R-modules and a retract of R-algebras. By the previous lemma and proposition, the pushout is isomorphic under A to the bar construction $\beta^R(\mathbb{T}CM, \mathbb{T}M, A)$. All of the degeneracy operators of $\beta^R_*(\mathbb{T}CM, \mathbb{T}M, A)$ are inclusion of wedge summands of R-modules, and it follows that $\beta^R_*(\mathbb{T}CM, \mathbb{T}M, A)$ is proper in the sense of X.2.2. This implies that the map from the zero skeleton $\mathbb{T}CM \amalg A$ into $\beta^R(\mathbb{T}CM, \mathbb{T}M, A)$ is a cofibration, and the conclusion follows. □

We shall also need the following elementary complement.

LEMMA 3.10. *Let $\{A_i\}$ be a sequence of maps of R-algebras that are cofibrations of spectra. Then the underlying spectrum of the colimit of the sequence computed in the category of R-algebras is the colimit of the sequence computed in the category of spectra.*

PROOF. The colimit in the category of spectra computes the colimit in the category of R-modules and satisfies

$$(\mathrm{colim}\ A_i) \wedge_R (\mathrm{colim}\ A_i) \cong \mathrm{colim}\,(A_i \wedge_R A_i).$$

Therefore the spectrum level colimit inherits an R-algebra structure from the A_i, and the universal property in the category of R-algebras follows from the universal property in the category of R-modules. □

We must still prove Propositions 3.2 and 3.3. Let $s\mathscr{C}$ denote the category of simplicial objects in a category $\mathscr{C}$.

PROOF OF PROPOSITION 3.2. For a space Y, let $\mathscr{U}(\Delta_*, Y)$ be the evident simplicial space with q-simplices $\mathscr{U}(\Delta_q, Y)$. This functor of Y is right adjoint to geometric realization,

$$\mathscr{U}(|X_*|, Y) \cong s\mathscr{U}(X_*, \mathscr{U}(\Delta_*, Y)). \tag{3.11}$$

Similarly, for an object F of $\mathscr{E}$, let $F_{\mathscr{E}}(\Delta_*, F)$ be the evident simplicial object of $\mathscr{E}$ with q-simplices $F_{\mathscr{E}}(\Delta_q, F)$. This functor of F is right adjoint to the internal geometric realization,

$$\mathscr{E}(|E_*|_{\mathscr{E}}, F) \cong s\mathscr{E}(E_*, F_{\mathscr{E}}(\Delta_*, F)). \tag{3.12}$$

These adjunctions, together with tensor and cotensor adjunctions, give the chain of natural isomorphisms

$$\begin{aligned}\mathscr{E}(E \otimes_{\mathscr{E}} |X_*|, F) &\cong \mathscr{U}(|X_*|, \mathscr{E}(E, F)) \\ &\cong s\mathscr{U}(X_*, \mathscr{U}(\Delta_*, \mathscr{E}(E, F)) \\ &\cong s\mathscr{U}(X_*, \mathscr{E}(E, F_{\mathscr{E}}(\Delta_*, F)) \\ &\cong s\mathscr{E}(E \otimes_{\mathscr{E}} X_*, F_{\mathscr{E}}(\Delta_*, F)) \\ &\cong \mathscr{E}(|E \otimes_{\mathscr{E}} X_*|_{\mathscr{E}}, F).\end{aligned}$$

The conclusion follows. □

PROOF OF PROPOSITION 3.3. Our interest is in the examples $\mathscr{E} = \mathscr{A}_R$ and $\mathscr{E} = \mathscr{C}\mathscr{A}_R$, but the argument works more generally. In fact, it applies whenever realizations $|E_*|$ inherit structure present in $\mathscr{E}$, with the induced structure "arising pointwise". To explain what this means, note that we have an adjunction like those of (3.11) and (3.12) for simplicial spectra K_* and spectra L, namely

$$\mathscr{S}(|K_*|, L) \cong s\mathscr{S}(K_*, F((\Delta_*)_+, L)), \tag{3.13}$$

where $F((\Delta_*)_+, L)$ has q-simplices $F((\Delta_q)_+, L)$. Now let E_* be a simplicial object of $\mathscr{E}$ and F be an object of $\mathscr{E}$. When $|E_*|$ is again an object of $\mathscr{E}$, we have the subspace

$$\mathscr{E}(|E_*|, F) \subset \mathscr{S}(|E_*|, F)$$

of maps in $\mathscr{E}$. We say that the induced structure on $|E_*|$ arises pointwise if this subspace coincides under the adjunction (3.13) with the subspace of $s\mathscr{S}(E_*, F((\Delta_*)_+, F))$ consisting of those points $f = \{f_q\}$ such that the adjoint $\tilde{f}_q : E_q \wedge (\Delta_q)_+ \longrightarrow F$ of $f_q : E_q \longrightarrow F((\Delta_q)_+, F))$ restricts to a map $E_q \longrightarrow F$ in $\mathscr{E}$ on the copy of E_q in $E_q \wedge (\Delta_q)_+$ determined by each point of Δ_q. By Proposition 2.11 and Remark 2.12, such a map $\tilde{f}_q$ extends uniquely to a map $\tilde{g}_q : E_q \otimes_{\mathscr{E}} \Delta_q \longrightarrow F$ in $\mathscr{E}$. In turn, under the tensor-cotensor adjunction, $\tilde{g}_q$ corresponds to a map $g_q : E_q \longrightarrow F_{\mathscr{E}}(\Delta_q, F)$ in $\mathscr{E}$. The function $\{f_q\} \longrightarrow \{g_q\}$ determines an adjunction

$$\mathscr{E}(|E_*|, F) \cong s\mathscr{E}(E_*, F_{\mathscr{E}}(\Delta_*, F)). \tag{3.14}$$

Comparison of (3.12) and (3.14) gives the conclusion. An alternative argument based on the properties of the monads $\mathbb{T}$ and $\mathbb{P}$ is also possible. The adjunctions above can be used to check that

$$\mathbb{T}|A_*| \cong |\mathbb{T}(A_*)|_{\mathscr{A}_R}.$$

The functor $\mathbb{T}$ commutes with $|\cdot|$ on simplicial R-modules, the functor $|\cdot|_{\mathscr{A}_R}$ preserves coequalizers, and a comparison of coequalizer diagrams gives the result. □

4. Model categories of ring, module, and algebra spectra

We shall prove that our various categories of structured spectra admit model structures. A more general, axiomatic, framework is possible; compare Blanc [5]. We assume familiarity with the language of model categories, by which we understand closed model categories in Quillen's original sense [57]. A good exposition is given in [18]. We explain our results in this section and prove them in the next.

In this paper, cofibrations and fibrations in any of our categories mean maps that satisfy the homotopy extension property (HEP) or covering homotopy property (CHP) in that category. Cofibrations in this sense will play a central role in the work of the next section. It is a pity that the language of model categories has, in the literature, been superimposed on the classical language, with resulting ambiguity. We shall use q-cofibrations and q-fibrations for the model theoretic terms.

In all of our model categories, the weak equivalences in the model sense will be those maps in the category which are weak equivalences of underlying spectra. We say that the weak equivalences are created in $\mathscr{S}$. Observe that a retract of a weak equivalence is a weak equivalence. Recall that a q-fibration or q-cofibration in a model category is said to be acyclic if it is a weak equivalence.

Implicitly or explicitly, we must constantly think in terms of diagrams

$$\begin{array}{ccc} E & \xrightarrow{\alpha} & X \\ {\scriptstyle i}\downarrow & \overset{g}{\nearrow} & \downarrow{\scriptstyle p} \\ F & \xrightarrow[\beta]{} & Y, \end{array}$$

where the square is given to be commutative and we seek a lift g that makes both triangles commute. We say that i has the left lifting property (LLP) with respect to a class of morphisms $\mathscr{P}$ if there exists such a lift g for any square in which $p \in \mathscr{P}$. We say that p satisfies the right lifting property (RLP) with respect to a class of morphisms $\mathscr{I}$ if there exists such a lift g for any square in which $i \in \mathscr{I}$.

For example, a Serre fibration of spectra is a map that satisfies the CHP with repect to the set of "cone spectra"

$$\{\Sigma_q^\infty CS^n \mid q \geq 0 \text{ and } n \geq 0\}.$$

This means that it is a map that satisfies the RLP with respect to the set of inclusions

$$i_0 : \Sigma_q^\infty CS^n \longrightarrow \Sigma_q^\infty CS^n \wedge I_+.$$

Again, a retract of a Serre fibration is a Serre fibration. The q-fibrations in $\mathscr{S}$ will be the Serre fibrations.

The following definition will allow us to give succinct statements of our results.

DEFINITION 4.1. Let $\mathscr{C}$ be a model category with a forgetful functor to $\mathscr{S}$ that creates weak equivalences and let $\mathscr{E}$ be a category with a forgetful functor to $\mathscr{C}$. We say that $\mathscr{C}$ creates a model structure in $\mathscr{E}$ if $\mathscr{E}$ is a model category whose weak equivalences are created in $\mathscr{S}$ and whose q-fibrations are created in

$\mathscr{C}$. That is, a map in $\mathscr{E}$ is a q-fibration if it is a q-fibration when regarded as a map in $\mathscr{C}$. The q-cofibrations in $\mathscr{E}$ must then be those maps which satisfy the left lifting property with respect to the acyclic q-fibrations.

Our categories are enriched, and our model structures will reflect this. Quillen defined the notion of a simplicial model category in [57, II§2], and the appropriate topological analogue of his definition reads as follows.

DEFINITION 4.2. A model category $\mathscr{E}$ is topological if it is topologically complete and cocomplete and if, for any q-cofibration $i : E \longrightarrow F$ and q-fibration $p : X \longrightarrow Y$, the induced map

$$(i^*, p_*) : \mathscr{E}(F, X) \longrightarrow \mathscr{E}(E, X) \times_{\mathscr{E}(E,Y)} \mathscr{E}(F, Y) \tag{4.3}$$

is a Serre fibration of spaces which is acyclic if either i or p is acyclic.

THEOREM 4.4. *The category $\mathscr{S}$ is a topological model category with respect to the weak equivalences and Serre fibrations. If $\mathbb{T} : \mathscr{S} \longrightarrow \mathscr{S}$ is a continuous monad that preserves reflexive coequalizers and satisfies the "Cofibration Hypothesis", then $\mathscr{S}$ creates a topological model structure in $\mathscr{S}[\mathbb{T}]$.*

We think of the first statement as the specialization to the identity monad of the second. We shall specify the "Cofibration Hypothesis" shortly. It will obviously be satisfied by the identity monad and by the monad $\mathbb{L}$, and arguments like those of the previous section verify it for the monads $\mathbb{TL}$ and $\mathbb{PL}$ that define A_∞ and E_∞ ring spectra.

COROLLARY 4.5. *The categories of $\mathbb{L}$-spectra and of A_∞ and E_∞ ring spectra are topological model categories.*

Of course, we are far more interested in our categories of modules and algebras. The crux of the proof of Theorem 4.4 is the adjunction

$$\mathscr{S}[\mathbb{T}](\mathbb{T}X, A) \cong \mathscr{S}(X, A)$$

for spectra X and $\mathbb{T}$-algebras A. By the adjunction, the q-fibrations in $\mathscr{S}[\mathbb{T}]$ are the maps that satisfy the RLP with respect to the set of inclusions

$$\mathbb{T}i_0 : \mathbb{T}\Sigma_q^\infty CS^n \longrightarrow \mathbb{T}\Sigma_q^\infty CS^n \wedge I_+.$$

That is, they satisfy the CHP with respect the set of "cone $\mathbb{T}$-algebras" $\mathbb{T}\Sigma_q^\infty CS^n$. These maps deserve to be called Serre fibrations of $\mathbb{T}$-algebras.

Similarly, we define a Serre fibration of S-modules to be a map that satisfies the CHP with respect to the "cone S-modules" $S \wedge_{\mathscr{L}} \mathbb{L}\Sigma_q^\infty CS^n$. For S-modules, the adjunction above must be replaced by the adjunction

$$\mathscr{M}_S(S \wedge_{\mathscr{L}} \mathbb{L}X, M) \cong \mathscr{S}(X, F_{\mathscr{L}}(S, M))$$

that we obtain by composing the first adjunction of II.2.2 with the freeness adjunction for the monad $\mathbb{L}$. Thus, when interpreting Definition 4.1 for S-modules, we must change our forgetful functor from the obvious one to the functor $F_{\mathscr{L}}(S, -)$. Since $F_{\mathscr{L}}(S, M)$ is naturally weakly equivalent to M, by I.8.7, the weak equivalences are unchanged. However, the q-fibrations are changed.

THEOREM 4.6. *The category $\mathscr{M}_S$ is a topological model category with weak equivalences created in $\mathscr{S}$. Its q-fibrations are the Serre fibrations of S-modules, which are the maps $f : M \longrightarrow N$ of S-modules such that*

$$F(\mathrm{id}, f) : F_{\mathscr{L}}(S, M) \longrightarrow F_{\mathscr{L}}(S, N)$$

is a Serre fibration of spectra.

Although the functor $\mathbb{T}X = S \wedge_{\mathscr{L}} \mathbb{L}X$ from spectra to S-modules is not a monad, the proof of Theorem 4.4 nevertheless applies. To understand this, we think in terms of the "mirror image" category $\mathscr{M}^S$ of counital $\mathbb{L}$-spectra specified in II.2.1. By II.2.7 and composition (see II.6.1), we have a continuous monad $F_{\mathscr{L}}(S, \mathbb{L}(-))$ on $\mathscr{S}$ whose algebras are the counital $\mathbb{L}$-spectra. We have a topological equivalence of categories $\mathscr{M}^S \longrightarrow \mathscr{M}_S$ that carries N to $S \wedge_{\mathscr{L}} N$. By II.2.5, $S \wedge_{\mathscr{L}} F_{\mathscr{L}}(S, M)$ is naturally isomorphic to $S \wedge_{\mathscr{L}} M$ for any $\mathbb{L}$-spectrum M. Thus the monad that defines counital $\mathbb{L}$-spectra is transported under the equivalence to the functor $\mathbb{T}$ relevant to the construction of the model structure on $\mathscr{M}_S$. The equivalence has the effect of changing the forgetful functor.

The proof of Theorem 4.4 will apply equally well if we change our ground category to $\mathscr{M}_S$.

THEOREM 4.7. *If $\mathbb{T} : \mathscr{M}_S \longrightarrow \mathscr{M}_S$ is a continuous monad that preserves reflexive coequalizers and satisfies the "Cofibration Hypothesis", then $\mathscr{M}_S$ creates a topological model structure in $\mathscr{M}_S[\mathbb{T}]$.*

Of course, the description of the q-fibrations as maps f such that $F_{\mathscr{L}}(S, f)$ is a Serre fibration persists. Again, the Cofibration Hypothesis will be specified shortly and holds in our examples.

COROLLARY 4.8. *The categories of S-algebras, commutative S-algebras, and modules over an S-algebra R are topological model categories.*

Now that we have a model structure on $\mathscr{M}_R$, we can generalize Theorem 4.7 by changing its ground category to $\mathscr{M}_R$.

THEOREM 4.9. *Let R be a commutative S-algebra. If $\mathbb{T} : \mathscr{M}_R \longrightarrow \mathscr{M}_R$ is a continuous monad that preserves reflexive coequalizers and satisfies the "Cofibration Hypothesis", then $\mathscr{M}_R$ creates a topological model structure in $\mathscr{M}_R[\mathbb{T}]$.*

COROLLARY 4.10. *The categories of algebras and commutative algebras over a commutative S-algebra R are topological model categories.*

In fact, Theorems 4.7 and 4.9 both apply, and they give the same model structures since they give the same q-fibrations and weak equivalences. We prefer to think of the model structure as created in $\mathscr{M}_R$, since that makes visible more information about the q-cofibrations. While the model category theory dictates what the q-cofibrations must be, the proofs of the theorems will lead to explicit descriptions.

DEFINITION 4.11. Let $\mathbb{T}$ be a monad in $\mathscr{S}$ as in Theorem 4.7. A relative cell $\mathbb{T}$-algebra Y under a $\mathbb{T}$-algebra X is a $\mathbb{T}$-algebra $Y = \operatorname{colim} Y_n$, where $Y_0 = X$ and Y_{n+1} is obtained from Y_n as the pushout of a sum of attaching maps $\mathbb{T}S^q \longrightarrow Y_n$ along the coproduct of the natural maps $\mathbb{T}S^q \longrightarrow \mathbb{T}CS^q$. When X is an initial $\mathbb{T}$-algebra, we say that Y is a cell $\mathbb{T}$-algebra. Relative and absolute cell $\mathbb{T}$-algebras are defined in precisely the same way for a monad $\mathbb{T}$ in $\mathscr{M}_R$ as in Theorem 4.9, except that the sphere spectra S^q are replaced by the sphere R-modules S^q_R.

REMARK 4.12. The functor $\mathbb{T} : \mathscr{S} \longrightarrow \mathscr{S}[\mathbb{T}]$, being a left adjoint, preserves coproducts. Thus, when attaching a coproduct of cells $\mathbb{T}CS^q$ to Y_n to obtain Y_{n+1}, we are considering a pushout in $\mathscr{S}[\mathbb{T}]$ of the general form

$$\begin{array}{ccc} \mathbb{T}E & \longrightarrow & A \\ \downarrow & & \downarrow{\scriptstyle i} \\ \mathbb{T}CE & \longrightarrow & B, \end{array} \tag{4.13}$$

where E is a wedge of spheres, and similarly when the ground category is $\mathscr{M}_S$ or $\mathscr{M}_R$.

The Cofibration Hypothesis is just the minimum condition necessary to obtain homotopical control over these pushout diagrams and their colimits. It holds in our examples by Proposition 3.9 and Lemma 3.10.

COFIBRATION HYPOTHESIS. The map i in any pushout of the form (4.13) is a cofibration of spectra (for Theorem 4.4) or of S-modules (for Theorem 4.7) or of R-modules (for Theorem 4.9). The underlying spectrum of the $\mathbb{T}$-algebra colimit of a sequence of cofibrations of $\mathbb{T}$-algebras is their colimit as a sequence of maps of spectra.

Actually, for the model structure in Theorems 4.7 and 4.9, we only need the maps i to be cofibrations of spectra, or even just spacewise closed inclusions of spectra. However, the stronger R-module cofibration condition holds in practice and is important in the applications.

THEOREM 4.14. *Under the hypotheses of Theorems 4.4, 4.7, and 4.9, a map of $\mathbb{T}$-algebras is a q-cofibration if and only if it is a retract of a relative cell $\mathbb{T}$-algebra. Moreover, any q-cofibration is a cofibration of underlying spectra (in Theorem 4.4) or of underlying S-modules (in Theorem 4.7) or of underlying R-modules (in Theorem 4.9).*

By the Cofibration Hypothesis, the second statement will follow from the first. In all of our categories of $\mathbb{T}$-algebras, the trivial spectrum is a terminal object and every $\mathbb{T}$-algebra is q-fibrant. By the previous result, a $\mathbb{T}$-algebra is q-cofibrant if and only if it is a retract of a cell $\mathbb{T}$-algebra. Note in particular that the unit $R \longrightarrow A$ of a q-cofibrant R-algebra or commutative R-algebra is a cofibration of R-modules.

As in our discussion of Theorem 4.6, the proof of the previous theorem will apply to give the following expected conclusion.

THEOREM 4.15. *For an S-algebra R, such as $R = S$, a map of R-modules is a q-cofibration if and only if it is a retract of a relative cell R-module.*

Thus, in the case of R-modules, model category theory just brings us back to the cell theory that we took as our starting point. We can turn this around. We certainly want the weak equivalences and q-cofibrations in $\mathscr{M}_R$ to be the weak equivalences of underlying spectra and the retracts of relative cell R-modules. Since the weak equivalences and q-cofibrations determine the q-fibrations, we see that the q-fibrations specified in Theorem 4.6 are in fact forced on us by the cell theory that we began with.

Returning to the general context of Theorem 4.14, we also have that the natural notion of homotopy in any of our categories of $\mathbb{T}$-algebras, namely that discussed in Remark 2.12, agrees with the notion of homotopy that is dictated by our model category structures.

LEMMA 4.16. *If A is a q-cofibrant $\mathbb{T}$-algebra, then $A \otimes I$ is a cylinder object for A in the sense of Quillen. That is, the folding map* $\mathrm{id} + \mathrm{id} : A \amalg A \longrightarrow A$ *factors as the composite of a q-cofibration $A \amalg A \longrightarrow A \otimes I$ and a weak equivalence $A \otimes I \longrightarrow A$.*

5. The proofs of the model structure theorems

We must prove Theorems 4.4, 4.6, 4.7, 4.9, 4.14, and 4.15 and Lemma 4.16. For uniformity of treatment, let $\mathscr{C}$ be either $\mathscr{S}$ or $\mathscr{M}_R$ for a commutative S-algebra R. Logically, of course, we should treat the case $R = S$ before going on to the general case. Let $\mathbb{T}$ be a continuous monad in $\mathscr{C}$ that preserves reflexive coequalizers. By Proposition 2.10, we already know that $\mathscr{C}[\mathbb{T}]$ is complete, cocomplete, tensored and cotensored, and indeed has all indexed limits and colimits. It is clear that if $g \circ f$ is defined and two of f, g, and $g \circ f$ are weak equivalences, then so is the third. It is also clear that the collections of q-fibrations, q-cofibrations, and weak equivalences are closed under composition and retracts and contain all isomorphisms. It remains to prove that arbitrary maps factor appropriately and that the q-fibrations satisfy the right lifting property (RLP) with respect to the acyclic q-cofibrations. The essential point is that Quillen's "small object argument" applies to construct the required factorizations. A general version of Quillen's original argument is given in [18, §6], and we shall give a modified version of that argument.

DEFINITION 5.1. For the purposes of this section, define a finite pair of spectra to be a pair of the form $(\Sigma_q^\infty B, \Sigma_q^\infty A)$, where B is a finite based CW complex, A is a subcomplex, and $q \geq 0$. Define a finite pair of $\mathbb{L}$-spectra to be a pair obtained by applying $\mathbb{L}$ to a finite pair of spectra. Define a finite pair of R-modules to be a pair obtained by applying $\mathbb{F}_R$ to a finite pair of spectra.

As a matter of esoterica, we actually only need A, not B, to be finite in our arguments.

LEMMA 5.2. *Let $\mathscr{F}$ be a set of maps in $\mathscr{C}[\mathbb{T}]$, each of which is of the form $\mathbb{T}E \longrightarrow \mathbb{T}F$ for some finite pair (F, E) in $\mathscr{C}$. Then any map $f : X \longrightarrow Y$ in $\mathscr{C}[\mathbb{T}]$ factors as a composite*

$$X \xrightarrow{i} X' \xrightarrow{p} Y,$$

where p satisfies the RLP with respect to each map in $\mathscr{F}$ and i satisfies the LLP with respect to any map that satisfies the RLP with respect to each map in $\mathscr{F}$.

PROOF. Let $X = X_0$. We construct a commutative diagram

$$\begin{array}{ccccccccc} X_0 & \xrightarrow{i_0} & X_1 & \longrightarrow \cdots \longrightarrow & X_n & \xrightarrow{i_n} & X_{n+1} & \longrightarrow & \cdots \\ {\scriptstyle f=p_0}\downarrow & & \downarrow{\scriptstyle p_1} & & \downarrow{\scriptstyle p_n} & & \downarrow{\scriptstyle p_{n+1}} & & \\ Y & \xrightarrow[\text{id}]{} & Y & \longrightarrow \cdots \longrightarrow & Y & \xrightarrow[\text{id}]{} & Y & \longrightarrow & \cdots \end{array} \tag{5.3}$$

as follows. Suppose inductively that we have constructed p_n. Consider all maps from a map in $\mathscr{F}$ to p_n. Each such map is a commutative diagram of the form

$$\begin{array}{ccc} \mathbb{T}E & \xrightarrow{\alpha} & X_n \\ \downarrow & & \downarrow{\scriptstyle p_n} \\ \mathbb{T}F & \xrightarrow[\beta]{} & Y. \end{array} \tag{5.4}$$

Summing over such diagrams, we construct a pushout diagram of the form

$$\begin{array}{ccc} \coprod \mathbb{T}E & \xrightarrow{\sum \alpha} & X_n \\ \downarrow & & \downarrow{\scriptstyle i_n} \\ \coprod \mathbb{T}F & \longrightarrow & X_{n+1}. \end{array}$$

The maps β induce a map $p_{n+1} : X_{n+1} \longrightarrow Y$ such that $p_{n+1} \circ i_n = p_n$. Let $X' = \operatorname{colim} X_n$, let $i : X \longrightarrow X'$ be the canonical map, and let $p : X' \longrightarrow Y$ be obtained by passage to colimits from the p_n. Constructing lifts by passage to coproducts, pushouts, and colimits in $\mathscr{C}[\mathbb{T}]$, we see that each i_n and therefore also i satisfies the LLP with respect to maps that satisfy the RLP with respect to maps in $\mathscr{F}$. Assume given a commutative square

$$\begin{array}{ccc} \mathbb{T}E & \xrightarrow{\alpha'} & X' \\ {\scriptstyle i}\downarrow & \overset{g}{\nearrow} & \downarrow{\scriptstyle p} \\ \mathbb{T}F & \xrightarrow[\beta]{} & Y, \end{array}$$

where i is in $\mathscr{F}$. To verify that p satisfies the RLP with respect to i, we must construct a map g that makes the diagram commute. The Cofibration Hypothesis implies that X' is constructed as the colimit of a sequence of cofibrations of spectra. By [38, App.3.9], a cofibration of spectra is a spacewise closed inclusion.

Therefore, using that $\mathbb{T}$ is the free functor from $\mathscr{C}$ to $\mathscr{C}[\mathbb{T}]$, we see by III.1.7 that the natural map

$$(5.5) \qquad \operatorname{colim} \mathscr{C}[\mathbb{T}](\mathbb{T}E, X_n) \longrightarrow \mathscr{C}[\mathbb{T}](\mathbb{T}E, X')$$

is a bijection. This ensures that $\alpha' : \mathbb{T}E \longrightarrow X'$ factors through some X_n, giving us one of the commutative squares (5.4) used in the construction of X_{n+1}. By construction, there is a map $\mathbb{T}F \longrightarrow X_{n+1}$ whose composite with the natural map to X' gives a map g as required. □

LEMMA 5.6. *Any map $f : X \longrightarrow Y$ in $\mathscr{C}[\mathbb{T}]$ factors as $p \circ i$, where i is an acyclic q-cofibration that satisfies the LLP with respect to any q-fibration and p is a q-fibration.*

PROOF. Let $\mathscr{F}$ be the set of pairs obtained by letting (B, A) in Definition 5.1 run through all pairs of spaces $(CS^n \wedge I_+, CS^n \wedge \{0\}_+)$, $n \geq 0$. By chasing through adjunctions and using the definition of q-fibrations and q-cofibrations, we see that a map is a q-fibration if and only if it satisfies the RLP with respect to every map in $\mathscr{F}$ and that every map in $\mathscr{F}$ is a q-cofibration, of course an acyclic one. Note the relevance of the first adjunction of II.2.2 when $\mathscr{C} = \mathscr{M}_S$: this is where the definition of q-fibrations in Theorem 4.6 is forced on us. Now use Lemma 5.2 to factor f. Then that lemma says that p is a q-fibration and that i satisfies the LLP with respect to all q-fibrations. In particular, i is a q-cofibration. We use the cylinders $(-) \otimes I$ to define homotopies in the category $\mathscr{C}[\mathbb{T}]$, as discussed in Remark 2.12. Then the free functor $\mathbb{T}$ and the adjoint forgetful functor preserve homotopies. A formal argument shows that each i_n is the inclusion of a deformation retraction of $\mathbb{T}$-algebras, and it follows that i is also a deformation retraction. Therefore i is an acyclic q-cofibration. □

LEMMA 5.7. *The q-fibrations satisfy the RLP with respect to the acyclic q-cofibrations.*

PROOF. This is formal. Let $f : E \longrightarrow F$ be any acyclic q-cofibration. We must show that f satisfies the LLP with respect to q-fibrations. By the previous lemma, we may factor f as $f = p \circ i$, where $i : E \longrightarrow E'$ is an acyclic q-cofibration that does satisfy the LLP with respect to q-fibrations and $p : E' \longrightarrow F$ is a q-fibration. Since f and i are weak equivalences, so is p. Since f satisfies the LLP with respect to acyclic q-fibrations, there exists $g : F \longrightarrow E'$ such that $g \circ f = i$ and $p \circ g = \mathrm{id}_F$. Clearly p and g, together with the identity map on E, express f as a retract of i. Since i satisfies the LLP with respect to q-fibrations, so does f. □

LEMMA 5.8. *Any map $f : X \longrightarrow Y$ in $\mathscr{S}[\mathbb{T}]$ factors as $p \circ i$, where i is a q-cofibration and p is an acyclic q-fibration.*

PROOF. This is another application of Lemma 5.2. Let $\mathscr{A}\mathscr{F}$ be the set of pairs obtained by letting (B, A) in Definition 5.1 run through all pairs of spaces (CS^n, S^n), $n \geq 0$. By tracing through adjunctions again, we see that a map of $\mathbb{T}$-algebras is an acyclic q-fibration if and only if it satisfies the RLP with respect to all maps in $\mathscr{A}\mathscr{F}$ and that each map in $\mathscr{A}\mathscr{F}$ is thus a q-cofibration. In the factorization $f = p \circ i$ that we now obtain from Lemma 5.2, that lemma says that p is an acyclic q-fibration and i is a q-cofibration. $\square$

We must still prove that $\mathscr{C}[\mathbb{T}]$ is topological, in the sense of Definition 4.2. As in [57, SM7(a), p.II.2.3], the description of q-cofibrations as retracts of relative cell $\mathbb{T}$-algebras implies that we need only check that the map (4.3) is a Serre fibration when $i : E \longrightarrow F$ is in the set $\mathscr{F}$ defined in the proof of Lemma 5.6 and an acyclic Serre fibration when $i : E \longrightarrow F$ is in the set $\mathscr{A}\mathscr{F}$ defined in the proof of Lemma 5.8. The freeness adjunction for the monad $\mathbb{T}$ reduces this to the case of spectra or of R-modules, and further adjunctions then reduce it to its known space level analogue. This completes the proofs of Theorems 4.4, 4.6, 4.7, and 4.9.

PROOF OF THEOREM 4.14. As in the proof of Lemmas 5.2 and 5.8, a relative cell $\mathbb{T}$-algebra $E \longrightarrow E'$ satisfies the LLP with respect to the acyclic q-fibrations and is thus a q-cofibration. Let $f : E \longrightarrow F$ be a q-cofibration. The proof of Lemma 5.8 gives a factorization of f as the composite of a relative cell $\mathbb{T}$-algebra $i : E \longrightarrow E'$ and an acyclic q-fibration $p : E' \longrightarrow F$. As in the proof of Lemma 5.7, there exists $g : F \longrightarrow E'$ such that $g \circ f = i$ and $p \circ g = \mathrm{id}_F$, and p and g express f as a retract of i. $\square$

PROOF OF LEMMA 4.16. We may assume without loss of generality that A is a cell $\mathbb{T}$-algebra. Write $A = \operatorname{colim} A_n$, where $A_0 = \mathbb{T}(*)$ and, for $n > 0$, $A_n = \mathbb{T}(CE_n) \amalg_{\mathbb{T}(E_n)} A_{n-1}$ for some wedge E_n of sphere objects of $\mathscr{C}$. Let $B_n = A \amalg_{A_n} (A_n \otimes I) \amalg_{A_n} A$. Since the functor $(-) \otimes I$ commutes with colimits, we have

$$A \otimes I \cong A \amalg_A (A \otimes I) \amalg_A A \cong \operatorname{colim} B_n.$$

Clearly $B_0 \cong A \amalg A$. We have $\mathbb{T}(E_n \wedge I_+) \cong \mathbb{T}E_n \otimes I$, and, for $n > 0$, B_n is constructed from B_{n-1} by a pushout diagram of the form

$$\begin{array}{ccc} \mathbb{T}(E_n \wedge I_+ \cup CE_n \wedge (\partial I)_+) & \longrightarrow & B_{n-1} \\ \big\downarrow & & \big\downarrow \\ \mathbb{T}(CE_n \wedge I_+) & \longrightarrow & B_n \end{array}$$

It follows that $A \otimes I$ is the colimit of a sequence of inclusions of cell $\mathbb{T}$-algebras relative to $A \amalg A$. The canonical weak equivalences $A_n \otimes I \longrightarrow A_n$ induce the canonical weak equivalence $A \otimes I \longrightarrow A$ on passage to colimits. $\square$

The reader should be convinced that the construction of model structures is a nearly formal consequence of the monadic descriptions of our various notions of structured ring, module, and algebra spectra.

REMARK 5.9. In [35], categories of A_∞ and E_∞ k-algebras and their modules were defined, and derived categories of modules were constructed, using a cell theory based on "sphere and cone modules". Replacing the ground category $\mathscr{S}$ with the ground category $\mathscr{M}_k$ of differential graded k-modules, the arguments of this section apply to give model structures to the analogous algebraic categories of modules. Unlike the treatments in [57, 5, 18], with this approach there is not the slightest reason to restrict attention to bounded below k-modules.

6. The underlying R-modules of q-cofibrant R-algebras

Let R be a fixed q-cofibrant commutative S-algebra. We study the underlying R-modules of q-cofibrant R-algebras and commutative R-algebras. The main point is to prove that the point-set level iterated smash products of q-cofibrant R-algebras represent their smash product in the derived category $\mathscr{D}_R$.

We begin with the simpler non-commutative case, and we do not need R to be q-cofibrant in the following two results. Recall from III.7.3 that smash products of cell R-modules are cell R-modules.

PROPOSITION 6.1. *Let A and B be R-algebras that are cell R-modules relative to R. Then their coproduct $A \amalg B$ is a cell R-module relative to R. In more detail, $A \amalg B$ is the colimit of an expanding sequence of relative cell R-modules $\{C_n\}$ such that $C_0 = R$ and, for $n \geq 1$, C_n/C_{n-1} is the wedge of the two monomial word modules of length n in A/R and B/R.*

PROOF. The monomial word modules in R-modules M and N are the smash products

$$M \wedge_R N \wedge_R M \wedge_R \cdots \qquad \text{and} \qquad N \wedge_R M \wedge_R N \wedge_R \cdots .$$

By II.7.4, we see that $A \amalg B$ is constructed via a coequalizer diagram in $\mathscr{M}_R$

$$\mathbb{T}(\mathbb{T}A \vee \mathbb{T}B) \rightrightarrows \mathbb{T}(A \vee B) \longrightarrow A \amalg B.$$

Writing out the source and target of the pair of parallel arrows as wedges of smash products and restricting to those wedge summands with at most n smash factors, we define C_n to be the coequalizer of the resulting restricted parallel pair of arrows. Clearly there result compatible maps $C_n \longrightarrow C_{n+1}$ and $C_n \longrightarrow A \amalg B$ such that $A \amalg B$ is the colimit of the C_n. In view of the use of the action maps $\mathbb{T}A \longrightarrow A$ and $\mathbb{T}B \longrightarrow B$ in II.7.4, we see that the wedge of the monomial words in A and B of length at most n maps onto C_n. That is, elements of word monomials involving $A \wedge_R A$ or $B \wedge_R B$ are identified in the coequalizer with elements of word monomials of lower length.

Let U_n be the coproduct in the category of R-modules under R of the two monomial words in A and B of length n, so that the copies of R in these R-modules under R are identified. Then U_n is a relative cell R-module. Let V_n be the union of the subcomplexes of U_n that are obtained by replacing any one A or B in either of the monomial words by its submodule R. The isomorphisms $R \wedge_R A \longrightarrow A$ and $R \wedge_R B \longrightarrow B$ induce a map $V_n \longrightarrow C_n$. Inspection of the restricted coequalizer diagrams (and comparison with algebra for intuition) shows that there result pushout diagrams of R-modules

$$\begin{array}{ccc} V_n & \longrightarrow & C_n \\ \downarrow & & \downarrow \\ U_{n+1} & \longrightarrow & C_{n+1}. \end{array}$$

Inductively, the C_n and $A \amalg B$ are cell R-modules relative to R. $\square$

THEOREM 6.2. *If A is a q-cofibrant R-algebra, then A is a retract of a cell R-module relative to R. Thus the unit $R \longrightarrow A$ is a q-cofibration of R-modules.*

PROOF. If M is a cell R-module, then M^j is a cell R-module for $j \geq 1$ and $(\mathbb{T}M, R)$ is a relative cell R-module. Moreover, since $M \longrightarrow CM$ is cellular, $\mathbb{T}M \longrightarrow \mathbb{T}CM$ is the inclusion of a subcomplex in a relative cell R-module. Now suppose that (A, R) is a relative cell R-module and that we have a pushout diagram of R-algebras

$$\begin{array}{ccc} \mathbb{T}M & \longrightarrow & A \\ \downarrow & & \downarrow \\ \mathbb{T}CM & \longrightarrow & B. \end{array}$$

As in the proof of Proposition 3.9, B is isomorphic to the geometric realization of a simplicial R-module that is proper because its degeneracies are given by inclusions of wedge summands. The previous proposition implies that its R-module of p-simplices is a cell R-module relative to R. Moreover, the face and degeneracy maps are sequentially cellular. Therefore, by X.2.7, (B, R) is isomorphic to a relative cell R-module, and A is a subcomplex. By passage to colimits, any cell R-algebra is a relative cell R-module. The conclusion follows from Theorem 4.14. $\square$

In the commutative case, the argument fails because we must pass to orbits over actions of symmetric groups. Tracing the proof of III.7.3 back to that of I.6.1, we see that it depends on the homeomorphism $\mathscr{L}(j) \cong \mathscr{L}(1)$ induced by a linear isomorphism $f : U^j \longrightarrow U$. Since this homeomorphism is not Σ_j-equivariant, we cannot deduce that symmetric powers of CW $\mathbb{L}$-spectra are, or even have the homotopy types of, CW $\mathbb{L}$-spectra, although they do have the homotopy types of CW spectra. For this reason, we cannot conclude that the symmetric power M^j/Σ_j of a cell R-module M has the homotopy type of a cell R-module; we refer the reader to III.5.1 for an analysis of the homotopy type of its underlying spectrum. We get around this problem by use of the following result, which gives canonical CW S-module approximations of smash products of "extended powers" of CW spectra. Here the jth extended power of X is defined to be

$$D_jX = (\mathbb{L}X)^j/\Sigma_j \cong \mathscr{L}(j) \ltimes_{\Sigma_j} X^j.$$

We adopt the convention that $D_0X = S$.

THEOREM 6.3. *Let $\{X_1, \dots, X_n\}$ be CW spectra, let $j_i \geq 0$, and consider the following commutative diagram of $\mathbb{L}$-spectra:*

$$\begin{array}{ccc} \bigwedge_{\mathscr{L}}(S \wedge_{\mathscr{L}} \mathbb{L}D_{j_i}X_i) & \xrightarrow{\wedge_{\mathscr{L}}\lambda} & \bigwedge_{\mathscr{L}} \mathbb{L}D_{j_i}X_i \\ {\scriptstyle \wedge_{\mathscr{L}}(\mathrm{id} \wedge_{\mathscr{L}} \xi)} \downarrow & & \downarrow {\scriptstyle \wedge_{\mathscr{L}}\xi} \\ \bigwedge_{\mathscr{L}}(S \wedge_{\mathscr{L}} D_{j_i}X_i) & \xrightarrow[\wedge_{\mathscr{L}}\lambda]{} & \bigwedge_{\mathscr{L}} D_{j_i}X_i. \end{array}$$

All spectra in the diagram have the homotopy types of CW-spectra, all maps in the diagram are homotopy equivalences of spectra, and $\wedge_S(S\wedge_{\mathscr{L}}\mathbb{L}D_{j_i}X_i)$ *has the homotopy type of a CW S-module.*

PROOF. By [38, VI.5.2 or VIII.2.4], X^j has the homotopy type of a Σ_j-CW spectrum indexed on U^j. By XI.1.7, $\mathscr{L}(j)$ has the homotopy type of a Σ_j-CW complex. Therefore, by the equivariant form of I.2.6, $\mathscr{L}(j)\ltimes X^j$ has the homotopy type of a Σ_j-CW spectrum indexed on U. Thus, by [38, I.5.6], D_jX has the homotopy type of a CW-spectrum. By I.4.6 and II.1.9, $\mathbb{L}D_jX$ has the homotopy type of a CW $\mathbb{L}$-spectrum and $S\wedge_{\mathscr{L}}\mathbb{L}D_jX$ has the homotopy type of a CW S-module. These conclusions pass to smash products by I.6.1 and III.7.3. The top horizontal arrow is a homotopy equivalence of $\mathbb{L}$-spectra by I.4.6 and I.8.5, and the bottom horizontal arrow is a homotopy equivalence of spectra by XI.2.5. We claim that the right vertical arrow and therefore the left vertical arrow are also homotopy equivalences of spectra. Indeed, if all $j_i\geq 1$, then use of I.5.4 and I.5.6 shows that the right vertical arrow is isomorphic to

$$\begin{array}{c}(\mathscr{L}(n)\times\mathscr{L}(j_1)\times\cdots\times\mathscr{L}(j_n))\ltimes_{\Sigma_{j_1}\times\cdots\times\Sigma_{j_n}}(X_1^{j_1}\wedge\cdots\wedge X_n^{j_n})\\ \Big\downarrow{\scriptstyle\gamma\ltimes\mathrm{id}}\\ \mathscr{L}(j_1+\cdots+j_n)\ltimes_{\Sigma_{j_1}\times\cdots\times\Sigma_{j_n}}(X_1^{j_1}\wedge\cdots\wedge X_n^{j_n}).\end{array}$$

Since γ is a $(\Sigma_{j_1}\times\cdots\times\Sigma_{j_n})$-equivariant homotopy equivalence, the map before passage to orbits is an equivariant homotopy equivalence by the equivariant version of I.2.5. If any $j_i=0$, then use of I.6.1 reduces us to the case when a single $j_i=0$, and in this case the conclusion follows from I.8.6. □

Now return to the study of our given q-cofibrant commutative S-algebra R.

DEFINITION 6.4. Define $\mathscr{E}_R$ to be the collection of R-modules of the form

$$R\wedge_S(S\wedge_{\mathscr{L}}D_jX),$$

where X is any spectrum of the homotopy type of a CW-spectrum and $j\geq 0$. Define $\bar{\mathscr{E}}_R$ to be the closure of $\mathscr{E}_R$ under finite $\wedge_R$-products, wedges, pushouts along cofibrations, colimits of countable sequences of cofibrations, and homotopy equivalences, where all of these operations are taken in the category of R-modules. That is, if $\{M_1,\dots,M_n\}\subset\bar{\mathscr{E}}_R$, then $M_1\wedge_R\cdots\wedge_R M_n\in\bar{\mathscr{E}}_R$, and so forth.

Observe that $\bar{\mathscr{E}}_R$ contains all R-modules of the homotopy types of cell R-modules, that being the collection that would be obtained if we only allowed $j=1$ in our initial class. One point of the definition is the following observation. Its proof is just like that of Theorem 6.2, and we shall say more about the commutative case shortly.

THEOREM 6.5. *The underlying R-module of a q-cofibrant R-algebra or commutative R-algebra A is in* $\bar{\mathscr{E}}_R$.

Another point is the following reassuring consequence of Theorem 6.3 and the definition.

PROPOSITION 6.6. *The underlying spectrum of an R-module in $\bar{\mathscr{E}}_R$ has the homotopy type of a CW-spectrum.*

These lead to the main point, which is that we have control of the behavior of derived smash product of R-modules that are in $\bar{\mathscr{E}}_R$.

THEOREM 6.7. *Let R be a q-cofibrant commutative S-algebra. Choose a cell R-module ΓM and a weak equivalence of R-modules $\gamma : \Gamma M \longrightarrow M$ for each $M \in \bar{\mathscr{E}}_R$. Then, for any finite subset $\{M_1, \dots, M_n\}$ of $\bar{\mathscr{E}}_R$,*

$$\gamma \wedge_R \cdots \wedge_R \gamma : \Gamma M_1 \wedge_R \cdots \wedge_R \Gamma M_n \longrightarrow M_1 \wedge_R \cdots \wedge_R M_n$$

is a weak equivalence of R-modules. That is, the derived smash product of the M_i in the category $\mathscr{D}_R$ is represented by their point-set level smash product.

PROOF. When $R = S$ and each M_i is in $\mathscr{E}_S$, Theorem 6.3 gives the conclusion. The conclusion for general M_i follows by standard commutation formulas relating smash products to the chosen operations. For general R and $M_i = R \wedge_S N_i$, where $N_i \in \mathscr{E}_S$ has CW S-approximation ΓN_i, $R \wedge_S \Gamma N_i$ is a CW R-approximation of M_i. Here we have the identification

$$(R \wedge_S N_1) \wedge_R \cdots \wedge_R (R \wedge_S N_n) \cong R \wedge_S (N_1 \wedge_S \cdots \wedge_S N_n),$$

and similarly and compatibly for the ΓN_i. By Theorem 6.5, R is in $\bar{\mathscr{E}}_S$, hence the result for S implies the result for these M_i. The result for general M_i follows as in the case $R = S$. □

Observe that the γ and their smash products are necessarily homotopy equivalences of underlying spectra, since these are CW homotopy types.

REMARK 6.8. In any model category, the coproduct of q-cofibrant objects is q-cofibrant. In particular, if A and B are q-cofibrant commutative R-algebras, then so is $A \wedge_R B$.

7. q-Cofibrations and weak equivalences; cofibrations

Again, let R be a fixed q-cofibrant commutative S-algebra. We here prove several useful lemmas concerning the relationship between q-cofibrations and weak equivalences. We also prove a key technical lemma about cofibrations. Recall the bar construction $\beta^R(A', A, A'')$ from Definition 3.5. The following result is a direct consequence of Proposition 3.7 and Lemma 4.16.

LEMMA 7.1. *If $A \longrightarrow A'$ and $A \longrightarrow A''$ are maps of q-cofibrant R-algebras, then $\beta^R(A', A, A'')$ is a q-cofibrant R-algebra over $A' \amalg_A A''$, and the natural map $A' \amalg A'' \longrightarrow \beta^R(A', A, A'')$ is a q-cofibration. The same statement holds for commutative R-algebras.*

This result gains force from the following two, which give good homotopical behavior of the functor $\beta^R(A', A, A'')$ and allow us to transport that behavior to the functor $A' \amalg_A A''$. Remember that $A' \amalg_A A'' = A' \wedge_A A''$ in the commutative case.

LEMMA 7.2. *If $A \longrightarrow A'$ is a map of q-cofibrant R-algebras, then the functor $\beta^R(A', A, -)$ preserves weak equivalences between q-cofibrant R-algebras. The same statement holds for commutative R-algebras.*

PROOF. This is not obvious since the given weak equivalence of R-algebras will be a homotopy equivalence of R-algebras but not necessarily of R-algebras under A. The functor β^R is obtained by passage to geometric realization from a functor β^R_* to simplicial R-modules. Since coproducts of weak equivalences between q-cofibrant objects are weak equivalences in any model category, it is clear from Definition 3.5 that each functor $\beta^R_q(A', A-)$ preserves weak equivalences. By Lemma 7.4 below, $\beta_*(A', A, A'')$ is a proper simplicial R-module, and the conclusion follows from X.2.4. □

LEMMA 7.3. *If $A \longrightarrow A'$ is a q-cofibration of q-cofibrant R-algebras and A'' is a q-cofibrant R-algebra, then the natural map $\psi : \beta^R(A', A, A'') \longrightarrow A' \amalg_A A''$ is a weak equivalence of R-algebras. The same statement holds for commutative R-algebras.*

PROOF. Recall from Proposition 3.7 that we can identify $\beta^R(A', A, A'')$ with the double mapping cylinder $M(A', A, A'')$ as an R-algebra over $A' \amalg_A A''$. Without loss of generality, we may assume that $A \longrightarrow A'$ is the inclusion of a relative cell R-algebra. Write $A' = \operatorname{colim} A'_n$ where $A'_0 = A$ and $A'_n = \mathbb{T}(CE_n) \amalg_{\mathbb{T}E_n} A'_{n-1}$ for some wedge of sphere R-modules E_n. Then each map $A'_n \amalg_A A'' \longrightarrow A'_{n+1} \amalg_A A''$ and $M(A'_n, A, A'') \longrightarrow M(A'_{n+1}, A, A'')$ is a q-cofibration of R-algebras and therefore a cofibration of R-modules. Thus it suffices to show that each map $\psi : M(A'_n, A, A'') \longrightarrow A'_n \amalg_A A''$ is a weak equivalence. This is clear for $n = 0$; assume inductively that it is true for $n - 1$. We have the following commutative diagram, in which the horizontal arrows are the evident isomorphisms:

$$\begin{array}{ccc}
M(A'_n, A, A'') & \xrightarrow{\cong} & \mathbb{T}(CE_n) \amalg_{\mathbb{T}E_n} M(A'_{n-1}, A, A'') \\
\downarrow{\scriptstyle \psi} & & \downarrow{\scriptstyle \mathrm{id} \amalg \psi} \\
A'_n \amalg_A A'' & \xrightarrow[\cong]{} & \mathbb{T}(CE_n) \amalg_{\mathbb{T}E_n} A'_{n-1} \amalg_A A''.
\end{array}$$

We also have the following commutative diagram:

$$\begin{array}{ccc}
M(\mathbb{T}(CE_n), \mathbb{T}E_n, M(A'_{n-1}, A, A'')) & \xrightarrow{\psi} & \mathbb{T}(CE_n) \amalg_{\mathbb{T}E_n} M(A'_{n-1}, A, A'') \\
\downarrow{\scriptstyle M(\mathrm{id},\mathrm{id},\psi)} & & \downarrow{\scriptstyle \mathrm{id} \amalg \psi} \\
M(\mathbb{T}(CE_n), \mathbb{T}E_n, A'_{n-1} \amalg_A A'') & \xrightarrow[\psi]{} & \mathbb{T}(CE_n) \amalg_{\mathbb{T}E_n} A'_{n-1} \amalg_A A''.
\end{array}$$

Here, by Proposition 3.8, the horizontal maps ψ are homotopic to isomorphisms and are thus weak equivalences. The map $M(\mathrm{id}, \mathrm{id}, \psi)$ is a weak equivalence by the previous lemma. Therefore $\mathrm{id} \amalg \psi$ is a weak equivalence. The conclusion follows. □

This has the following important consequence concerning cobase change.

PROPOSITION 7.4. *If $A \longrightarrow A'$ is a q-cofibration of q-cofibrant R-algebras, then the functor $A' \amalg_A (-)$ preserves weak equivalences between q-cofibrant R-algebras. The same statement holds for commutative R-algebras.*

In fact, in the commutative case, we need only assume given a weak equivalence between R-algebras that are q-cofibrant as S-algebras, since the construction of $A \wedge_A A''$ does not depend on R.

We conclude with the following lemma on cofibrations. It will imply that the simplicial R-modules used in the following chapters are proper, in the sense of X.2.2. As explained at the start of X§2, we abuse language by writing about unions, images, and inclusions when we should be writing more precisely about maps from a suitable coend to A^q. The abuse is justified by the conclusion, since a cofibration of spectra is a spacewise closed inclusion [38, I.8.1].

LEMMA 7.5. *Let A be a q-cofibrant R-algebra or a q-cofibrant commutative R-algebra. Let $sA^q \subset A^q$ be the "union of the images" of the maps*

$$s_i = (\mathrm{id})^i \wedge \eta \wedge (\mathrm{id})^{q-i} : A^{q-1} \longrightarrow A^q.$$

Then the "inclusion" $sA^q \subset A^q$ is a cofibration of R-modules. In particular, the unit $\eta : R \longrightarrow A$ is a cofibration of R-modules.

PROOF. In the non-commutative case, (A, R) is a relative cell R-module. Its qth smash power inherits such a structure, by III.7.3, and sA^q is a subcomplex. In the commutative case, we can apply the same brief argument, once we observe that (A, R) is a suitably general kind of relative cell R-module. Thus we consider generalized relative cell R-modules that are constructed with (cell, sphere) pairs replaced by pairs of the form $(N \wedge B^q_+, N \wedge S^{q-1}_+)$, where N runs through all finite smash products over R of R-modules of the form $(S^n_R)^j/\Sigma_j$ or $(CS^n_R)^j/\Sigma_j$ for integers n and for $j \geq 1$. Here the (B^q, S^{q-1}) are ordinary space level (cell, sphere) pairs. Observe that these R-modules are finite colimits of compact R-modules, so that III.1.7 applies to them. Let $\mathbb{P}$ be the monad in the category of R-modules that defines commutative R-algebras. Obviously $\mathbb{P}M$ and $\mathbb{P}CM$ are relative cell R-modules in this generalized sense when M is a wedge of sphere R-modules. Equally obviously, the smash product over R of two such generalized relative cell R-modules is another such. Suppose that A is such a commutative R-algebra and consider a pushout diagram

$$\begin{array}{ccc} \mathbb{P}M & \longrightarrow & A \\ \downarrow & & \downarrow \\ \mathbb{P}CM & \longrightarrow & B. \end{array}$$

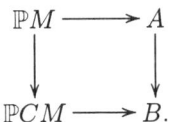

As explained in 3.5–3.9, B is isomorphic to the geometric realization of the proper simplicial R-module $\beta^R_*(\mathbb{P}CM, \mathbb{P}M, A)$. Remember that, here in the commutative case, the coproduct used in VII§3 is the smash product over R. We may construct the geometric realization by first using degeneracy identifications, which serve simply to eliminate redundant wedge summands from the relevant coend, and then face identifications; compare X.2.6. The latter identifications can be expressed by pushout diagrams of R-modules

$$\begin{array}{ccc} \beta_q^R(\mathbb{P}CM, \mathbb{P}M, A) \wedge \partial\Delta_{q+} & \xrightarrow{g} & F_{q-1}\beta^R(\mathbb{P}CM, \mathbb{P}M, A) \\ \downarrow & & \downarrow \\ \beta_q^R(\mathbb{P}CM, \mathbb{P}M, A) \wedge \Delta_{q+} & \longrightarrow & F_q\beta^R(\mathbb{P}CM, \mathbb{P}M, A). \end{array}$$

Of course, we think of $(\Delta_q, \partial\Delta_q)$ as a model for (B^q, S^{q-1}). Proceeding inductively and using III.1.7 and the proof of III.2.2, we can make g a sequentially cellular map and deduce that the qth filtration is a generalized relative cell R-module. By passage to colimits, so is any commutative cell R-algebra. The same holds for smash powers. The inclusion of a subcomplex is a cofibration, by reduction to the obvious case of cell pairs, and the conclusion follows. □

CHAPTER VIII

Bousfield localizations of R-modules and algebras

We study Bousfield localizations in this chapter. For any S-algebra R and cell R-module E, we show that $\mathscr{M}_R$ admits a new model category structure in which the weak equivalences are the E-equivalences. With this model structure, a factorization of the trivial map $M \longrightarrow *$ as the composite of an E-acyclic E-cofibration and an E-fibration constructs the localization of M at E.

Restricting to a q-cofibrant commutative S-algebra R, we combine formal constructions with the homotopical analysis of the previous chapter to prove that localizations at E of cell R-algebras can be constructed as cell R-algebras, and similarly for commutative cell R-algebras. Of course, this applies quite generally since any R-algebra is weakly equivalent to a cell R-algebra. That is, we can conclude that Bousfield localizations of R-algebras and commutative R-algebras are again such. In the case $R = S$, Hopkins and McClure had an unpublished argument, sketched in e-mails to us, that Bousfield localizations of E_∞ ring spectra are E_∞ ring spectra.

We deduce that the derived category of E-local R-modules is equivalent to the full subcategory of the derived category of R_E-modules whose objects are those R_E-modules that are E-local as R-modules. In particular our new derived categories of S_E-modules are intrinsically important to a complete understanding of the classical Bousfield localizations of spectra.

As a simple direct application, we deduce that KO and KU are commutative ko and ku-algebras since they are Bousfield localizations of ko and ku obtained by inverting the Bott elements. By neglect of structure, they are therefore commutative S-algebras. This solves a problem that was first studied in McClure's 1978 PhD thesis.

We refer the reader to [26] for a discussion of the special cases of Bousfield localization that give localizations and completions of R-modules at ideals of the coefficient ring R_*.

1. Bousfield localizations of R-modules

Let R be an S-algebra, such as S itself, and suppose given a cell R-module E. We shall construct Bousfield localizations of R-modules at E. The treatment is based on Bousfield's papers [11, 12], but in the latter he was handicapped by working in a primitive category of spectra that did not admit a model category structure.

A map $f : M \longrightarrow N$ of R-modules is said to be an E-equivalence if the induced map

$$\mathrm{id} \wedge_R f : E \wedge_R M \longrightarrow E \wedge_R N$$

is a weak equivalence. Homologically, we should call such maps E_*^R-equivalences, and we shall often refer to them as E-acyclic maps. An R-module W is said to be E-acyclic if $E \wedge_R W \simeq *$, and a map f is E-acyclic if and only if its cofiber is E-acyclic. We say that an R-module L is E-local if $f^* : \mathscr{D}_R(N, L) \longrightarrow \mathscr{D}_R(M, L)$ is an isomorphism for any E-equivalence f or, equivalently, if $\mathscr{D}_R(W, L) = 0$ for any E-acyclic R-module W. Since this is a derived category criterion, it suffices to test it when W is a cell R-module. A localization of M at E is a map $\lambda : M \longrightarrow M_E$ such that λ is an E-equivalence and M_E is E-local. Of course, the formal properties of such localizations discussed in [11, 12] carry over verbatim to the present context. We shall construct a model structure on $\mathscr{M}_R$ that implies the existence of E-localizations of R-modules.

THEOREM 1.1. *The category $\mathscr{M}_R$ admits a new structure as a topological model category in which the weak equivalences are the E-equivalences and the cofibrations are the q-cofibrations in the model structure already constructed. The fibrations in the new model structure are the maps that satisfy the right lifting property with respect to the E-acyclic q-cofibrations.*

Although the theorem gives the best way to think about the new model structure, it will be convenient to construct it in a way that parallels the proofs in VII§5. To that end, we give apparently different definitions of E-fibrations and E-cofibrations.

DEFINITION 1.2. A map $f : M \longrightarrow N$ is an E-fibration if it has the right lifting property with respect to the E-acyclic inclusions of subcomplexes in cell R-modules. A map $f : M \longrightarrow N$ is an E-cofibration if it satisfies the left lifting property with respect to the E-acyclic E-fibrations.

The following comparisons will emerge during our proof of Theorem 1.1.

LEMMA 1.3. *A map is an E-cofibration if and only if it is a q-cofibration.*

LEMMA 1.4. *A map is an E-fibration if and only if it satisfies the RLP with respect to the E-acyclic q-cofibrations.*

An R-module L is said to be E-fibrant if the unique map $L \longrightarrow *$ is an E-fibration.

PROPOSITION 1.5. *An R-module is E-fibrant if and only if it is E-local.*

PROOF. The argument is the same as that of [11, 3.5]. As in [11, 3.6], one checks that the class of E-equivalences in the derived category $\mathscr{D}_R$ admits a calculus of left fractions. This implies [11, 2.5] that an R-module L is E-local if and only if $f^* : \mathscr{D}_R(N, L) \longrightarrow \mathscr{D}_R(M, L)$ is a surjection for any E-equivalence $f : M \longrightarrow N$. Since we are working in derived categories, there is no loss of generality to assume that f is the inclusion of a subcomplex in a cell R-module. If L is E-fibrant, the RLP already gives surjectivity on the point-set level, hence on the level of homotopy classes. If L is E-local, we have surjectivity on the level of homotopy classes and deduce it on the point-set level by use of HEP. □

THEOREM 1.6. *Every R-module M admits a localization $\lambda : M \longrightarrow M_E$.*

PROOF. We may factor the trivial map $M \longrightarrow *$ as the composite of an E-acyclic E-cofibration $\lambda : M \longrightarrow M_E$ and an E-fibration $M_E \longrightarrow *$. □

Localizations of the underlying spectra of R-modules at spectra can be recovered as special cases of our new localizations of R-modules at R-modules. Therefore, up to equivalence, the localization of an R-module at a spectrum can be constructed as a map of R-modules.

PROPOSITION 1.7. *Let K be a CW-spectrum and let E be the R-module $\mathbb{F}_R K$. Regarded as a map of spectra, a localization $\lambda : M \longrightarrow M_E$ of an R-module M at E is a localization of M at K.*

PROOF. By IV.1.9, we have $K_*(M) \cong E^R_*(M)$ for R-modules M. Therefore an E-equivalence of R-modules is a K-equivalence of spectra. If W is a K-acyclic spectrum, then $\mathbb{F}_R W$ is an E-acyclic R-module since $E \wedge_R \mathbb{F}_R W$ is equivalent to $\mathbb{F}_R(K \wedge W)$. Therefore, if N is an E-local R-module, then $[W, N] \cong [\mathbb{F}_R W, N]_R = 0$ and N is a K-local spectrum. The conclusion follows. □

The argument generalizes to show that, for an R-algebra A, the localization of an A-module at an R-module E can be constructed as a map of A-modules.

PROPOSITION 1.8. *Let A be a q-cofibrant commutative R-algebra, let E be a cell R-module and let F be the A-module $A \wedge_R E$. Regarded as a map of R-modules, a localization $\lambda : M \longrightarrow M_F$ of an A-module M at F is a localization of M at E.*

We prove Theorem 1.1 in the rest of the section. Of course, $\mathscr{M}_R$ is topologically complete and cocomplete. It is clear that retracts of E-equivalences, E cofibrations, and E-fibrations are again such (and similarly with cofibrations and fibrations as in the statement of the theorem). The following result motivates our definition of E-fibrations in terms of inclusions of subcomplexes rather than general q-cofibrations. Let $\#X$ denote the cardinality of the set of cells of a cell R-module X and let c be a fixed infinite cardinal that is at least the cardinality of $E^R_*(S_R)$. Long exact sequences and the commutation of homology with colimits imply that if $X \longrightarrow Y$ is the inclusion of a subcomplex in a cell R-module Y such that $\#Y \leq c$, then the cardinality of $E^R_*(Y/X)$ is at most c. Let $\mathscr{T}$ be the set of E-acyclic inclusions of subcomplexes in cell R-modules Y such that $\#Y \leq c$. Then $\mathscr{T}$ is a test set for E-fibrations.

LEMMA 1.9. *A map $f : M \longrightarrow N$ is an E-fibration if and only p has the RLP with respect to maps in $\mathscr{T}$.*

PROOF. Arguing as in [11, 11.2,11.3], we see that for any proper E-acyclic inclusion $X \longrightarrow Y$ of a subcomplex in a cell R-module Y, there is a subcomplex $X' \subset Y$ such that $\#X' \leq c$, X' is not contained in X, and $X' \cap X \longrightarrow X'$ is E-acyclic. We construct X' as the union of a sequence of subcomplexes X_n of Y such that $\#X_n \leq c$, X_n is not contained in X, and the map

$$E^R_*(X_n/X_n \cap X) \longrightarrow E^R_*(X_{n+1}/X_{n+1} \cap X)$$

is zero. The fact that homology commutes with colimits and that $E^R_*(Y/X) = 0$ allows us to perform the inductive step by adjoining a finite subcomplex of Y to X_n to kill each element of $E^R_*(X_n/X_n \cap X)$. We conclude that if f has the RLP with respect to maps in $\mathscr{T}$, then it has the RLP with respect to $X \longrightarrow X' \cup X$ since it has the RLP with respect to $X' \cap X \longrightarrow X'$. By transfinite induction, it follows that f has the RLP with respect to $X \longrightarrow Y$. $\square$

LEMMA 1.10. *Any map $f : M \longrightarrow N$ factors as a composite*

$$M \xrightarrow{i} M' \xrightarrow{p} N,$$

where p is an E-fibration and i is an E-acyclic q-cofibration that satisfies the LLP with respect to E-fibrations.

PROOF. The construction is exactly like that in the proof of VII.5.2, with $\mathscr{T}$ here playing the role of $\mathscr{F}$ there. However, since we do not have compactness, we must perform the construction transfinitely. We carry the construction through to the least ordinal of cardinality greater than c. We can then use set theory rather than VII.5.5 to ensure the requisite factorization α' in the cited proof. The resulting map p is certainly an E-fibration. The construction by successive pushouts of wedges of maps in $\mathscr{T}$ for ordinals n that are successors and by passage to colimits for limit ordinals shows that the constructed map i is E-acyclic and satisfies the LLP with respect to the E-fibrations. To see that i, despite its transfinite construction, is a q-cofibration, we must specify a sequential filtration inductively. Given the sequential filtration on M_n, we obtain a sequential filtration on M_{n+1} by using III.2.3 to arrange that the pushout that constructs M_{n+1} is a diagram of sequentially cellular maps. If n is a limit ordinal and we have compatible sequential filtrations on the M_m for $m < n$, then each cell of M_n has a preassigned filtration and we take the qth filtration of M_n to be the union of the qth filtrations of the M_m for $m < n$. $\square$

REMARK 1.11. The previous proof uses that if $i : X \longrightarrow Y$ is an E-acyclic q-cofibration and $f : X \longrightarrow M$ is any map, then the pushout $j : M \longrightarrow M \cup_X Y$ is E-acyclic. This holds because i and j are cofibrations of R-modules with isomorphic quotients $Y/X \cong (M \cup_X Y)/M$.

LEMMA 1.12. *The following conditions on a map $f : M \longrightarrow N$ are equivalent.*

(i) *f is an E-acyclic E-fibration.*
(ii) *f is an E-acyclic map that satisfies the RLP with respect to all q-cofibrations.*

(iii) f is an acyclic q-fibration.

PROOF. Obviously (ii) implies (i) and (iii) implies (ii). Assume (i). Clearly f is a q-fibration, and we must prove that it is a weak equivalence. By factoring a weak equivalence from a cell R-module to M as the composite of an acyclic q-cofibration and a q-fibration, we can construct an acyclic q-fibration $p : M' \longrightarrow M$, where M' is a cell R-module. By VII.5.8, we may factor $f \circ p$ as the composite of a q-cofibration $i : M' \longrightarrow N'$ and an acyclic q-fibration $p' : N' \longrightarrow N$. By III.2.6 and the proof of VII.5.2, we can arrange that N' is a cell R-module that contains M' as a subcomplex. Since $f \circ p$ is E-acyclic, so is i. Summarizing, we have the diagram

$$\begin{array}{ccccc} M' & =\!=\!= & M' & \xrightarrow{p} & M \\ {\scriptstyle i}\downarrow & {\scriptstyle r}\nearrow & \downarrow{\scriptstyle f\circ p} & & \downarrow{\scriptstyle f} \\ N' & \xrightarrow{p'} & N & =\!=\!= & N. \end{array}$$

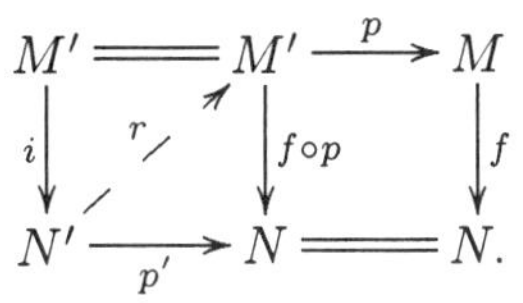

Since $f \circ p$ is an E-fibration and i is an E-acyclic inclusion of a subcomplex in a cell R-module, there exists a lift r. This expresses $f \circ p$ as a retract of the weak equivalence p'. Therefore $f \circ p$ and f are weak equivalences. □

Observe that Lemma 1.3 is an immediate consequence of Lemma 1.12.

PROOF OF LEMMA 1.4. It suffices to show that any E-acyclic q-cofibration $i : X \longrightarrow Y$ satisfies the LLP with respect to all E-fibrations. Factor i as in Lemma 1.10 and consider the resulting diagram

$$\begin{array}{ccc} X & \xrightarrow{j} & X' \\ {\scriptstyle i}\downarrow & {\scriptstyle \iota}\nearrow & \downarrow{\scriptstyle p} \\ Y & =\!=\!= & Y, \end{array}$$

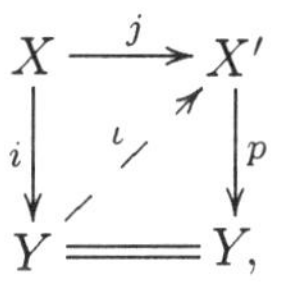

where p is an E-fibration and j is an E-acyclic q-cofibration that satisfies the LLP with respect to the E-fibrations. Clearly p is E-acyclic, hence Lemma 1.12 implies that it satisfies the RLP with respect to the cofibration i. There results a lift ι, and ι and p show that i is a retract of j. Since j satisfies the LLP with respect to all E-fibrations, so does i. □

PROOF OF THEOREM 1.1. We have proven one of the factorization axioms in Lemma 1.10, and the remaining axioms for a model structure are now direct consequences of the corresponding axioms for the original model structure on $\mathscr{M}_R$. □

2. Bousfield localizations of R-algebras

In this section, we restrict R to be a q-cofibrant commutative S-algebra and let E be a cell R-module. We shall prove the following pair of theorems.

THEOREM 2.1. *For a cell R-algebra A, the localization $\lambda : A \longrightarrow A_E$ can be constructed as the inclusion of a subcomplex in a cell R-algebra A_E. Moreover, if $f : A \longrightarrow B$ is a map of R-algebras into an E-local R-algebra B, then f lifts to a map of R-algebras $\tilde{f} : A_E \longrightarrow B$ such that $\tilde{f} \circ \lambda = f$, and $\tilde{f}$ is unique up to homotopy through maps of R-algebras. If f is an E-equivalence, then $\tilde{f}$ is a weak equivalence.*

THEOREM 2.2. *For a cell commutative R-algebra A, the localization $\lambda : A \longrightarrow A_E$ can be constructed as the inclusion of a subcomplex in a cell commutative R-algebra A_E. Moreover, if $f : A \longrightarrow B$ is a map of R-algebras into an E-local commutative R-algebra, then f lifts to a map of R-algebras $\tilde{f} : A_E \longrightarrow B$ such that $\tilde{f} \circ \lambda = f$, and $\tilde{f}$ is unique up to homotopy through maps of commutative R-algebras. If f is an E-equivalence, then $\tilde{f}$ is a weak equivalence.*

PROOFS. The idea is to replace the category $\mathscr{M}_R$ by either the category $\mathscr{A}_R$ or $\mathscr{C}\mathscr{A}_R$ in the work of the previous section. Most arguments go through with little change, the crucial exception being the part of the proof of Lemma 1.10 that is singled out in Remark 1.11. The problem there is that, to prove the analogue of the cited lemma in full generality, we would have to allow A to be an arbitrary R-algebra or commutative R-algebra. However, to keep homotopical control, we need A to be a cell R-algebra. This is enough to prove our theorems, although we do not actually obtain new model structures on $\mathscr{A}_R$ and $\mathscr{C}\mathscr{A}_R$.

For definiteness, consider the non-commutative case. Proceeding as in the proofs of VII.5.2 and Lemma 1.10, we let $A_0 = A$ and construct a transfinite sequence

$$A_0 \xrightarrow{i_0} A_1 \longrightarrow \cdots \longrightarrow A_n \xrightarrow{i_n} A_{n+1} \longrightarrow \cdots \tag{2.3}$$

as follows. Suppose inductively that we have constructed A_n. Consider all diagrams of R-modules

$$Y \xleftarrow{i} X \xrightarrow{\alpha} A_n, \tag{2.4}$$

where i is in $\mathscr{T}$. Using the free R-algebra functor $\mathbb{T}$ on R-modules, we take the sum over such diagrams of the adjoint maps $\tilde{\alpha} : \mathbb{T}X \longrightarrow A_n$ and construct the pushout diagram of R-algebras

$$\begin{array}{ccc} \coprod \mathbb{T}X & \xrightarrow{\sum \tilde{\alpha}} & A_n \\ \downarrow & & \downarrow{\scriptstyle i_n} \\ \coprod \mathbb{T}Y & \longrightarrow & A_{n+1}. \end{array} \tag{2.5}$$

If n is a limit ordinal and A_m has been constructed for $m < n$, we let A_n be the colimit of the A_m. We take A_E to be the colimit up to the least ordinal of cardinality greater than c. Then any map of R-modules from a cell R-module X with $\#X \leq c$ into A_E factors through some A_n, and we let $\lambda : A \longrightarrow A_E$

be the canonical map. Regarded as an R-module, A_E is E-fibrant and therefore E-local. To see this, consider a diagram of R-modules

$$\begin{array}{ccc} X & \xrightarrow{\alpha'} & A_E, \\ {\scriptstyle i}\downarrow & \overset{g}{\nearrow} & \\ Y & & \end{array}$$

where i is in $\mathscr{T}$. We must construct a map g that makes the diagram commute. Since α' factors through some A_n, we have one of the diagrams (2.4) used in the construction of A_{n+1}. By construction, there is a map $\mathbb{T}Y \longrightarrow A_{n+1}$ the adjoint of whose composite with the natural map to A_E is a map g as required. As in Lemmas 1.10 and 1.12, we can use the cell R-algebra analogue of III.2.6 to arrange that $\lambda : A \longrightarrow A_E$ is the inclusion of a subcomplex in a cell R-algebra.

We must prove that λ is an E-equivalence. By the commutation of homology with directed colimits, only the pushout maps $A_n \longrightarrow A_{n+1}$ are at issue. Observe that $\amalg \mathbb{T}X \cong \mathbb{T}(\vee X)$. Lemma 2.6 below shows that the left vertical arrow in (2.5) is an E-equivalence. In the commutative case, we must replace $\mathbb{T}$ by $\mathbb{P}$, and here Lemma 2.7 shows that the left vertical arrow in the commutative analogue of (2.5) is an E-equivalence. Finally, in both cases, Lemma 2.8 shows that the right vertical arrow $A_n \longrightarrow A_{n+1}$ is an E-equivalence.

We prove the lifting statement for a map $f : A \longrightarrow B$ by inductively lifting f to maps $f_n : A_n \longrightarrow B$. For the inductive step when $f_n : A_n \longrightarrow B$ is given, we apply the fact that B is E-fibrant to lift the evident composites

$$X \longrightarrow \mathbb{T}X \longrightarrow A_n \longrightarrow B$$

to maps of R-modules $Y \longrightarrow B$ and then apply freeness to obtain maps of R-algebras $\mathbb{T}Y \longrightarrow B$ that lift the maps of R-algebras $\mathbb{T}X \longrightarrow B$. Passage to pushouts then gives the required map $f_{n+1} : A_{n+1} \longrightarrow B$. We construct $\tilde{f}_n$ by passage to colimits when n is a limit ordinal. It is clear that $\tilde{f}$ must be an E-equivalence and therefore a weak equivalence if f is an E-equivalence.

To see the uniqueness of $\tilde{f}$ up to homotopy through maps of R-algebras, it suffices by VII.2.11 to construct a homotopy $A \otimes I \longrightarrow B$ between two such maps. We observe that $A_E \otimes I$ can be built inductively as the union of its subalgebras

$$(A \otimes I)_n = (A \otimes \{0\}) \amalg_{A_n} (A_n \otimes I) \amalg_{A_n} (A \otimes \{1\}),$$

where $(A \otimes I)_{n+1}$ is obtained from $(A \otimes I)_n$ by adding a cell

$$(\mathbb{T}(Y \wedge I_+), \mathbb{T}(Y \vee_X (X \wedge I_+) \vee_X Y))$$

for each cell $(\mathbb{T}Y, \mathbb{T}X)$ used in forming A_{n+1} from A_n. Since the map

$$Y \vee_X (X \wedge I_+) \vee_X Y \longrightarrow Y \wedge I_+$$

is an E-acyclic E-cofibration, we can construct the required homotopy one such cell at a time. $\square$

LEMMA 2.6. *If $f : M \longrightarrow N$ is an E-equivalence of cell R-modules, then $\mathbb{T}f : \mathbb{T}M \longrightarrow \mathbb{T}N$ is an E-equivalence of R-modules.*

PROOF. If $f : M \longrightarrow N$ and $f' : M' \longrightarrow N'$ are E-equivalences, then, factoring

$$\mathrm{id} \wedge_R f \wedge_R f' : E \wedge_R M \wedge_R M' \longrightarrow E \wedge_R N \wedge_R N'$$

as the composite $(\mathrm{id} \wedge_R f \wedge_R \mathrm{id})(\mathrm{id} \wedge_R \mathrm{id} \wedge_R f')$ and using the commutativity and associativity of $\wedge_R$ and the fact that all R-modules in sight are cell R-modules, we see that $\mathrm{id} \wedge_R f \wedge_R f'$ is an equivalence and thus that $f \wedge_R f'$ is an E-equivalence. Inductively, $f^j : M^j \longrightarrow N^j$ is an E-equivalence for all $j \geq 0$. □

LEMMA 2.7. *If $f : M \longrightarrow N$ is an E-equivalence of cell R-modules, then $\mathbb{P}f : \mathbb{P}M \longrightarrow \mathbb{P}N$ is an E-equivalence of R-modules.*

PROOF. We must show that $f^j/\Sigma_j : M^j/\Sigma_j \longrightarrow N^j/\Sigma_j$ is an E-equivalence for all $j \geq 0$. By III.5.1, this will hold if

$$\mathrm{id} \wedge f^j : (E\Sigma_j)_+ \wedge_{\Sigma_j} M^j \longrightarrow (E\Sigma_j)_+ \wedge_{\Sigma_j} N^j$$

is an E-equivalence. By the previous proof, we have an E-equivalence before passage to orbits. Using the skeletal filtration of $E\Sigma_j$, we may set up a natural spectral sequence

$$H_*(\Sigma_j; E^R_*(M^j)) \Longrightarrow E^R_*((E\Sigma_j)_+ \wedge_{\Sigma_j} M^j)$$

and so deduce the conclusion. □

LEMMA 2.8. *Suppose given a pushout diagram of R-algebras*

$$\begin{array}{ccc} A & \xrightarrow{f} & C \\ {\scriptstyle i}\downarrow & & \downarrow{\scriptstyle j} \\ B & \xrightarrow[g]{} & D, \end{array}$$

where i is an E-acyclic inclusion of a subcomplex in a cell R-algebra and C is a cell R-algebra. Then j is E-acyclic. The same conclusion holds in the case of commutative R-algebras.

PROOF. Recall the bar construction $\beta^R(B, A, C)$ of VII.3.5. By definition or by VII.3.7, we may interpret $\beta^R(B, A, C)$ as the homotopy pushout of i and f. Since i is a cofibration of R-algebras, the natural map $\beta^R(B, A, C) \longrightarrow D$ is a homotopy equivalence of R-algebras under C. Moreover, the map $C \longrightarrow \beta^R(B, A, C)$ factors as the composite of a map $\eta : C \longrightarrow \beta^R(A, A, C)$ and the map

$$\beta^R(i, \mathrm{id}, \mathrm{id}) : \beta^R(A, A, C) \longrightarrow \beta^R(B, A, C).$$

Here η is a homotopy equivalence of R-modules by XII.1.2 and X.1.2. The map $\beta^R(i, \mathrm{id}, \mathrm{id})$ is the geometric realization of a map of proper simplicial R-modules, where properness is defined in X.2.2. Properness holds in the case of R-algebras since, by VII.6.2, the inclusions of degeneracy sub R-modules are inclusions of subcomplexes in relative cell R-modules. It holds in the case of commutative R-algebras by VII.7.5. The smash product over R with E commutes with geometric realization by X.1.4. Since i is an E-equivalence, we find in the R-algebra case that the map $E \wedge_R \beta^R_q(i, \mathrm{id}, \mathrm{id})$ on q-simplices is a homotopy equivalence for each q because it is a weak equivalence between relative cell R-modules. In the commutative case, this map is the smash product over R of the weak equivalence $E \wedge_R A \longrightarrow E \wedge_R B$ with the identity map on $A^q \wedge_R C$ and is therefore a weak equivalence by VII.6.7. In either case, we conclude from X.2.4 that $\beta^R(i, \mathrm{id}, \mathrm{id})$ is an E-equivalence. □

3. Categories of local modules

Again, let R be a q-cofibrant commutative S-algebra and E be a cell R-module. let R_E be a q-cofibrant commutative R-algebra whose unit is a localization of R at E. The fact that Bousfield localization preserves R-algebras and commutative R-algebras gives a powerful tool for the construction of new R-algebras and opens up a new approach to the study of Bousfield localizations.

To see the latter, let us compare the derived category $\mathscr{D}_{R_E}$ to the stable homotopy category $\mathscr{D}_R[E^{-1}]$ associated to the model structure on $\mathscr{M}_R$ determined by E. Here $\mathscr{D}_R[E^{-1}]$ is obtained from $\mathscr{D}_R$ by inverting the E-equivalences and is equivalent to the full subcategory of $\mathscr{D}_R$ whose objects are the E-local R-modules. Observe that, for a cell R-module M, we have the canonical E-equivalence

$$\xi = \eta \wedge \mathrm{id} : M \cong R \wedge_R M \longrightarrow R_E \wedge_R M.$$

The following observation is the same as in the classical case.

LEMMA 3.1. *If M is a finite cell R-module, then $R_E \wedge_R M$ is E-local and therefore ξ is the localization of M at E.*

PROOF. If W is an E-acyclic R-module, then

$$\mathscr{D}_R(W, R_E \wedge_R M) \cong \mathscr{D}_R(W \wedge_R D_R M, R_E) = 0$$

since $W \wedge_R D_R M$ is E-acyclic and R_E is E-local. □

We say that localization at E is smashing if, for all cell R-modules M, $R_E \wedge_R M$ is E-local and therefore ζ is the localization of M at E. The following observation is due to Wolbert [72].

PROPOSITION 3.2 (WOLBERT). *If localization at E is smashing, then the categories $\mathscr{D}_R[E^{-1}]$ and $\mathscr{D}_{R_E}$ are equivalent.*

These categories are closely related even when localization at E is not smashing, as the following elaboration of Wolbert's result shows.

THEOREM 3.3. *The following three categories are equivalent.*

(i) *The category $\mathscr{D}_R[E^{-1}]$ of E-local R-modules.*
(ii) *The full subcategory $\mathscr{D}_{R_E}[E^{-1}]$ of $\mathscr{D}_{R_E}$ whose objects are the R_E-modules that are E-local as R-modules.*
(iii) *The category $\mathscr{D}_{R_E}[(R_E \wedge_R E)^{-1}]$ of $(R_E \wedge_R E)$-local R_E-modules.*

This implies that the question of whether or not localization at E is smashing is a question about the category of R_E-modules, and it leads to the following factorization of the localization functor. In the classical case $R = S$, this shows that our new commutative S-algebras S_E and their categories of modules are intrinsic to the theory of Bousfield localization.

THEOREM 3.4. *The E-localization functor $\mathscr{D}_R \longrightarrow \mathscr{D}_R[E^{-1}]$ is equivalent to the composite of the extension of scalars functor*

$$R_E \wedge_R (-) : \mathscr{D}_R \longrightarrow \mathscr{D}_{R_E}$$

and the $(R_E \wedge_R E)$-localization functor

$$(-)_{R_E \wedge_R E} : \mathscr{D}_{R_E} \longrightarrow \mathscr{D}_{R_E}[(R_E \wedge_R E)^{-1}].$$

COROLLARY 3.5. *Localization at E is smashing if and only if all R_E-modules are E-local as R-modules, so that*

$$\mathscr{D}_R[E^{-1}] \approx \mathscr{D}_{R_E} \approx \mathscr{D}_{R_E}[(R_E \wedge_R E)^{-1}].$$

The proofs of the results above are based on the following generalization of III.4.4 and a special case of III.4.1 from the ground category S to the ground category R. The proofs are the same as those of the cited results.

PROPOSITION 3.6. *Let A be an R-algebra with unit $\eta : R \longrightarrow A$. The forgetful functor $\eta^* : \mathscr{D}_A \longrightarrow \mathscr{D}_R$ has the functor $A \wedge_R (-) : \mathscr{D}_R \longrightarrow \mathscr{D}_A$ as left adjoint, and it also has a right adjoint $\eta^\# : \mathscr{D}_R \longrightarrow \mathscr{D}_A$.*

PROOF OF THEOREM 3.3. We apply the previous result to $\eta : R \longrightarrow R_E$, obtaining

$$\mathscr{D}_{R_E}(N, \eta^\# M) \cong \mathscr{D}_R(\eta^* N, M)$$

and

$$\mathscr{D}_{R_E}(R_E \wedge_R M, N) \cong \mathscr{D}_R(M, \eta^* N)$$

for R-modules M and R_E-modules N. We claim that the functors η^* and $\eta^\#$ of the first adjunction restrict to give inverse adjoint equivalences between $\mathscr{D}_R[E^{-1}]$ and $\mathscr{D}_{R_E}[E^{-1}]$, and we also claim that an R_E-module N is $(R_E \wedge_R E)$-acyclic if and only if $\eta^* N$ is E-acyclic. These claims will give the conclusion.

If W is an E-acyclic R-module, then $R_E \wedge_R W$ is an $(R_E \wedge_R E)$-acyclic R_E-module since

$$(R_E \wedge_R W) \wedge_{R_E} (R_E \wedge_R E) \cong R_E \wedge_R (W \wedge_R E) \simeq *.$$

Therefore, by the second adjunction, if N is an $(R_E \wedge_R E)$-local R_E-module, then $\eta^* N$ is an E-local R-module.

If V is an $(R_E \wedge_R E)$-acyclic R_E-module, then $\eta^* V$ is an E-acyclic R-module since

$$(\eta^* V) \wedge_R E \cong \eta^*(V \wedge_{R_E} (R_E \wedge_R E)) \simeq *.$$

Therefore, by the first adjunction, if M is an E-local R-module, then $\eta^\# M$ is an $(R_E \wedge_R E)$-local R_E-module and thus $\eta^* \eta^\# M$ is again an E-local R-module.

We claim that if if M is E-local, then the counit $\varepsilon : \eta^*\eta^\# M \longrightarrow M$ of the first adjunction is a weak equivalence of R-modules. To see this, consider the commutative diagrams:

$$\begin{array}{ccc} & \mathscr{D}_{R_E}(R_E \wedge_R S^q, \eta^\# M) & \\ {}^{\eta^*}\swarrow & & \searrow{}^{\cong} \\ \mathscr{D}_R(\eta^*(R_E \wedge_R S^q), \eta^*\eta^\# M) & \xrightarrow{\quad \xi^* \quad} & \mathscr{D}_R(S^q, \eta^*\eta^\# M) \\ {\scriptstyle \varepsilon_*}\downarrow & & \downarrow{\scriptstyle \varepsilon_*} \\ \mathscr{D}_R(\eta^*(R_E \wedge_R S^q), M) & \xrightarrow{\quad \xi^* \quad} & \mathscr{D}_R(S^q, M). \end{array}$$

The left vertical composite $\varepsilon_* \circ \eta^*$ is an instance of the first adjunction. The right diagonal is an instance of the second adjunction. The horizontal arrows are induced by $\xi : S^q \longrightarrow R_E \wedge_R S^q$ and are isomorphisms since ξ is localization at E and M and $\eta^*\eta^\# M$ are E-local. Therefore the maps ε_* in the diagram are isomorphisms and ε is a weak equivalence.

If N is an R_E-module such that $\eta^* N$ is E-local, then the unit $\zeta : N \longrightarrow \eta^\#\eta^* N$ of the first adjunction is a weak equivalence since the composite

$$\eta^* N \xrightarrow{\eta^*\zeta} \eta^*\eta^\#\eta^* N \xrightarrow{\varepsilon} \eta^* N$$

is the identity and ε is a weak equivalence. Since $\eta^\#\eta^* N$ is $(R_E \wedge_R E)$-local, this also implies that N is $(R_E \wedge_R E)$-local and so completes the proof. $\square$

PROOF OF THEOREM 3.4. Any E-local R-module is isomorphic in $\mathscr{D}_R$ to one of the form $\eta^* N$, where N is an R_E-module that is E-local as an R-module. Therefore the E-localization of any R-module M is given by a map $\lambda : M \longrightarrow M_E$, where M_E is an R_E-module. Such a map λ factors uniquely through a map $\tilde{\lambda} : R_E \wedge_R M \longrightarrow M_E$ in $\mathscr{D}_{R_E}$. Clearly $\tilde{\lambda}$ is an E-equivalence of R-modules and therefore an $(R_E \wedge_R E)$-equivalence of R_E-modules. This proves the claimed factorization of localization at E. $\square$

PROOF OF COROLLARY 3.5. Localization at E is smashing if and only if all R-modules of the form $R_E \wedge_R M$ for an R-module M are E-local. In this case, if M is an R_E-module, then, as an R-module, M is a retract of $R_E \wedge_R M$ and is therefore also E-local. $\square$

4. Periodicity and K-theory

We illustrate the constructive power of our results on R-algebras by showing that the algebraic localizations of R considered in Chapter V take R to commutative R-algebras on the point set level and not just on the homotopical level studied in V.2.3. Thus let X be a set of elements of R_* and consider $R[X^{-1}]$. We saw in V.2.3 that $R[X^{-1}]$ is a commutative and associative R-ring spectrum whose product is an equivalence under R. For an R-module M, we have the canonical map

$$\lambda = \lambda \wedge_R \mathrm{id} : M \cong R \wedge_R M \longrightarrow R[X^{-1}] \wedge_R M.$$

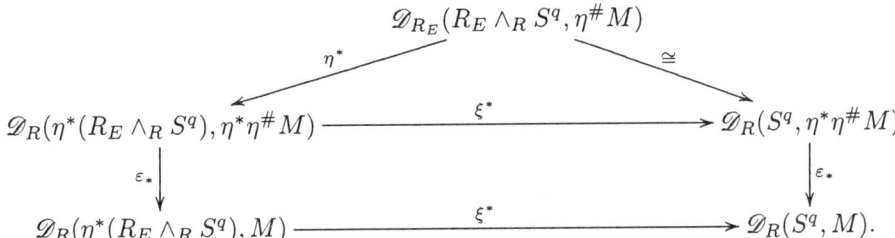

PROPOSITION 4.1. *For any R-module M, λ is the Bousfield localization of M at $R[X^{-1}]$.*

PROOF. Upon smashing with $R[X^{-1}]$, λ becomes an equivalence with inverse given by the product on $R[X^{-1}]$. Thus λ is an $R[X^{-1}]$-equivalence. By a standard argument, $R[X^{-1}] \wedge_R M$ is $R[X^{-1}]$-local since it is an $R[X^{-1}]$-module spectrum. $\square$

So far we have been working homotopically, in the derived category $\mathscr{D}_R$. Theorem 2.1 allows us to translate to the point-set level: R is a cell R-algebra, so its localization at $E = R[X^{-1}]$ can be constructed as a map of R-algebras. We are entitled to the following conclusion.

THEOREM 4.2. *The localization $R \longrightarrow R[X^{-1}]$ can be constructed as the unit of a cell R-algebra.*

By multiplicative infinite loop space theory [50] and our model category structure on the category of S-algebras, the spectra ko and ku that represent real and complex connective K-theory can be taken to be q-cofibrant commutative S-algebras. It is standard (see e.g. [48, II§3]) that the spectra that represent periodic K-theory can be reconstructed up to homotopy by inverting the Bott element $\beta_O \in \pi_8(ko)$ or $\beta_U \in \pi_2(ku)$. That is,

$$KO \simeq ko[\beta_O^{-1}] \qquad \text{and} \qquad KU \simeq ku[\beta_U^{-1}].$$

We are entitled to the following result as a special case of the previous one.

THEOREM 4.3. *The spectra KO and KU can be constructed as commutative ko and ku-algebras.*

Restricting the unit maps $ko \longrightarrow KO$ and $ku \longrightarrow KU$ along the unit maps $S \longrightarrow ko$ and $S \longrightarrow ku$, we see that KO and KU are commutative S-algebras. McClure studied the problem of obtaining such a structure in his thesis. He proved that KO and KU are H_∞ ring spectra, this being a weakened up-to-homotopy analogue of the notion of an E_∞ ring spectrum, with some additional structure; see [14, VII§7]. More recently, in unpublished work, he returned to the problem and proved that the completion of KU at a prime p is an E_∞ ring spectrum. Of course, this also follows from our work since completion at p is another example of a Bousfield localization.

Wolbert [72] has studied the algebraic structure of the derived categories of modules over the connective and periodic versions of the real and complex K-theory S-algebras.

REMARK 4.4. For finite groups G, Theorem 4.3 applies with the same proof to construct the periodic spectra KO_G and KU_G of equivariant K-theory as commutative ko_G and ku_G-algebras. As explained in [27], this leads to an elegant proof of the Atiyah-Segal completion theorem in equivariant K-cohomology and of its analogue for equivariant K-homology.

CHAPTER IX

Topological Hochschild homology and cohomology

As another application of our theory, we construct the topological Hochschild homology of R-algebras with coefficients in bimodules. The relevance to THH of a theory such as ours has long been known. McClure and Staffeldt gave a good introduction of ideas in [55, §3], and in fact our paper provides the foundations that were optimistically assumed in theirs (with reference to a four author paper in preparation that will never exist). Our generalization of Bökstedt's topological Hochschild homology [8] is under active investigation by a number of people, and we shall just lay the foundations.

Actually, in Sections 1 and 2, we give two different constructions. First, for a q-cofibrant commutative S-algebra R, a q-cofibrant R-algebra A, and an (A, A)-bimodule M, we define $THH^R(A; M)$ to be the derived smash product $M \wedge_{A^e} A$, exactly as in algebra (for flat algebras over rings). With this definition, we prove that algebraic Hochschild homology can be realized as the homotopy groups of the topological Hochschild homology of suitable Eilenberg-Mac Lane spectra, and we construct spectral sequences for the calculation of the homotopy and homology groups of $THH^R(A; M)$ in general.

Second, we define $thh^R(A; M)$ by mimicking the standard complex for the computation of algebraic Hochschild homology and we prove that, when M is a cell A^e-module, $thh^R(A; M)$ and $THH^R(A; M)$ are equivalent. When $M = A$, the resulting construction $thh^R(A)$ has exceptionally nice formal properties. For example, it is immediate from the construction that $thh^R(A)$ is a commutative R-algebra when A is. While A is not equivalent to a cell A^e-module, we shall use our standing hypotheses that R and A are q-cofibrant to prove that $THH^R(A)$ is equivalent to $thh^R(A)$.

Further formal properties of $thh^R(A)$ are explained in the brief Section 3, which contains the results of the recent paper [54] of McClure, Schwänzl, and Vogt. They exploit the tensor structure of the category of commutative S-algebras to give a conceptual description of $thh^R(A)$ as $A \otimes S^1$ when A is a commutative R-algebra. Their paper was based on the now obsolete definitions of a preliminary draft of this paper, and we have since found simpler ways to carry out their intriguing application of our theory.

Bökstedt's original definition of topological Hochschild homology [8] (see also

[10]) was based on "functors with smash products", abbreviated FSP's. These are defined in elementary spacewise terms, and Mandell has recently made rigorous an insight of Jeff Smith that shows how to compare FSP's and S-algebras. He has also proven that our construction of $thh^S(A;M)$ is equivalent to Bökstedt's. However, Bökstedt's construction is intrinsically restricted to the "absolute case" $R = S$. There is no previous construction of the general relative version $thh^R(A;M)$ or of topological Hochschild cohomology. At this writing, passage from topological Hochschild homology to topological cyclic homology, which is vital to the applications to algebraic K-theory, relies on Bökstedt's definition; see for example Hesselholt and Madsen [30]. However, it seems possible that a variant of our construction will allow a rederivation and relative generalization of topological cyclic homology: this is work in progress.

1. Topological Hochschild homology: first definition

We assume given an algebra A over a commutative S-algebra R and an (A,A)-bimodule M. We here define the topological Hochschild homology and cohomology spectra $THH^R(A;M)$ and $THH_R(A;M)$. We mimic the conceptual definition of Hochschild homology and cohomology given by Cartan-Eilenberg [15, IX§§3-4]. In the next section, we give an alternative construction that mimics Hochschild's original definition in terms of the standard complex [31] and compare definitions.

We are only interested in relative (A,A)-bimodules, that is, those for which the induced left and right actions of R agree under transposition of M and R, and we define the enveloping R-algebra of A by

$$A^e = A \wedge_R A^{op}.$$

Then an (A,A)-bimodule M can be viewed as either a left or a right A^e-module. We usually view A itself as a left A^e-module and our general bimodule M as a right or a left A^e-module, whichever is convenient. If A is commutative, then $A^e = A \wedge_R A$, the product $A^e \longrightarrow A$ is a map of R-algebras, and A can be viewed as an (A^e, A)-bimodule.

We assume once and for all that our given commutative S-algebra R is q-cofibrant in the model category of commutative S-algebras and that A is q-cofibrant in the model category of R-algebras or of commutative R-algebras. There is no loss of generality in these assumptions since we could replace any given pair (A,R) by a weakly equivalent pair that satisfies our assumptions. By VII.6.7, these assumptions ensure that if $\gamma : \Gamma A \longrightarrow A$ is a weak equivalence of R-modules, where ΓA is a cell R-module, then

$$\gamma \wedge \gamma : \Gamma A \wedge_R \Gamma A \longrightarrow A \wedge_R A$$

is a weak equivalence of R-modules. Thus the underlying R-module of A^e represents $A \wedge_R A$ in the derived category $\mathscr{D}_R$.

DEFINITION 1.1. Working in derived categories, define topological Hochschild homology and cohomology with values in $\mathscr{D}_R$ by

$$THH^R(A;M) = M \wedge_{A^e} A \qquad \text{and} \qquad THH_R(A;M) = F_{A^e}(A,M).$$

If A is commutative, then these functors take values in the derived category $\mathscr{D}_{A^e}$. On passage to homotopy groups, define

$$THH_*^R(A;M) = \mathrm{Tor}_*^{A^e}(M,A) \quad \text{and} \quad THH_R^*(A;M) = \mathrm{Ext}_{A^e}^*(A,M).$$

When $M = A$, we delete it from the notations.

Since we are working in derived categories, we are implicitly taking M to be a cell A^e-module in the definition of $THH^R(A;M)$ and approximating A by a weakly equivalent cell A^e-module in the definition of $THH_R(A;M)$. When A is commutative, we have the following observation, which will be amplified in the next section.

PROPOSITION 1.2. *If A is a commutative R-algebra, then $THH^R(A)$ is isomorphic in $\mathscr{D}_{A^e}$ to a commutative A^e-algebra.*

PROOF. By VII.6.8, A^e is a q-cofibrant commutative R-algebra since A is assumed to be one, and A is clearly a commutative A^e-algebra. Let $\Psi A \longrightarrow A$ be a weak equivalence of A^e-algebras from a q-cofibrant commutative A^e-algebra ΨA to A. Then, by VII.6.5 and VII.6.7, the commutative A^e-algebra $\Psi A \wedge_{A^e} \Psi A$ is isomorphic in $\mathscr{D}_{A^e}$ to $THH^R(A)$. □

The module structures on $THH^R(A;M)$ have the following standard implication.

PROPOSITION 1.3. *If either R or A is the Eilenberg-Mac Lane spectrum of a commutative ring, then $THH^R(A;M)$ is a product of Eilenberg-Mac Lane spectra.*

As we have said, there is no analogue in the literature of our $THH^R(A;M)$ except in the absolute case $R = S$, and there is no analogue of our $THH_R(A;M)$ even then. The relationship between algebraic and topological Hochschild homology becomes far more transparent when one works in full generality. To describe this relationship, we must first fix notations for algebraic Hochschild homology and cohomology. For a commutative graded ring R_*, a graded R_*-algebra A_* that is flat as an R_*-module, and a graded (A_*, A_*)-bimodule M_*, we write

$$HH_{p,q}^{R_*}(A_*;M_*) = \mathrm{Tor}_{p,q}^{(A_*)^e}(M_*,A_*)$$

and

$$HH_{R^*}^{p,q}(A^*;M^*) = \mathrm{Ext}_{(A^*)^e}^{p,q}(A^*,M^*),$$

where p is the homological degree and q is the internal degree. When $M_* = A_*$, we delete it from the notation.

Observe that there is an evident epimorphism

$$\iota : M_* \longrightarrow HH_{0,*}^{R_*}(A_*;M_*) \tag{1.4}$$

and that ι is an isomorphism if the left and right action of A on M are related by $\xi_\ell = \xi_r \circ \tau$. If A_* is commutative, then $HH_{*,*}^{R_*}(A_*)$ is a graded A_*-algebra and ι is a ring homomorphism; see [15, XI§6].

There is also a map

$$(1.5) \qquad \sigma : A_* \longrightarrow HH^{R_*}_{1,*}(A_*)$$

that sends an element a to the 1-cycle $1 \otimes a$ in the standard complex, and σ is a derivation if A is commutative.

Specialization of IV.4.1 gives the following result. Observe that $(A^{op})_* = (A_*)^{op}$.

THEOREM 1.6. *There are spectral sequences of the form*

$$E^2_{p,q} = \mathrm{Tor}^{R_*}_{p,q}(A_*, A^{op}_*) \Longrightarrow (A^e)_{p+q},$$

$$E^2_{p,q} = \mathrm{Tor}^{(A^e)_*}_{p,q}(M_*, A_*) \Longrightarrow THH^R_{p+q}(A; M),$$

and

$$E^{p,q}_2 = \mathrm{Ext}^{p,q}_{(A^e)_*}(A^*, M^*) \Longrightarrow THH^{p+q}_R(A; M).$$

If A_* is a flat R_*-module, so that the first spectral sequence collapses, then the initial terms of the second and third spectral sequences are, respectively,

$$HH^{R_*}_{*,*}(A_*; M_*) \quad \text{and} \quad HH^{*,*}_{R^*}(A^*; M^*).$$

This is of negligible use in the absolute case $R = S$, where the flatness hypothesis is unrealistic. However, in the relative case, it implies that algebraic Hochschild homology and cohomology are special cases of topological Hochschild homology and cohomology.

THEOREM 1.7. *Let R be a (discrete, ungraded) commutative ring, let A be an R-flat R-algebra, and let M be an (A, A)-bimodule. Then*

$$HH^R_*(A; M) \cong THH^{HR}_*(HA; HM)$$

and

$$HH^*_R(A; M) \cong THH^*_{HR}(HA; HM).$$

If A is commutative, then $HH^R_*(A) \cong THH^{HR}_*(HA)$ as A-algebras.

PROOF. By VII.1.3 and the naturality of multiplicative infinite loop space theory [50], we can construct HA as an HR-algebra, commutative if A is so. The results of IV§2 construct HM as an (HA, HA)-bimodule. Thus the statement makes sense. The spectral sequences collapse since their internal gradings are concentrated in degree zero. We will prove the last statement below. □

We concentrate on homology henceforward. In the absolute case $R = S$, it is natural to approach $THH^S_*(A; M)$ by first determining the ordinary homology of $THH^S(A; M)$, using the case $E = H\mathbb{F}_p$ of the following spectral sequence, and then using the Adams spectral sequence. A spectral sequence like the following one was first obtained by Bökstedt [9]. An interesting case, essentially $THH(ku)$, was worked out by McClure and Staffeldt [55], who assumed without proof that the second spectral sequence in the following theorem could be constructed. The flatness hypotheses required when E is only a commutative ring spectrum are still unrealistic in the absolute case, but the situation is saved by the lack of need for such hypotheses when E is a commutative S-algebra, such as $H\mathbb{F}_p$.

Observe that there is a natural map

$$(1.8)\qquad \zeta = \mathrm{id}\wedge\phi : M \cong M \wedge_{A^e} A^e \longrightarrow M \wedge_{A^e} A = THH_R(A;M).$$

THEOREM 1.9. *Let E be a commutative ring spectrum. If $E_*(R)$ is a flat R_*-module, or if E is a commutative S-algebra, there is a spectral sequence of differential $E_*(R)$-modules of the form*

$$E^2_{p,q} = \mathrm{Tor}^{E_*R}_{p,q}(E_*A, E_*(A^{op})) \Longrightarrow E_{p+q}(A^e).$$

If $E_*(A^e)$ is a flat $(A^e)_*$-module, or if E is a commutative S-algebra, there is a spectral sequence of differential $E_*(R)$-modules of the form

$$E^2_{p,q} = \mathrm{Tor}^{E_*(A^e)}_{p,q}(E_*(M), E_*(A)) \Longrightarrow E_{p+q}(THH^R(A;M)).$$

In either case, if $E_*(A)$ is $E_*(R)$-flat, so that

$$E^2_{p,q} = HH^{E_*(R)}_{*,*}(E_*(A); E_*(M))$$

in the second spectral sequence, then the composite

$$E_*(M) \xrightarrow{\ \iota\ } E^2_{0,*} \longrightarrow E^\infty_{0,*} \xrightarrow{\ \subset\ } E_*(THH^R(A;M))$$

coincides with $\zeta_* : E_*(M) \longrightarrow E_*(THH^R(A;M))$.

PROOF. When E is just a commutative ring spectrum, both spectral sequences are immediate from IV.6.2. When E is a commutative S-algebra, both spectral sequences are immediate from IV.6.4. The statement about ζ is clear from the discussion of the edge homomorphism in IV§5. □

Applied to Eilenberg-Mac Lane spectra, the following complement implies the last statement of Theorem 1.7. Clearly Proposition 1.2 implies that if A is a commutative R-algebra, then $THH^R(A)$ is a commutative R-ring spectrum. Moreover, by Corollary 3.8 below, there is then a natural map (in $\mathscr{D}_R$)

$$(1.10)\qquad \omega : A \wedge S^1_+ \longrightarrow THH^R(A).$$

PROPOSITION 1.11. *Let A be a commutative R-algebra and assume sufficient hypotheses that Theorem 1.9 gives a spectral sequence*

$$E^2_{p,q} = HH^{E_*(R)}_{p,q}(E_*(A)) \Longrightarrow E_{p+q}(THH^R(A)).$$

Then this is a spectral sequence of differential $E_*(A)$-algebras such that E^2 has the standard product in Hochshild homology. Moreover, the composite

$$E_*(A) \xrightarrow{\ \sigma\ } E^2_{1,*} \longrightarrow E^\infty_{1,*} \xrightarrow{\ \subset\ } E_*(THH^R(A))/\operatorname{im}\zeta_*$$

coincides with the restriction of

$$(\omega)_* : E_*(A \wedge S^1_+) \longrightarrow E_*(THH^R(A))/\operatorname{im}\zeta_*$$

to the wedge summand ΣA.

PROOF. We may use the standard complex for the computation of algebraic Hochschild homology in the construction of the spectral sequences in IV§5. We can then construct a pairing of the resolutions constructed there that realizes the standard product and so deduce a pairing of spectral sequences. We omit details, since the result will be more transparent with the alternative construction of the spectral sequence (albeit under different hypotheses) in the next section. For the last statement, under the usual stable splitting of S^1_+ as $S^0 \vee S^1$, the restriction of ω to the wedge summand A coincides with ζ. Under the equivalence with the standard complex description of THH in the next section, ω lands in simplicial filtration one. Thinking of ΣA as $\Sigma(\mathbb{F}S \wedge_{A^e} A)$, where $\mathbb{F}$ is the free A^e-module functor, we find that the restriction of ω to ΣA provides a factorization through $A \wedge_{A^e} A$ of the first stage of the inductive construction of the spectral sequence given in IV§5. Again, this will be more transparent with the alternative construction of the spectral sequence. □

REMARK 1.12. We have given Theorem 1.9 in the form appropriate to classical stable homotopy theory. It is perhaps more natural to give a version that makes sense from the point of view of the multiplicative homology theories E^R_* on R-modules of IV.1.7, where E is a commutative R-algebra. The ground ring in this context is $E_* = E^R_*(R)$. We leave the details to the reader. The essential point is the relative case of III.3.10.

2. Topological Hochschild homology: second definition

We again assume given a q-cofibrant commutative S-algebra R, a q-cofibrant R-algebra or q-cofibrant commutative R-algebra A, and an (A, A)-bimodule M. Write A^p for the p-fold $\wedge_R$-power of A, and let

$$\phi : A \wedge_R A \longrightarrow A \quad \text{and} \quad \eta : R \longrightarrow A$$

be the product and unit of A. Let

$$\xi_\ell : A \wedge_R M \longrightarrow M \quad \text{and} \quad \xi_r : M \wedge_R A \longrightarrow M$$

be the left and right action of A on M. We have canonical cyclic permutation isomorphisms

$$\tau : M \wedge_R A^p \wedge_R A \longrightarrow A \wedge_R M \wedge_R A^p.$$

The following definition precisely mimics the definition of the standard complex for the computation of Hochschild homology, as given in [15, p. 175]. The topological analogue of passage from a simplicial k-module to a chain complex of k-modules is passage from a simplicial spectrum E_* to its geometric realization $|E_*|$.

DEFINITION 2.1. Define a simplicial R-module $thh^R(A; M)_*$ as follows. Its R-module of p-simplices is $M \wedge_R A^p$. Its face and degeneracy operators are

$$d_i = \begin{cases} \xi_r \wedge (\mathrm{id})^{p-1} & \text{if } i = 0 \\ \mathrm{id} \wedge (\mathrm{id})^{i-1} \wedge \phi \wedge (\mathrm{id})^{p-i-1} & \text{if } 1 \le i < p \\ (\xi_\ell \wedge (\mathrm{id})^{p-1}) \circ \tau & \text{if } i = p \end{cases}$$

and $s_i = \mathrm{id} \wedge (\mathrm{id})^i \wedge \eta \wedge (\mathrm{id})^{p-i}$. Define

$$thh^R(A;M) = |thh^R(A;M)_*|.$$

When $M = A$, we delete it from the notation, writing $thh^R(A)_*$ and $|thh^R(A)_*|$.

Since geometric realization converts simplicial R-modules to R-modules, by X.1.5, $thh^R(A;M)$ and $thh^R(A)$ are R-modules. Observe that the maps

$$\zeta_p = \mathrm{id} \wedge \eta^p : M \cong M \wedge_R R^p \longrightarrow M \wedge_R A^p$$

specify a map of simplicial R-modules from the constant simplicial R-module $\underline{M}$ to $thh^R(A;M)_*$; it induces a natural map of R-modules

$$\zeta = |\zeta_*| : M \longrightarrow thh^R(A;M).$$

Inspection of the simplicial structure shows that, if A is commutative, there is a natural map of R-modules

$$\omega : A \wedge S^1_+ \longrightarrow thh^R(A)$$

with image in the simplicial 1-skeleton; see (3.2) and Corollary 3.8 below. Moreover, we then have the following observation.

PROPOSITION 2.2. *Let A be a commutative R-algebra. Then $thh^R(A)$ is naturally a commutative A-algebra with unit $\zeta : A \longrightarrow thh^R(A)$, and $thh^R(A;M)$ is a $thh^R(A)$-module. By neglect of structure, $thh^R(A)$ is a commutative R-algebra.*

PROOF. Clearly $thh^R(A)_*$ is a simplicial commutative R-algebra, $thh^R(A;M)_*$ is a simplicial $thh^R(A)_*$-module, and $\zeta_* : \underline{A} \longrightarrow thh^R(A)_*$ is a map of simplicial commutative R-algebras. Since all structure in sight is preserved by geometric realization, by X.1.5, this implies the result. □

The definition of $THH^R(A;M)$ was homotopical and led directly to spectral sequences for its calculational study. The definition of $thh^R(A;M)$ is formal and algebraic. We must establish a connection between these two definitions.

The starting point is the relative two-sided bar construction $B^R(M,A,N)$, which is defined for a commutative S-algebra R, an R-algebra A, and right and left A-modules M and N. The definition is the same as that of $B(M,R,N) = B^S(M,R,N)$ in IV.7.2, except that smash products over S are replaced by smash products over R. By XII.1.2 and X.1.2, there is a natural map

$$\psi : B^R(A,A,N) \longrightarrow N$$

of A-modules that is a homotopy equivalence of R-modules. More generally, by use of the product on A and its action on the given modules, we obtain a natural map of R-modules

$$\psi : B^R(M,A,N) \longrightarrow M \wedge_A N.$$

The proof of the following result is the same as that of its special case IV.7.5.

PROPOSITION 2.3. *For a cell A-module M and an A-module N,*

$$\psi : B^R(M,A,N) \longrightarrow M \wedge_A N$$

is a weak equivalence of R-modules.

The relevance of the bar construction to thh is shown by the following observation, which is the same as in algebra. We agree to write

$$B^R(A) = B^R(A, A, A).$$

Observe that $B^R(A)$ is an (A, A)-bimodule; on the simplicial level, $B_0^R(A) = A^e$.

LEMMA 2.4. *For (A, A)-bimodules M, there is a natural isomorphism*

$$thh^R(A; M) \cong M \wedge_{A^e} B^R(A).$$

PROOF. If $\underline{M}$ is the constant simplicial (A, A)-bimodule at M, then

$$M \wedge_{A^e} B^R(A) \cong |\underline{M} \wedge_{A^e} B_*^R(A, A, A)|.$$

We have canonical isomorphisms

$$M \wedge_R A^p \cong M \wedge_{A^e} (A^e \wedge_R A^p) \cong M \wedge_{A^e} (A \wedge_R A^p \wedge_R A)$$

given by the permutation of $A^{op} = A$ past A^p. Inspection shows that these commute with the face and degeneracy operations and so induce the stated isomorphism. □

Since the natural map $\psi : B^R(A) \longrightarrow A$ of (A, A)-bimodules is a homotopy equivalence of R-modules, this has the following immediate consequence, by III.3.8.

PROPOSITION 2.5. *For cell A^e-modules M, the natural map*

$$thh^R(A; M) \cong M \wedge_{A^e} B^R(A) \xrightarrow{\mathrm{id} \wedge \psi} M \wedge_{A^e} A = THH^R(A; M)$$

is a weak equivalences of R-modules, or of A^e-modules if A is commutative.

While we are perfectly happy, indeed forced, to assume that M is a cell A^e-module in our derived category level definition of THH, we are mainly interested in the case $M = A$ of our point-set level construction thh, and A is not of the A^e-homotopy type of a cell A^e-module except in trivial cases. However, we have the following result.

THEOREM 2.6. *Let $\gamma : M \longrightarrow A$ be a weak equivalence of A^e-modules, where M is a cell A^e-module. Then the map*

$$thh^R(\mathrm{id}; \gamma) : thh^R(A; M) \longrightarrow thh^R(A; A) = thh^R(A)$$

is a weak equivalence of R-modules, or of A^e-modules if A is commutative. Therefore $THH^R(A; M)$ is weakly equivalent to $thh^R(A)$.

PROOF. With the notation of VII.6.4, it is clear from VII.6.5 that M and A are both in $\bar{\mathscr{E}}_R$, and it follows from VII.6.7 that the map $thh_p^R(\mathrm{id}; \gamma)$ of p-simplices is a weak equivalence for each p. By Proposition 2.8 below, the simplicial R-modules $B_*^R(A)$ and $thh_*^R(A)$ are proper, in the sense of X.2.2, and X.2.4 gives the conclusion. □

Combining the previous two results, we arrive at the conclusion we really want.

THEOREM 2.7. *If R is a q-cofibrant commutative S-algebra and A is either a q-cofibrant R-algebra or a q-cofibrant commutative R-algebra, then $thh(A)$ and $THH(A)$ are canonically isomorphic in the derived category $\mathscr{D}_R$.*

PROPOSITION 2.8. *For right and left A-modules M and N, $B^R_*(M, A, N)$ is a proper simplicial R-module. For an (A, A)-bimodule M, $thh^R_*(A; M)$ is a proper simplicial R-module.*

PROOF. The condition of being proper involves only the degeneracy and not the face operators of a simplicial R-module. In our cases, the degeneracies are obtained from the unit $\eta : R \longrightarrow A$ and, since smashing over R with M and N in the first statement and with M in the second preserves cofibrations of R-modules, the result in both cases is an immediate consequence of VII.7.5. □

Returning to the study of spectral sequences in the previous section, we find that use of the standard complex gives us spectral sequences under different flatness hypotheses, just as in IV§7. We consider the spectral sequence derived in X.2.9 from the simplicial filtration of $thh^R(A; M)$. For simplicity, we restrict attention to the absolute case $R = S$.

THEOREM 2.9. *Let E be a commutative ring spectrum, let A be an S-algebra, and let M be a cell A^e-module. If $E_*(A)$ is E_*-flat, then there is a spectral sequence of the form*

$$E^2_{p,q} = HH^{E_*}_{p,q}(E_*(A); E_*(M)) \Longrightarrow E_{p+q}(thh^S(A; M)).$$

The composite

$$E_*(M) \xrightarrow{\iota} E^2_{0,*} \longrightarrow E^\infty_{0,*} \xrightarrow{\subset} E_*(thh^S(A; M))$$

coincides with $\zeta_* : E_*(M) \longrightarrow E_*(thh^S(A; M))$. If A is a commutative S-algebra then the spectral sequence

$$E^2_{p,q} = HH^{E_*}_{p,q}(E_*(A)) \Longrightarrow E_{p+q}(thh^S(A))$$

is a spectral sequence of differential $E_*(A)$-algebras, and the composite

$$E_*(A) \xrightarrow{\sigma} E^2_{1,*} \longrightarrow E^\infty_{1,*} \xrightarrow{\subset} E_*(thh^S(A))/\operatorname{im}\zeta_*$$

coincides with the restriction of

$$(\omega)_* : E_*(A \wedge S^1_+) \longrightarrow E_*(thh^S(A))/\operatorname{im}\zeta_*$$

to the wedge summand ΣA.

PROOF. Using VII.6.2 or VII.6.7 and our standing q-cofibrancy hypothesis to see that our point-set level constructions can be used to compute derived smash products, we see that the E^1 term is exactly the standard complex for the computation of the algebraic Hochschild homology groups in the E^2 term. The standard product on the standard complex is realized on E^1, and the rest is clear. □

3. The isomorphism between $thh^R(A)$ and $A \otimes S^1$

We here explain a reinterpretation of the definition of $thh^R(A)$ that was observed by McClure, Schwänzl, and Vogt [54]. Recall that the category $\mathscr{C}\mathscr{A}_R$ of commutative R-algebras $\mathscr{C}\mathscr{A}_R$ is tensored and cotensored over the category $\mathscr{U}$ of unbased topological spaces, so that we have adjunction homeomorphisms

$$\mathscr{C}\mathscr{A}_R(A \otimes X, B) \cong \mathscr{U}(X, \mathscr{C}\mathscr{A}_R(A, B)) \cong \mathscr{C}\mathscr{A}_R(A, F(X_+, B)). \tag{3.1}$$

As in VII.3.7, we easily obtain an identification of simplicial commutative R-algebras

$$thh^R(A)_* \cong A \otimes S^1_*. \tag{3.2}$$

by writing out the standard simplicial set S^1_* whose realization is the circle and comparing faces and degeneracies. We give a slightly different proof of the following result. We think of S^1 as the unit complex numbers.

THEOREM 3.3 (MCCLURE, SCHWÄNZL, VOGT). *For commutative R-algebras A, there is a natural isomorphism of commutative R-algebras*

$$thh^R(A) \cong A \otimes S^1.$$

The product of $thh^R(A)$ is induced by the codiagonal $S^1 \amalg S^1 \longrightarrow S^1$. The unit $\zeta : A \longrightarrow thh^R(A)$ is induced by the inclusion $\{1\} \longrightarrow S^1$.

PROOF. We may identify S^1 with the pushout in the diagram

$$\begin{array}{ccc} \partial I & \longrightarrow & I \\ \downarrow & & \downarrow \\ \{pt\} & \longrightarrow & S^1. \end{array}$$

We arrange our identification to map $\{pt\}$ to $\{1\}$. Applying the functor $A \otimes (-)$, we obtain a pushout diagram

$$\begin{array}{ccc} A \otimes \partial I & \longrightarrow & A \otimes I \\ \downarrow & & \downarrow \\ A \otimes \{pt\} & \longrightarrow & A \otimes S^1. \end{array}$$

By VII.3.7, we have $B^R(A) \cong A \otimes I$. By Lemma 2.4, $thh^R(A) \cong A \wedge_{A^e} B^R(A)$. This gives the isomorphism $thh^R(A) \cong A \otimes S^1$ by a comparison of pushouts. The statement about the product follows from the isomorphism of coproducts

$$(A \otimes_R S^1) \wedge_R (A \otimes_R S^1) \cong A \otimes (S^1 \amalg S^1).$$

Since ζ is determined by $\{pt\} \longrightarrow S^1$, the last statement is clear. $\square$

COROLLARY 3.4. *The pinch map $S^1 \longrightarrow S^1 \vee S^1$ and trivial map $S^1 \longrightarrow *$ induce a (homotopy) coassociative and counital coproduct and counit*

$$\psi : thh^R(A) \longrightarrow thh^R(A) \wedge_A thh^R(A) \quad \textit{and} \quad \varepsilon : thh^R(A) \longrightarrow A$$

that make $thh^R(A)$ a homotopical Hopf A-algebra.

PROOF. Since $S^1 \vee S^1$ is the pushout of $S^1 \longleftarrow * \longrightarrow S^1$ and the functor $A \otimes (-)$ preserves pushouts, we see from VII.1.6 that

$$thh^R(A) \wedge_A thh^R(A) \cong thh^R(A) \otimes (S^1 \vee S^1).$$

The rest is clear. □

The next few corollaries are based on the case $X = S^1$ of the adjunctions (3.1). Of course, left adjoints preserve colimits.

COROLLARY 3.5. *The functor $thh^R(A)$ preserves colimits in A.*

The word "naive" in the following corollary refers to the fact that we only consider spectra indexed on universes with trivial S^1-actions here; genuine S^1-spectra must be indexed on universes that contain all representations of S^1. In the naive sense, we define commutative S^1-S-algebras, and so on, simply by requiring S^1 to act compatibly with all structure in sight. We think of S^1 as acting trivially on R and A.

COROLLARY 3.6. *$thh^R(A)$ is a naive commutative S^1-R-algebra. If B is a naive commutative S^1-R-algebra and $f : A \longrightarrow B$ is a map of commutative R-algebras, then there is a unique map of naive commutative S^1-R-algebras $\tilde{f} : thh^R(A) \longrightarrow B$ such that $\tilde{f} \circ \zeta = f$.*

PROOF. The product on S^1 gives a map

$$\alpha : (A \otimes S^1) \otimes S^1 \cong A \otimes (S^1 \times S^1) \longrightarrow A \otimes S^1.$$

Its adjoint $S^1 \longrightarrow \mathscr{C}\mathscr{A}_R(thh^R(A), thh^R(A))$ gives actions by elements of S^1 via maps of commutative R-algebras, with the requisite continuity, and the adjunction immediately implies the universal property. □

For an integer r, define $\phi^r : S^1 \longrightarrow S^1$ by

$$\phi^r(e^{2\pi i t}) = e^{2\pi i r t}.$$

It is immediate that these induce power operations Φ^r on $thh^R(A)$.

COROLLARY 3.7. *There are natural maps of commutative R-algebras*

$$\Phi^r : thh^R(A) \longrightarrow thh^R(A)$$

such that

$$\Phi^0 = \zeta\varepsilon, \quad \Phi^1 = \mathrm{id}, \quad \Phi^r \circ \Phi^s = \Phi^{rs},$$

and the following diagrams commute:

$$\begin{array}{ccc} thh^R(A) \otimes S^1 & \xrightarrow{\alpha} & thh^R(A) \\ \downarrow{\scriptstyle \Phi^r \otimes \phi^s} & & \downarrow{\scriptstyle \Phi^{r+s}} \\ thh^R(A) \otimes S^1 & \xrightarrow[\alpha]{} & thh^R(A). \end{array}$$

COROLLARY 3.8. *There is a natural S^1-equivariant map of R-modules*

$$\omega : A \wedge S^1_+ \longrightarrow thh^R(A)$$

such that if B is a commutative R-algebra and $f : A \wedge S^1_+ \longrightarrow B$ is a map of spectra such that the composite $f \circ (\mathrm{id} \wedge i_x) : A \longrightarrow B$, $i_x : \{x\}_+ \subset S^1_+$, is a map of R-algebras for each $x \in S^1$, then f uniquely determines a map of R-algebras $\tilde{f} : thh^R(A) \longrightarrow B$ such that $f = \tilde{f} \circ \omega$. Moreover, ω is obtained by passage to geometric realization from the natural map of simplicial spectra

$$\omega_* : A \wedge (S^1_*)_+ \longrightarrow A \otimes S^1_*.$$

PROOF. This is immediate from VII.2.11. Its transitivity diagram and a naturality diagram imply the S^1-equivariance. □

The image of ω lies in the simplicial 1-skeleton. The intuition is that the rest of $thh^R(A)$ freely builds up the R-algebra structure.

REMARK 3.9. When A is an R-algebra, inspection of the simplicial structure shows that $thh^R(A)_*$ is a cyclic spectrum, and it follows exactly as for cyclic spaces [17] that $thh^R(A)$ is a naive S^1-R-module. We believe that a variant of our construction of $thh^R(A)$ will yield a genuine S^1-spectrum that is cyclotomic in the sense of Hesselholt and Madsen [30, 1.2]. If so, this variant should lead to a relative generalization of topological cyclic homology. We intend to return to this elsewhere.

CHAPTER X

Some basic constructions on spectra

We have used geometric realization of simplicial spectra in several places, and we shall make further use of it to prove some of our earlier claims. We study it in the first two sections, concentrating on formal properties in Section 1 and on homotopical properties in Section 2. We then use geometric realization to define homotopy colimits in Section 3. All of the basic definitions and much of the work in the first three sections carries over to any of our model categories of module, ring, and algebra categories, as we have already indicated in VII§3. We prefer to be more concrete in this service chapter. It will be evident that geometric realization in the category of $\mathbb{L}$-spectra or the category of R-modules for an S-algebra R is given by geometric realization in the underlying category of spectra.

After discussing various special kinds of prespectra in Section 4, we use homotopy colimits to construct the "cylinder functor" in Section 5. This functor converts spectra to weakly equivalent Σ-cofibrant spectra while preserving ring, module, and algebra structures.

Much of the material of this chapter has long been known to the authors, and to others, but little if any of it has appeared in the literature.

1. The geometric realization of simplicial spectra

We first recall from [43] the definition of a coend, or tensor product of functors. Let Λ be any small category and let $\mathscr{C}$ be any category that has all (small) colimits. Write $\vee$ for the categorical coproduct in $\mathscr{C}$. Suppose given a functor

$$F : \Lambda^{op} \times \Lambda \longrightarrow \mathscr{C}.$$

Define the coend of F, denoted

$$\int^{\Lambda} F(n,n)$$

to be the coequalizer of the pair of maps

$$\bigvee_{\phi : m \longrightarrow n} F(n,m) \underset{f}{\overset{e}{\rightrightarrows}} \bigvee_n F(n,n)$$

where the restriction of e to the ϕth summand is $F(\phi, \mathrm{id})$ and the restriction of f to the ϕth summand is $F(\mathrm{id}, \phi)$. Using equalizers, we obtain the dual notion of the end of F, denoted $\int_\Lambda F(n,n)$.

Now recall that a simplicial object in any category $\mathscr{C}$ is a contravariant functor from the simplicial category Δ to $\mathscr{C}$. We have the classical geometric realization functor, denoted $|X_*|$, from simplicial spaces to spaces, and we need to extend it to the level of spectra. We shall begin with a spectrum level definition, and we shall then show how to interpret it in terms of the space level definition. This will allow us to deduce many properties of the spectrum level functor simply by quoting standard properties of the space level functor. Recall that, using the usual face and degeneracy maps, we obtain a covariant functor from Δ to spaces that sends q to the standard topological q-simplex Δ_q.

DEFINITION 1.1. Let K_* be a simplicial spectrum. Define its geometric realization to be the coend

$$|K_*| = \int^\Delta K_q \wedge (\Delta_q)_+.$$

Of course, the functor $\Delta^{op} \times \Delta \longrightarrow \mathscr{S}$ that is implicit in the definition sends (p,q) to $K_p \wedge (\Delta_q)_+$. The geometric realization of a simplicial space X_* is defined similarly:

$$|X_*| = \int^\Delta X_q \times \Delta_q.$$

If X_* is a simplicial based space, so that all its face and degeneracy maps are basepoint-preserving, then all points of each subspace $* \times \Delta_q$ are identified to the point $(*, 1) \in X_0 \times \Delta_0$ in the construction of $|X_*|$, hence

$$|X_*| = \int^\Delta X_q \wedge (\Delta_q)_+.$$

This places the two definitions in the same form. Actually, as with any categorical colimit, the geometric realization of a simplicial spectrum is obtained by applying the spectrification functor L to the spacewise geometric realization of its underlying simplicial prespectrum. In more detail, for a simplicial prespectrum K_*, we have simplicial based spaces $K_*(V)$ for indexing spaces V. Their geometric realizations form a prespectrum with structural maps induced from those of K_* as the composites

$$\Sigma^{W-V}|K_*(V)| \cong |\Sigma^{W-V}K_*(V)| \longrightarrow |K_*(W)|.$$

If K_* is a simplicial spectrum, then $|K_*|$ is obtained by applying L to this prespectrum.

As we explain in the proof, the following result has two different senses, and it is correct in both senses.

PROPOSITION 1.2. *The geometric realization functor from simplicial spectra to spectra preserves homotopies.*

PROOF. As on the space level [45, §11], we have two kinds of homotopy between simplicial maps, and geometric realization carries both to homotopies of the usual sort. One kind is just a simplicial map with domain of the form $K_* \wedge I_+$, and geometric realization preserves this kind of homotopy by part (ii) of the next proposition applied to the constant simplicial space at I_+. The other is the combinatorial kind of homotopy that makes sense for maps between simplicial objects in any category [45, 9.1]. It can be viewed as a simplicial map with domain of the form $K_* \wedge \Delta_*[1]_+$, where $\Delta_*[1]$ is the standard simplicial 1-simplex viewed as a discrete simplicial space, hence part (ii) of the next proposition also applies to show that geometric realization preserves this kind of homotopy. □

On the level of based spaces, it is standard that geometric realization commutes with wedges and products and therefore with smash products. This easily implies a direct proof of (ii) and (iv) of the following result, and (i) can be viewed as a special case of (ii). Recall that functors on $\mathscr{C}$ are extended termwise to simplicial objects in $\mathscr{C}$; for example $(J_* \wedge K_*)_q = J_q \wedge K_q$ for simplicial spectra J_* and K_*.

PROPOSITION 1.3. *Geometric realization enjoys the following properties.*

(i) *For simplicial based spaces X_*, there is a natural isomorphism*

$$\Sigma^\infty |X_*| \cong |\Sigma^\infty X_*|.$$

(ii) *For simplicial based spaces X_* and simplicial spectra K_*, there is a natural isomorphism*

$$|K_* \wedge X_*| \cong |K_*| \wedge |X_*|.$$

(iii) *For simplicial spectra K_* indexed on U and spaces A over $\mathscr{I}(U, U')$, there is a natural isomorphism*

$$|A \ltimes K_*| \cong A \ltimes |K_*|.$$

(iv) *For simplicial spectra $|J_*|$ and $|K_*|$, there is a natural isomorphism*

$$|J_* \wedge K_*| \cong |J_*| \wedge |K_*|,$$

where external smash products are understood.

(v) *For simplicial spectra K_*, there are natural isomorphisms*

$$|\mathbb{B}K_*| \cong \mathbb{B}|K_*| \quad \text{and} \quad |\mathbb{C}K_*| \cong \mathbb{C}|K_*|,$$

and similarly for the monads $\mathbb{B}[1]$ and $\mathbb{C}[1]$ and for the corresponding reduced monads that were defined in II§§4–5.

PROOF. Part (i-iii) hold since left adjoints commute with colimits. Parts (iv) and (v) follow from parts (i)-(iii) and a Fubini theorem for iterated coends. □

PROPOSITION 1.4. *For simplicial $\mathbb{L}$-spectra K_* and L_*, there is a natural isomorphism*

$$|K_*| \wedge_{\mathscr{L}} |L_*| \cong |K_* \wedge_{\mathscr{L}} L_*|.$$

For an S-algebra R, such as $R = S$, and simplicial R-modules M_* and N_*, there is a natural isomorphism

$$|M_*| \wedge_R |N_*| \cong |M_* \wedge_R N_*|.$$

PROOF. The first isomorphism is immediate from the previous proposition, and it directly implies the second when $R = S$. In view of the coequalizer description of smash products over R, the case of general R follows by the commutation of coequalizers with geometric realization. □

PROPOSITION 1.5. *The geometric realization of a simplicial A_∞ or E_∞ ring spectrum is an A_∞ or E_∞ ring spectrum. For a commutative S-algebra R, such as $R = S$, the geometric realization of a simplicial (commutative) R-algebra is a (commutative) R-algebra. The analogous preservation properties hold for modules.*

2. Homotopical and homological properties of realization

To discuss the behavior of geometric realization with respect to equivalences and CW homotopy types, and to obtain useful spectral sequences from its canonical filtration, we need the following technical definition.

DEFINITION 2.1. Let K_* be a simplicial spectrum and let $sK_q \subset K_q$ be the "union" of the subspectra s_jK_{q-1}, $0 \le j < q$. Say that K_* is proper if the "inclusion" $sK_q \longrightarrow K_q$ is a cofibration for each $q \ge 0$.

The term "union" must be interpreted in terms of appropriate pushout diagrams. The corresponding "inclusions" must be interpreted, a priori, in terms of associated maps. However, a cofibration of spectra is a spacewise closed inclusion [38, I.8.1]. Rigorous notation would make this section unreadable, so we shall use notations as if we were dealing with simplicial spaces, leaving the translation to rigorous categorical language to the reader. One way to be precise about the degeneracy subspectrum sK_q and its associated map to K_q is to interpret the latter as the following map of coends

$$\int^{D_{q-1}} K_p \wedge D(\underline{q},\underline{p})_+ \longrightarrow \int^{D_q} K_p \wedge D(\underline{q},\underline{p})_+ \cong K_q.$$

Here D is the subcategory of Δ consisting of the monotonic surjections (which index the degeneracy and identity maps), and D_q is its full subcategory of objects $\underline{i}$ with $0 \le i \le q$. The isomorphism is an application of Yoneda's lemma. With this interpretation, we can generalize the context to $\mathbb{L}$-spectra or to R-modules for a fixed S-algebra R. Recall that colimits in the categories of $\mathbb{L}$-spectra and of R-modules are created in the category of spectra.

DEFINITION 2.2. A simplicial $\mathbb{L}$-spectrum is proper if the canonical map of $\mathbb{L}$-spectra $sK_q \longrightarrow K_q$ is a cofibration for each $q \ge 0$. A simplicial R-module is proper if the canonical map of R-modules $sK_q \longrightarrow K_q$ is a cofibration for each $q \ge 0$.

Since mapping cylinders of R-modules are created in $\mathscr{S}$, a cofibration of $\mathbb{L}$-spectra or of R-modules is a cofibration of spectra, but not conversely. Note that the inclusion $M \longrightarrow N$ of a relative cell R-module (N, M) is a cofibration, by HELP, and that VII.4.14 gives that q-cofibrations of R-algebras and of commutative R-algebras are cofibrations of R-modules.

LEMMA 2.3. *Let $i : A \longrightarrow X$ be a cofibration of spectra, $\mathbb{L}$-spectra, or R-modules. Then*

$$j = i \wedge \mathrm{id} : (A \wedge \Delta_{q+}) \cup (X \wedge \partial\Delta_{q+}) \longrightarrow X \wedge \Delta_{q+}$$

is a cofibration of spectra, $\mathbb{L}$-spectra, or R-modules. Therefore, if K_* is a proper simplicial spectrum, $\mathbb{L}$-spectrum, or R-module, then the inclusion

$$(sK_q \wedge \Delta_{q+}) \cup (K_q \wedge \partial\Delta_{q+}) \longrightarrow K_q \wedge \Delta_{q+}$$

is a cofibration for each $q \geq 1$.

PROOF. With the usual conventions on products and smash products of pairs, we are given that

$$(X, A) \wedge (I_+, \{0\}_+)$$

is a retract pair, and we must deduce that

$$(X, A) \wedge (\Delta_{q+}, \partial\Delta_{q+}) \wedge (I_+, \{0\}_+)$$

is a retract pair. There is a homeomorphism of pairs

$$(\Delta_q, \partial\Delta_q) \times (I, \{0\}) \cong (\Delta_q \times I, \Delta_q \times \{0\}).$$

In based notation, this clearly implies a homeomorphism of pairs

$$(\Delta_{q+}, \partial\Delta_{q+}) \wedge (I_+, \{0\}_+) \cong (I_+, \{0\}_+) \wedge \Delta_{q+}.$$

The conclusion follows upon smashing with (X, A) and using the given retraction. $\square$

Similarly, the theorems to follow are valid with essentially identical proofs in the contexts of spectra, $\mathbb{L}$-spectra, and R-modules.

THEOREM 2.4. *Let $f_* : K_* \longrightarrow K'_*$ be a map of proper simplicial spectra, $\mathbb{L}$-spectra, or simplicial R-modules.*

(i) *If each $f_q : K_q \longrightarrow K'_q$ is a homotopy equivalence, then so is $|f_*| : |K_*| \longrightarrow |K'_*|$.*
(ii) *In the $\mathbb{L}$-spectrum case, and therefore also in the R-module case, if each $f_q : K_q \longrightarrow K'_q$ is a weak equivalence, then so is $|f_*| : |K_*| \longrightarrow |K'_*|$; in the spectrum case, this holds if all given and constructed spectra are tame.*

PROOF. The proof is precisely parallel to the proof of the space level analogues [46, A.4]. The essential point is just the gluing theorem to the effect that a pushout of (weak) equivalences is a (weak) equivalence when corresponding legs of the given pushout diagrams are cofibrations. For tame spectra, the weak version of this statement is a consequence of I.3.5. For $\mathbb{L}$-spectra, the weak version is a consequence of I.6.5.

For $0 \leq k < q$, let $s^k K_q$ be the union over $0 \leq j \leq k$ of the subspectra $s_j K_{q-1}$ of K_q. We claim first that the inclusion $s^{k-1}K_q \longrightarrow s^k K_q$ is a cofibration for $0 < k < q$. This holds vacuously if $q = 1$. We assume it for $q - 1$ and deduce it for q. Since a composite of cofibrations is a cofibration and K_* is proper, the left vertical inclusion in the following commutative diagram is a cofibration:

$$\begin{array}{ccccc} s^{k-1}K_{q-1} & \longrightarrow & s^{k-1}K_q \cap s_k K_{q-1} & \longrightarrow & s^{k-1}K_q \\ \downarrow & & \downarrow & & \downarrow \\ K_{q-1} & \longrightarrow & s_k K_{q-1} & \longrightarrow & s^k K_q. \end{array} \tag{2.5}$$

The left horizontal arrows are induced by s_k and are isomorphisms with inverses induced by d_{k+1}, and the right square is a pushout. Therefore the middle and right vertical arrows are also cofibrations. This proves our claim.

Since $s_0 : K_{q-1} \longrightarrow s^0 K_q$ is an isomorphism, we find by induction on q and, for fixed q, by induction on k that $f_q : s^k K_q \longrightarrow s^k K'_q$ is a (weak) equivalence for each k and q. In particular, $f_q : sK_q \longrightarrow sK'_q$ is a (weak) equivalence for each q. As usual, $|K_*|$ is filtered, and we have successive pushouts

$$\begin{array}{ccccc} sK_q \wedge \partial\Delta_{q+} & \longrightarrow & K_q \wedge \partial\Delta_{q+} & & \\ \downarrow & & \downarrow & & \\ sK_q \wedge \Delta_{q+} & \longrightarrow & (sK_q \wedge \Delta_{q+}) \cup (K_q \wedge \partial\Delta_{q+}) & \xrightarrow{\;g\;} & F_{q-1}|K_*| \\ & & \downarrow & & \downarrow \\ & & K_q \wedge \Delta_{q+} & \longrightarrow & F_q|K_*|. \end{array} \tag{2.6}$$

Here the restrictions of the map g to $sK_q \wedge \Delta_{q+}$ and $K_q \wedge \partial\Delta_{q+}$ are dictated by the coequalizer description of $|K_*|$. The vertical arrows are cofibrations, the bottom middle one by Lemma 2.3. Therefore the restrictions $|f_*| : F_q|K_*| \longrightarrow F_q|K'_*|$ are (weak) equivalences by successive applications of the gluing theorem, and $|f_*|$ is a (weak) equivalence by passage to colimits over q, using III.1.7 in the weak case. □

THEOREM 2.7. *Let K_* be a simplicial spectrum, $\mathbb{L}$-spectrum, or R-module.*

(i) *If each K_q is a cell object, each degeneracy operator is the inclusion of a subcomplex, and each face operator is sequentially cellular, then $|K_*|$ is a cell object, and similarly for CW objects.*
(ii) *If K_* is proper and if each K_q has the homotopy type of a cell object, then so does $|K_*|$, and similarly for CW objects provided that, in the R-module case, R is connective.*

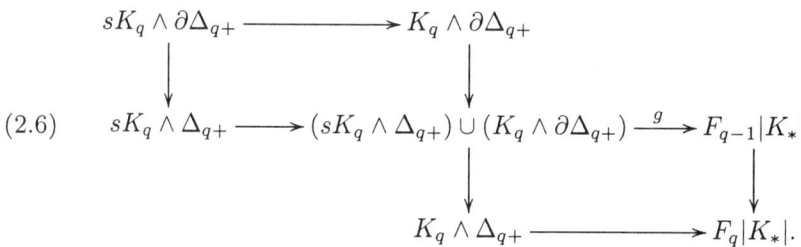

PROOF. The proofs follow the same outline as in the previous theorem, and part (i) is clear from the second pushout in (2.6). For the CW case of (ii), if each K_q has the homotopy type of a CW object, then, by induction on q and, for fixed q, by induction on k, (2.5) shows that each $s^k K_q$ has the homotopy type of a CW spectrum. By induction on q, (2.6) then shows that each $F_q|K_*|$ has the homotopy type of a CW spectrum. Therefore $|K_*|$ has the homotopy type of a CW spectrum.

The essential point is that if J, K, and L are CW homotopy types, then a pushout diagram obtained from a cofibration $J \longrightarrow K$ and a map $J \longrightarrow L$ is equivalent to a pushout diagram obtained from the inclusion of a subcomplex J' in a CW object K' and a cellular map from J' to a CW object L'. Since we are applying the cellular approximation theorem, we must assume that R is connective in the R-module case. By the gluing theorem, the pushout $K \cup_J L$ is therefore homotopy equivalent to the CW object $K' \cup_{J'} L'$. Since colimits are constructed from wedges and coequalizers and thus from wedges and pushouts, it follows that the colimit of a sequence of cofibrations of objects of the homotopy types of CW objects has the homotopy type of a CW object. The proof of the statement about cell objects is similar; here III.2.2 substitutes for the cellular approximation theorem. □

REMARK 2.8. A result similar to Theorem 2.4(i) was proven for simplicial LEC spectra (see §4) in [21]; in that context the argument proceeds by an immediate reduction to the space level analogue. However, that proof does not work to give Theorem 2.7(ii) and does not apply to the R-module setting.

The following is the spectrum level analogue of a frequently used space level spectral sequence. We have used it in our discussion of bar constructions and of topological Hochschild homology. For a spectrum E, we can apply E_q to the simplicial spectrum K_* to obtain a simplicial abelian group $E_q(K_*)$. Taking the homology of its normalized chain complex, we obtain groups $H_p(E_q(K_*))$, and we obtain the same groups if we take the homology of its unnormalized chain complex [44, 22.3]. If E is an R-module and K_* is a simplicial R-module, then, with the notation of IV.1.7, we obtain groups $H_p(E^R_q(K_*))$ the same way. If R is commutative, then each $H_p(E^R_*(K_*))$ is an R_*-module. If, further, E is a commutative R-ring spectrum, then $H_p(E^R_*(K_*))$ is an E_*-module.

THEOREM 2.9. *Let K_* be a proper simplicial spectrum and let E be any spectrum. There is a natural homological spectral sequence $\{E^r_{p,q}K_*\}$ such that*

$$E^2_{p,q}K_* = H_p(E_q(K_*))$$

and $\{E^r_{p,q}K_*\}$ converges strongly to $E_*(|K_*|)$. With E_* replaced by E^R_*, the same statement holds for a proper simplicial R-module K_* and an R-module E. Here, if R is commutative, then the spectral sequence is one of differential R_*-modules, and if E is a commutative R-ring spectrum, then the spectral sequence is one of differential E_*-modules.

PROOF. Let $F_p|K_*|$ be the image in $|K_*|$ of the wedge over $0 \leq q \leq p$ of the spectra $K_q \wedge (\Delta_{q+})$. Then the inclusions $F_{p-1}|K_*| \subset F_p|K_*|$ are cofibrations, and we have isomorphisms

$$F_p|K_*|/F_{p-1}|K_*| \cong (K_p/sK_p) \wedge (\Delta_p/\partial\Delta_p) \cong \Sigma^p(K_p/sK_p).$$

We apply E_* to obtain an exact couple, and thus a spectral sequence, with

$$E_{p+q}(F_p|K_*|/F_{p-1}|K_*|) = E^1_{p,q}K_*.$$

We now see that $E^1_{p,q}K_*$ is isomorphic to the p-chain group of the normalized chain complex of $E_q(K_*)$, and a diagram chase just like that of [45, p. 111] shows that d^1 agrees with the differential of this chain complex. This identifies E^2, and the convergence is standard. □

3. Homotopy colimits and limits

Homotopy colimits and limits of spectra do not appear explicitly in the literature. However, in view of our results on geometric realization, these constructions present no more difficulty in the category $\mathscr{S}$ of spectra than they do in the category $\mathscr{T}$ of based spaces. Of course, we are concerned with precise point-set level versions rather than with the cruder homotopical versions that are present in any Quillen model category. We record the definitions in this section. The same definitions apply in the category of R-modules for any S-algebra R.

Let $\mathscr{D}$ be any small category. Let $B_q(\mathscr{D})$ be the set of q-tuples $\underline{f} = (f_1, \dots, f_q)$ of composable arrows of $\mathscr{D}$, depicted

$$d_0 \xleftarrow{f_1} d_1 \xleftarrow{f_2} \cdots \xleftarrow{f_q} d_q, \tag{3.1}$$

and let $S(\underline{f}) = d_q$ and $T(\underline{f}) = d_0$ be the source and target of $\underline{f}$. We understand $B_0(\mathscr{D})$ to be the set $\mathscr{O}$ of objects of $\mathscr{D}$. With the usual faces and degeneracies, $B_*(\mathscr{D})$ is a simplicial set whose geometric realization is the classifying space $B(\mathscr{D})$.

We first specify homotopy colimits. A $\mathscr{D}$-shaped diagram of spectra is a covariant functor $D : \mathscr{D} \longrightarrow \mathscr{S}$. For any such D, there is a simplicial spectrum $B_*(*, \mathscr{D}, D)$, the space level analogue of which was specified in [47, §12]. (The left variable $*$ is a place holder.) The spectrum of q-simplices is the wedge over all $\underline{f} \in B_q(\mathscr{D})$ of the spectra $D(S(\underline{f}))$. The faces and degeneracies are the standard ones of the two-sided bar construction [47, §7]. Applied to $\underline{f}$ as in (3.1), the last face on $B_q(\mathscr{D})$ forgets f_q; the last face on the $\underline{f}$th wedge summand of $B_q(*, \mathscr{D}, D)$ is the map $D(f_q) : D(d_q) \longrightarrow D(d_{q-1})$. By definition, hocolim D is the geometric realization of this simplicial spectrum. Using the abbreviation $B(-) = |B_*(-)|$, we may write this as

$$\operatorname{hocolim} D = B(*, \mathscr{D}, D). \tag{3.2}$$

For example, if $\underline{K} : \mathscr{D} \longrightarrow \mathscr{S}$ is the constant functor at a spectrum K, then we see by inspection of definitions that

$$\operatorname{hocolim} \underline{K} \cong B(\mathscr{D})_+ \wedge K. \tag{3.3}$$

A map $f : D \longrightarrow D'$ of diagrams is a natural transformation of functors, and, since $B_*(*, \mathscr{D}, D)$ is clearly proper, Theorem 2.4 has the following immediate implication.

PROPOSITION 3.4. *If $f : D \longrightarrow D'$ is a map of diagrams such that each $f(d)$ is a homotopy equivalence, then* $\operatorname{hocolim} f : \operatorname{hocolim} D \longrightarrow \operatorname{hocolim} D'$ *is a homotopy equivalence. If all given and constructed spectra are tame, the same holds for weak equivalences.*

We shall need the following lemma on cofibrations. While we shall only use it in the context of based spaces, it works equally well in the context of spectra.

LEMMA 3.5. *Let $\mathscr{D}'$ be a subcategory of $\mathscr{D}$ such that any morphism of $\mathscr{D}$ with target in $\mathscr{D}'$ is a morphism of $\mathscr{D}'$. Let D be a functor from $\mathscr{D}$ to the category of spaces, based spaces, or spectra and let D' be the restriction of D to $\mathscr{D}'$. Then the induced map* $\operatorname{hocolim} D' \longrightarrow \operatorname{hocolim} D$ *is a cofibration.*

PROOF. We work with based spaces for definiteness, but the argument is the same in the other two cases. It suffices to construct a retraction

$$r : (\operatorname{hocolim} D) \wedge I_+ \longrightarrow (\operatorname{hocolim} D) \cup ((\operatorname{hocolim} D') \wedge I_+)$$

from the reduced cylinder to the reduced mapping cylinder of the inclusion. A point $z = |(\underline{f}, x), u| \wedge s$ of the domain is given by a composable tuple $\underline{f}$ of maps as in (3.1), a point $x \in D(d_q)$, a point $u \in \Delta_q$, and a point $s \in I$. There is an $i \geq 0$ such that the maps $f_1, \cdots, f_i$ are in $\mathscr{D} - \mathscr{D}'$ and the maps $f_{i+1}, \cdots, f_q$ are in $\mathscr{D}'$. If $i = 0$, define $r(z) = z$. If $i = q$, define $r(z)$ by replacing s by 0; that is, r is here just the retraction to the base of the cylinder. If $0 < i < q$, write $u = (tv, (1-t)w)$, where $v \in \Delta_{i-1}$, $w \in \Delta_{q-i}$, and $t \in I$. Then define

$$r(z) = \begin{cases} |(\underline{f}, x), (0, w)| \wedge (1-2t)s & \text{if } 0 \leq t \leq 1/2 \\ |(\underline{f}, x), ((2t-1)v, (2-2t)w)| \wedge 0 & \text{if } 1/2 \leq t \leq 1. \end{cases}$$

It is straightforward to check that r is a well-defined retraction. □

Although we shall not need it here, we take the opportunity to record our preferred definition of homotopy limits, which is precisely dual to the definition of homotopy colimits. We suppose given a contravariant functor $E : \mathscr{D} \longrightarrow \mathscr{S}$. We obtain a cosimplicial spectrum $C_*(E, \mathscr{D}, *)$, a two-sided cobar construction. Its spectrum of q-cosimplices is the product over all $\underline{f} \in B_q(\mathscr{D})$ of the spectra $E(T(\underline{f}))$. The cofaces and codegeneracies with target $C_q(E, \mathscr{D}, *)$ have $\underline{f}$th coordinate obtained by projecting onto the coordinate of the source that is indexed by the corresponding face or degeneracy applied to $\underline{f}$, and, for the zeroth coface, composing with the map $E(f_1) : E(d_0) \longrightarrow E(d_1)$.

We define the geometric realization or totalization "Tot K_*" of a cosimplicial spectrum K_* to be the end

$$\text{Tot } K_* = \int_{\Delta} F((\Delta_q)_+, K_q). \tag{3.6}$$

Here we are using the evident functor $\Delta^{op} \times \Delta \longrightarrow \mathscr{S}$ that sends (p, q) to $F((\Delta_p)_+, K_q)$.

We define the homotopy limit of a contravariant functor $E : \mathscr{D} \longrightarrow \mathscr{S}$ to be the totalization

$$\text{holim}\, E = \text{Tot}\ C_*(E, \mathscr{D}, *). \tag{3.7}$$

For example, if $\underline{K} : \mathscr{D} \longrightarrow \mathscr{S}$ is the constant functor at a spectrum K, then we see by use of adjunctions and inspection of definitions that

$$\text{holim}\, \underline{K} \cong F(B(\mathscr{D})_+, K). \tag{3.8}$$

The essential point is that the definition makes perfect sense with the precise point-set level definitions of product and function spectra given in [38, pp. 13, 17].

4. Σ-cofibrant, LEC, and CW prespectra

We here discuss several special types of prespectra that play an important technical role in point-set level studies in stable homotopy theory. We first put the notion of a Σ-cofibrant spectrum (from I.2.4) into perspective by recalling the following definitions from [38, I§8] and [21]. A space X is said to be LEC (locally equiconnected) if the inclusion of its diagonal subspace is an unbased cofibration; see e.g. Lewis [36] for a discussion of such spaces.

DEFINITION 4.1. A prespectrum D is said to be

(i) Σ-cofibrant if its structure maps

$$\sigma : \Sigma^{W-V} DV \longrightarrow DW$$

are based cofibrations.

(ii) an inclusion prespectrum if its adjoint structure maps

$$\tilde{\sigma} : DV \longrightarrow \Omega^{W-V} DW$$

are inclusions.

(iii) cofibrant if its adjoint structure maps $\tilde{\sigma}$ are based cofibrations.

(iv) LEC if it is Σ-cofibrant and each space DV is LEC.

(v) CW if it is LEC and each DV has the homotopy type of a CW complex.

A spectrum E is said to be Σ-cofibrant or LEC if it is isomorphic to LD for some Σ-cofibrant or LEC prespectrum D.

If E is a spectrum, then the maps $\tilde{\sigma}$ are homeomorphisms. Therefore, as a prespectrum, E is cofibrant, but it is not Σ-cofibrant (unless it is trivial). The first statement of the following result is clear. Although we have concentrated on Σ-cofibrant prespectra and spectra, the second statement, which is due to Lewis [36], gives one reason for interest in the LEC notion.

LEMMA 4.2. *A Σ-cofibrant prespectrum is an inclusion prespectrum. An LEC prespectrum is cofibrant.*

Our CW prespectra must not be confused with CW spectra; the latter are defined in terms of spectrum-level spheres and attaching maps. Since we are interested in notions that are appropriate for serious point-set level work and that admit usable equivariant generalizations, we have no interest in the old-fashioned and, to our minds, obsolete, notion of a CW prespectrum that requires CW complexes D_n and cellular structure maps $\Sigma D_n \longrightarrow D_{n+1}$. We have the following relations between CW prespectra and CW spectra [38, I.8.12-14].

THEOREM 4.3. *If D is a CW prespectrum, then LD has the homotopy type of a CW spectrum. If E is a CW spectrum, then each space EV has the homotopy type of a CW complex and E is homotopy equivalent to LD for some CW prespectrum D. Thus a spectrum has the homotopy type of a CW spectrum if and only if it has the homotopy type of LD for some CW prespectrum D.*

The first statement is an immediate consequence of the following description of spectra in terms of shift desuspensions of spaces [38, I.4.7]. The second is based on use of the cylinder construction defined in the next section.

PROPOSITION 4.4. *If D is an inclusion prespectrum, then*

$$LD \cong \operatorname{colim} \Sigma_V^\infty DV,$$

where the colimit is computed as the prespectrum level colimit of the maps

$$\Sigma_W^\infty \sigma : \Sigma_V^\infty DV \cong \Sigma_W^\infty \Sigma^{W-V} DV \longrightarrow \Sigma_W^\infty DW.$$

That is, the prespectrum level colimit is a spectrum that is isomorphic to LD.

In particular, if D is Σ-cofibrant, then LD is the colimit of shift desuspensions of space level based cofibrations. This makes the point-set level analysis of such spectra particularly convenient. The condition of being Σ-cofibrant is quite weak. It is clear from Theorem 4.3 that tame spectra, that is, spectra of the homotopy types of Σ-cofibrant spectra, are considerably more general than spectra of the homotopy types of CW spectra. The output spectra of the standard infinite loop space machines are Σ-cofibrant no matter what their input. The following closure properties of the category of Σ-cofibrant spectra are more directly relevant to us.

LEMMA 4.5. *The suspension and shift desuspension spectra of based spaces are Σ-cofibrant.*

PROOF. The prespectrum level structure maps of shift desuspensions are identity maps or the inclusions of basepoints (which are always based cofibrations) [38, I.4.1]. Explicitly, $\Sigma_V^\infty X = L\Pi_V^\infty X$ for an indexing space $V \subset U$, where

$$(\Pi_V^\infty X)(W) = \Sigma^{W-V} X \text{ if } W \supset V \quad \text{and} \quad (\Pi_V^\infty X)(W) = \{*\} \text{ otherwise.} \quad \square$$

PROPOSITION 4.6. *Suppose given compact spaces A_i, based cofibrations $A_i \longrightarrow B_i$, and indexing spaces V_i, where i runs through any indexing set. If*

$$\begin{array}{ccc} \bigvee_i \Sigma^\infty_{V_i} A_i & \longrightarrow & E \\ \downarrow & & \downarrow \\ \bigvee_i \Sigma^\infty_{V_i} B_i & \longrightarrow & F \end{array}$$

is a pushout of spectra and E is Σ-cofibrant, then F is Σ-cofibrant. If

$$\begin{array}{ccc} \bigvee_i \mathbb{L}\Sigma^\infty_{V_i} A_i & \longrightarrow & L \\ \downarrow & & \downarrow \\ \bigvee_i \mathbb{L}\Sigma^\infty_{V_i} B_i & \longrightarrow & M. \end{array}$$

is a pushout of $\mathbb{L}$-spectra and L is tame, then M is tame.

PROOF. Let $E = LD$, where D is a Σ-cofibrant prespectrum. Since E is a prespectrum level colimit, A_i is compact, and $\Sigma^\infty_{V_i}$ is adjoint to the V_ith space functor, we find that the given map $\Sigma^\infty_{V_i} A_i \longrightarrow E$ is induced by a map $A_i \longrightarrow \Omega^{W_i - V_i} DW_i$ for some $W_i \supset V_i$. There results a map of prespectra $\Pi^\infty_{W_i} \Sigma^{W_i - V_i} A_i \longrightarrow D$ that induces the given map of spectra under the isomorphism $\Sigma^\infty_{V_i} A_i \cong \Sigma^\infty_{W_i} \Sigma^{W_i - V_i} A_i$. This allows us to construct the pushout on the prespectrum level, where an inspection from the fact that the structure maps of the $\Pi^\infty_{W_i} \Sigma^{W_i - V_i} A_i$ and $\Pi^\infty_{W_i} \Sigma^{W_i - V_i} B_i$ are wedges of identity maps or inclusions of basepoints shows that the pushout is Σ-cofibrant. This proves the first statement. Since there is no claim about the action of $\mathbb{L}$, the second statement is an easy consequence, by comparisons of pushout diagrams and use of the fact that $\eta : \Sigma^\infty_V A \longrightarrow \mathbb{L}\Sigma^\infty_V A$ is a homotopy equivalence for any A and V. □

PROPOSITION 4.7. *The external smash product of two Σ-cofibrant spectra is Σ-cofibrant. The j-fold external smash power of a Σ-cofibrant spectrum is Σ-cofibrant as a Σ_j-spectrum.*

PROOF. The smash product $f \wedge g$ of based cofibrations is a based cofibration since it is the composite of based cofibrations $f \wedge \mathrm{id}$ and $\mathrm{id} \wedge g$. Indexing smash products on inner product spaces $V \oplus V'$, as we may, we see immediately that the smash product of Σ-cofibrant prespectra is a Σ-cofibrant prespectrum. Similarly, for j-fold smash powers, we may index on j-fold sums V^j and use the fact that the jth smash power of a based cofibration is a based Σ_j-cofibration to see that the j-fold smash power of a Σ-cofibrant prespectrum is a Σ-cofibrant Σ_j-prespectrum. By the following lemma, these prespectrum level observations imply the desired spectrum level conclusions. □

LEMMA 4.8. *For prespectra D and D', the units $D \longrightarrow \ell LD$ and $D' \longrightarrow \ell LD'$ of the spectrification adjunction induce an isomorphism of spectra*

$$L(D \wedge D') \longrightarrow L(\ell LD \wedge \ell LD').$$

PROOF. Factoring the map in question through $L(D \wedge \ell LD')$ in the evident way, we see that it suffices to prove that $L(D \wedge D') \longrightarrow L(D \wedge \ell LD')$ is an isomorphism. We have adjoint function prespectra and spectra [38, II.3.3] such that if E is a spectrum and D' is a prespectrum, then $F(D', \ell E)$ is a spectrum. Moreover, a glance at the cited definition shows that

$$F(D', \ell E) \cong \ell F(LD', E).$$

This isomorphism of right adjoints implies the desired isomorphism of left adjoints. □

Although the fact will not be used in our work, results like those above also apply to LEC spectra [21]: the suspension and shift desuspensions of LEC spaces are LEC spectra, and the smash product of LEC spectra is LEC.

For twisted half-smash products, we only have an up to homotopy version of Proposition 4.7.

PROPOSITION 4.9. *If $E \in \mathscr{S}U$ is Σ-cofibrant and A is a compact space over $\mathscr{I}(U, U')$, then $A \ltimes E$ is Σ-cofibrant. If $E \in \mathscr{S}U$ is tame and A is a space over $\mathscr{I}(U, U')$ that has the homotopy type of a colimit of a sequence of cofibrations between compact spaces, then $A \ltimes E$ is tame.*

PROOF. The first statement is immediate from the prespectrum level construction of twisted half-smash products in the Appendix (or [38, VI.2.7]). For the second statement, we may assume that $E = LD$, where D is a Σ-cofibrant prespectrum and, by I.2.5, we may also assume that $A = \operatorname{colim} A_i$ for a sequence of cofibrations $A_i \longrightarrow A_{i+1}$ between compact spaces A_i. Then [38, VI.2.5 and VI.2.18] give a concrete description of $A \ltimes E$ as $L(A \ltimes D)$, where the prespectrum $A \ltimes D$ is obviously Σ-cofibrant. □

For example, the second statement applies when A has the homotopy type of a CW complex with finite skeleta.

5. The cylinder construction

We show here that we can functorially replace an A_∞ or E_∞ ring spectrum R by a weakly equivalent Σ-cofibrant A_∞ or E_∞ ring spectrum KR, and similarly for modules. This replacement already works on the prespectrum level. We have used it in several technical proofs, and we shall use it again later. An iterated mapping cylinder functor K that sends prespectra to weakly equivalent Σ-cofibrant prespectra was constructed in [38, I.6.8]. We shall use the language of homotopy colimits to give a more conceptual version of the construction that allows us to prove that it preserves structured ring and module spectra.

CONSTRUCTION 5.1. Let D be a prespectrum indexed on U. Define KD as follows. For an indexing space V, let $\underline{V}$ be the category of subspaces $V' \subset V$ and inclusions. Define a functor D_V from $\underline{V}$ to the category of based spaces by letting $D_V(V') = \Sigma^{V-V'} DV'$. For an inclusion $V'' \longrightarrow V'$,

$$V - V'' = (V - V') \oplus (V' - V''),$$

and $\sigma : \Sigma^{V'-V''}DV'' \longrightarrow DV'$ induces $D_V(V'') \longrightarrow D_V(V')$. Define

$$(KD)(V) = \operatorname{hocolim} D_V.$$

An inclusion $i : V \longrightarrow W$ induces a functor $\underline{i} : \underline{V} \longrightarrow \underline{W}$, there is an evident isomorphism $\Sigma^{W-V}D_V \cong D_W\underline{i}$ of functors $\underline{V} \longrightarrow \underline{W}$, and the functor Σ^{W-V} commutes with homotopy colimits. Therefore $\underline{i}$ induces a map

$$\sigma : \Sigma^{W-V} \operatorname{hocolim} D_V \cong \operatorname{hocolim} \Sigma^{W-V} D_V \cong \operatorname{hocolim} D_W\underline{i} \longrightarrow \operatorname{hocolim} D_W,$$

and this map is a cofibration by Lemma 3.5. Thus, with these structural maps, KD is a Σ-cofibrant prespectrum. The structural maps $\sigma : D_V V' \longrightarrow DV$ specify a natural transformation to the constant functor at DV and so induce a map $r : (KD)(V) \longrightarrow DV$, and these maps r specify a map of prespectra. Regarding the object V as a trivial subcategory of $\underline{V}$, we obtain $j : DV \longrightarrow (KD)(V)$. Clearly $rj = \mathrm{id}$, and $jr \simeq \mathrm{id}$ via a canonical homotopy since V is a terminal object of $\underline{V}$. The maps j do not specify a map of prespectra, but they do specify a weak map, in the sense that $j\sigma \simeq \sigma\Sigma^{W-V}j : \Sigma^{W-V}DV \longrightarrow (KD)(W)$, via a canonical homotopy. Clearly K is functorial and homotopy-preserving, and r is natural.

The following example may be illuminating.

EXAMPLE 5.2. Let X be a based space and let D be the suspension prespectrum of X, so that $DV = \Sigma^V X$ and the structure maps $\sigma : \Sigma^{W-V}\Sigma^V X \longrightarrow \Sigma^W X$ are the evident identifications. Via these identifications, the functor D_V is isomorphic to the constant functor at $\Sigma^V X$, hence

$$(KD)(V) \cong B(\underline{V})_+ \wedge \Sigma^V X.$$

The structure maps of KD are induced by the cited identifications and the maps $B(\underline{i})$. In this case, we can use the initial objects $\{0\}$ of the $\underline{V}$ rather than the terminal objects V to obtain maps $\Sigma^V X \longrightarrow (KD)(V)$. Because the functors $\underline{i}$ preserve initial points, this gives a map of prespectra $\nu : D \longrightarrow KD$ such that $r\nu = \mathrm{id}$. We have simply fattened up the $\Sigma^V X$ via the compatible system of contractible spaces $B(\underline{V})$.

Construction 5.1 is a conceptual version of [38, I.6.8], and the discussion of "preternaturality" given in [38, I.7.5-I.7.7] applies to it. As usual, we extend the construction to spectra by setting $KE = LK\ell E$, and we then have a natural weak equivalence $r = Lr\ell : KE \longrightarrow E$. The following result implies the second statement of Theorem 4.3.

PROPOSITION 5.3. (i) *If each space DV has the homotopy type of a CW complex, then LKD has the homotopy type of a CW spectrum.*

(ii) *If E has the homotopy type of a CW spectrum, then KE has the homotopy type of a CW spectrum, hence $r : KE \longrightarrow E$ is a homotopy equivalence.*

PROOF. By Proposition 4.4, $LKD \cong \operatorname{colim} \Sigma_n^\infty (KD)_n$, $(KD)_n = (KD)(\mathbb{R}^n)$, where the colimit is taken over the cofibrations

$$\Sigma_n^\infty (KD)_n \cong \Sigma_{n+1}^\infty \Sigma (KD)_n \longrightarrow \Sigma_{n+1}^\infty (KD)_{n+1}.$$

The conclusion of (i) follows since the colimit of a sequence of cofibrations of spectra of the homotopy types of CW spectra has the homotopy type of a CW spectrum. By [38, I.8.14], each space EV of a CW spectrum E has the homotopy type of a CW complex. Thus (ii) follows from (i) and the Whitehead theorem. □

We must still discuss the behavior of the functor K with respect to smash products and twisted half-smash products.

PROPOSITION 5.4. *Let D and D' be prespectra indexed on U and U'. Then there is a natural unital, associative, and commutative system of isomorphisms*

$$\omega : KD \wedge KD' \longrightarrow K(D \wedge D')$$

over $D \wedge D'$, where external smash products are understood.

PROOF. Recall that the prespectrum level external smash product $D \wedge D'$ is naturally indexed on direct sums $V \oplus V'$ of indexing spaces V in U and V' in U'. Clearly the product category $\underline{V} \times \underline{V'}$ is isomorphic to a cofinal subcategory of $\underline{V \oplus V'}$, and we can restrict to this subcategory in our construction of $K(D \wedge D')$. By definition,

$$(D \wedge D')(V \oplus V') = DV \wedge D'V',$$

with the evident structural maps. Since homotopy colimits are two-sided bar constructions and geometric realization and simplicial bar constructions commute suitably with products, we obtain isomorphisms

$$(\operatorname{hocolim} D_V) \wedge (\operatorname{hocolim} D'_{V'}) \cong \operatorname{hocolim}(D_V \wedge D'_{V'}) \cong \operatorname{hocolim}(D \wedge D')_{V \oplus V'}$$

that are evidently compatible with the retractions to $DV \wedge D'V'$. The coherence statements are easily verified. For the unital condition, we allow $U = \{0\}$, in which case K is the identity functor; the space S^0 is the unit for the external smash product. □

Clearly this extends to j-fold external smash products, with all possible associativity and equivariance. We next consider changes of universe, preparatory to considering twisted half-smash products.

LEMMA 5.5. *Let $f : U \longrightarrow U'$ be a linear isometry. For prespectra D' indexed on U', Kf^*D' is isomorphic over f^*D' to f^*KD'. For spectra E indexed on U, there is a natural map $\omega : f_*KE \longrightarrow Kf_*E$ over f_*E.*

PROOF. For an indexing space V in U, f induces an isomorphism of categories $\underline{V} \longrightarrow \underline{f(V)}$. By definition, $(f^*D')(V) = D'f(V)$, with the evident structural maps. By inspection,

$$(Kf^*D')(V) = \operatorname{hocolim}(f^*D')_V \cong \operatorname{hocolim} D'_{f(V)} = (f^*KD')(V),$$

and these isomorphisms are compatible with the retractions to $(f^*D')(V)$. The functor f_* is left adjoint to f^* [38, p.58]. For a prespectrum D indexed on U, the unit $D \longrightarrow f^*f_*D$ of the adjunction induces a natural map

$$KD \longrightarrow Kf^*f_*D \cong f^*Kf_*D.$$

The adjoint of this map is a natural map $\phi : f_*KD \longrightarrow Kf_*D$ over f_*D. The spectrum level left adjoint to f^* is $f_*E = Lf_*\ell E$. The unit $D \longrightarrow \ell LD$ of the (L, ℓ) adjunction induces natural maps

$$Lf_*D \longrightarrow Lf_*\ell LD = f_*LD \quad \text{and} \quad LKD \longrightarrow LK\ell LD = KLD.$$

By [38, pp. 19, 58], the first of these is an isomorphism of spectra since $f^*\ell E = \ell f^*E$. Therefore ϕ specializes to give the required map

$$f_*KE = Lf_*\ell LK\ell E \cong Lf_*K\ell E \longrightarrow LKf_*\ell E \longrightarrow LK\ell Lf_*\ell E = Kf_*E. \quad \square$$

LEMMA 5.6. *For based spaces X, there is a natural map $\omega : \Sigma^\infty X \longrightarrow K\Sigma^\infty X$ such that $r \circ \omega = \mathrm{id}$.*

PROOF. We can obtain ω by applying the previous lemma to $i : \{0\} \longrightarrow U$ since, as noted in the proof of I.3.2, $i_*X = \Sigma^\infty X$. The map ω so obtained is the same as the map of spectra induced by the map ν of prespectra described in Example 5.2. $\square$

PROPOSITION 5.7. *Let $\alpha : A \longrightarrow \mathscr{I}(U, U')$ be a space over $\mathscr{I}(U, U')$. For spectra $E \in \mathscr{S}U$, there is a natural map*

$$\omega : A \ltimes KE \longrightarrow K(A \ltimes E)$$

over $A \ltimes E$.

PROOF. By A.5.4, a map of spectra $A \ltimes E \longrightarrow E'$ determines and is determined by maps of spectra $\alpha(a)_*E \longrightarrow E'$ for $a \in A$ that satisfy a certain continuity condition. In particular, the identity map of $A \ltimes E$ is determined by the evident maps $\iota(a) : \alpha(a)_*E \longrightarrow A \ltimes E$. Composing maps ω from Lemma 5.5 with maps $K\iota(a)$, we obtain maps

$$\alpha(a)_*KE \longrightarrow K(\alpha(a)_*E) \longrightarrow K(A \ltimes E).$$

It is not hard to trace through the definitions to check the required continuity condition, and it is clear by pointwise inspection that the resulting map ω covers the retractions to $A \ltimes E$, $r \circ \omega = \mathrm{id} \ltimes \omega$. $\square$

There are coherence diagrams that relate the maps ω of the proposition to the isomorphisms recorded in I.2.2. Putting these results together, using the definitions of $\mathbb{L}$-spectra and their smash product and its unit map (I.4.2, I.5.1, I.8.3) and the definition of the $\mathbb{L}$-spectrum structure on $\Sigma^\infty X$ (I.4.5), we arrive at the following conclusions.

THEOREM 5.8. *If N is an $\mathbb{L}$-spectrum with action ξ, then KN is an $\mathbb{L}$-spectrum with action the composite*

$$\mathscr{L}(1) \ltimes KN \xrightarrow{\omega} K(\mathscr{L}(1) \ltimes N) \xrightarrow{K\xi} KN.$$

Moreover, $r : KN \longrightarrow N$ is a map of $\mathbb{L}$-spectra. If X a based space, then $\omega : \Sigma^\infty X \longrightarrow K\Sigma^\infty X$ is a map of $\mathbb{L}$-spectra over $\Sigma^\infty X$. For $\mathbb{L}$-spectra M and N, there is a natural map of $\mathbb{L}$-spectra

$$\omega : KM \wedge_{\mathscr{L}} KN \longrightarrow K(M \wedge_{\mathscr{L}} N)$$

over $M \wedge_{\mathscr{L}} N$ such that the following unit, associativity, and commutativity diagrams commute:

$$\begin{array}{ccc} S \wedge_{\mathscr{L}} KN & \xrightarrow{\omega\wedge \mathrm{id}} & KS \wedge_{\mathscr{L}} KN \\ \downarrow{\scriptstyle\lambda} & & \downarrow{\scriptstyle\omega} \\ KN & \xleftarrow[K\lambda]{} & K(S \wedge_{\mathscr{L}} N), \end{array}$$

$$\begin{array}{ccc} KL \wedge_{\mathscr{L}} KM \wedge_{\mathscr{L}} KN & \xrightarrow{\omega\wedge \mathrm{id}} & K(L \wedge_{\mathscr{L}} M) \wedge_{\mathscr{L}} KN \\ \downarrow{\scriptstyle \mathrm{id}\wedge\omega} & & \downarrow{\scriptstyle\omega} \\ KL \wedge_{\mathscr{L}} K(M \wedge_{\mathscr{L}} N) & \xrightarrow[\omega]{} & K(L \wedge_{\mathscr{L}} M \wedge_{\mathscr{L}} N), \end{array}$$

and

$$\begin{array}{ccc} KM \wedge_{\mathscr{L}} KN & \xrightarrow{\tau} & KN \wedge_{\mathscr{L}} KM \\ \downarrow{\scriptstyle\omega} & & \downarrow{\scriptstyle\omega} \\ K(M \wedge_{\mathscr{L}} N) & \xrightarrow[K\tau]{} & K(M \wedge_{\mathscr{L}} N). \end{array}$$

THEOREM 5.9. *Let R be an A_∞ ring spectrum with unit η and product ϕ. Then KR is an A_∞ ring spectrum with unit and product the composites*

$$S \xrightarrow{\omega} KS \xrightarrow{K\eta} KR \quad \text{and} \quad KR \wedge_{\mathscr{L}} KR \xrightarrow{\omega} K(R \wedge_{\mathscr{L}} R) \xrightarrow{K\phi} KR.$$

Moreover, $r : KR \longrightarrow R$ is a map of A_∞ ring spectra. If R is an E_∞ ring spectrum, then so is KR. If M is an R-module (in the sense of II.3.3), then KM is a KR-module such that $r : KM \longrightarrow M$ is a map of KR-modules.

CHAPTER XI

Spaces of linear isometries and technical theorems

This chapter contains several deferred proofs concerning the structure of the linear isometries operad. They were used to build the foundations of Chapter I, but were not referred to thereafter.

1. Spaces of linear isometries

Many of our results depend on understanding the point-set topological and homotopical properties of spaces of linear isometries. We collect together the results that we need in this section and the next. However, we begin with a result on limits of cofibrations of unbased spaces. To prove it, we need the following generalization of the standard fact that a cofibration which is a homotopy equivalence is the inclusion of a strong deformation retract; it applies when the given map is also the map of total spaces of a pair of fibrations.

LEMMA 1.1. *Assume given a commutative diagram of spaces*

$$\begin{array}{ccc} A & \xrightarrow{i} & X \\ {\scriptstyle p}\downarrow & & \downarrow{\scriptstyle q} \\ B & \xrightarrow[j]{} & Y \end{array}$$

in which p and q are fibrations and i and j are cofibrations and homotopy equivalences. Assume given a map $\bar{r} : X \longrightarrow B$ such that $\bar{r} \circ i = p$ together with a homotopy $\bar{h} : q \simeq j \circ \bar{r}$ rel A. Then there is a map $r : X \longrightarrow A$ such that $r \circ i = \mathrm{id}$ and $p \circ r = \bar{r}$ together with a homotopy $h : \mathrm{id} \simeq ir$ rel A such that $q \circ h = \bar{h}$.

PROOF. Both (X, A) and $(X, A) \times (I, \partial I) = (X \times I, X \times \partial I \cup A \times I)$ are DR-pairs. The standard lifting property for fibrations and acyclic cofibrations gives r and h via the diagrams:

$$\begin{array}{ccc} A & \xrightarrow{\mathrm{id}} & A \\ {\scriptstyle i}\downarrow & \overset{r}{\nearrow} & \downarrow{\scriptstyle p} \\ X & \xrightarrow[\bar{r}]{} & B \end{array} \qquad \text{and} \qquad \begin{array}{ccc} X \times \partial I \cup A \times I & \xrightarrow{k} & X \\ \downarrow & \overset{h}{\nearrow} & \downarrow{\scriptstyle q} \\ X \times I & \xrightarrow[\bar{h}]{} & Y, \end{array}$$

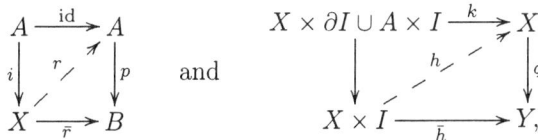

where $k(x,0,t) = x$, $k(x,1,t) = ir(x)$, and $k(a,s) = i(a)$. □

PROPOSITION 1.2. *For $n \geq 1$, assume given a commutative diagram of spaces*

$$\begin{array}{ccc} A_n & \xrightarrow{e_n} & X_n \\ {\scriptstyle p_n}\downarrow & & \downarrow{\scriptstyle q_n} \\ A_{n-1} & \xrightarrow[e_{n-1}]{} & X_{n-1} \end{array}$$

in which the p_n and q_n are fibrations and the e_n are cofibrations and homotopy equivalences. Then the induced map

$$e : A \equiv \lim A_i \longrightarrow \lim X_i \equiv X$$

is the inclusion of a strong deformation retract and a cofibration.

PROOF. Proceeding inductively, we use the lemma to construct retractions $r_n : X_n \longrightarrow A_n$ and homotopies $h_n : \mathrm{id} \simeq e_n \circ r_n$ rel A_n that are compatible with the given fibrations. The roles of $\bar{r}$ and $\bar{h}$ in the lemma are played by $r_{n-1} \circ q_n$ and $h_{n-1} \circ (q_n \times \mathrm{id})$. We obtain the retraction $r : X \longrightarrow A$ and homotopy $h : \mathrm{id} \simeq e \circ r$ rel A by passage to limits. By the standard (N)DR-pair criterion, to show that e is a cofibration, we need only construct a map $u : X \longrightarrow I$ such that $u^{-1}(0) = A$; of course, this is given by Urysohn's lemma if X is normal (e.g., metric). Since each (X_n, A_n) is a DR-pair, there are maps $v_n : X_n \longrightarrow I$ such that $v_n^{-1}(0) = A_n$. Let $u_n = v_n \circ \pi_n$, where $\pi_n : X \longrightarrow X_n$ is the projection, and define

$$u(x) = \sum_{n=1}^{\infty} \frac{1}{2^n} u_n(x).$$

Then u is continuous and $u(x) = 0$ if and only $\pi_n(x) \in A_n$ for each n. □

REMARK 1.3. The preceding may appear to be a model category result, but it depends on properties peculiar to the classical cofibrations of topological spaces.

Now Let U and U' be universes and write U and U' as the unions of expanding sequences of finite dimensional subspaces $\{V_n\}$ and $\{V_n'\}$, with the topology of the union. Thus a subset N of U is open if it intersects each V_n in an open subset. This topology is finer than the evident metric topology. If we identify U with $\mathbb{R}^\infty$ and think of $\mathbb{R}^\infty$ as a subset of the product of countably many copies of $\mathbb{R}$, then the intersection of $\mathbb{R}^\infty$ with the product of the intervals $(-1/q, 1/q)$ is an open neighborhood of zero which is not an open set in the metric topology.

For finite dimensional inner product spaces V and V', the space $\mathscr{I}(V, V')$ of linear isometries from V to V' is a smooth compact manifold. For a finite dimensional V, $\mathscr{I}(V, U')$ is the union of the $\mathscr{I}(V, V_n')$. It is homeomorphic to the Stiefel space of q-frames in U', where $q = \dim V$. As a union of smooth compact manifolds, it can be triangulated as a CW complex, and it is therefore paracompact.

The function space functor $\mathrm{map}(-, Y)$ converts colimits to limits, and it follows that $\mathscr{I}(U, U')$ is the inverse limit of the $\mathscr{I}(V_n, U')$. Each projection

$\mathscr{I}(V_{i+1}, V_j') \longrightarrow \mathscr{I}(V_i, V_j')$ is a bundle. By checking that the trivializations extend as j increases, we deduce that each projection $\mathscr{I}(V_{i+1}, U') \longrightarrow \mathscr{I}(V_i, U')$ is also a bundle.

Recall that a space X is LEC if the diagonal map $X \longrightarrow X \times X$ is a cofibration. It is standard that the inclusion $\{x\} \longrightarrow X$ is then a cofibration for all $x \in X$; that is, every point is a nondegenerate basepoint. In fact, more generally, the inclusion of a retract in an LEC space is a cofibration [36, 3.1].

PROPOSITION 1.4. *The space* $\mathscr{I}(U, U')$ *is LEC.*

PROOF. Any CW complex is LEC [36, 2.4], hence each $\mathscr{I}(V_i, U')$ is LEC. Since $\mathscr{I}(V_i, U')$ is also contractible [38, II.1.5] (or see the following lemma), its diagonal map is a cofibration and a homotopy equivalence. The diagonal map of $\mathscr{I}(U, U')$ is the inverse limit of the diagonal maps of the $\mathscr{I}(V_i, U')$. Now the conclusion is immediate from Proposition 1.2. □

Breaking with our rule of ignoring equivariant considerations, we prove the following result in full equivariant generality. As we have already used, $\mathscr{I}(U, U')$ is contractible. Thus, trivially, $\mathscr{I}(U, U')$ has the homotopy type of a CW complex. We record equivariant generalizations of these facts. We assume that some compact Lie group G acts on U and U'. Then G acts on $\mathscr{I}(U, U')$ by conjugation.

LEMMA 1.5. *For a G-space X, any two G-maps $f, g : X \longrightarrow \mathscr{I}(U, U')$ are homotopic.*

PROOF. Write $U' = U_1' \oplus U_2'$, where U_1' and U_2' are G-universes isomorphic to U'. Deformations of the identity on U' to isometries $U' \longrightarrow U_1'$ and $U' \longrightarrow U_2'$ show that f and g are homotopic to maps $f' : X \longrightarrow \mathscr{I}(U, U_1')$ and $g' : X \longrightarrow \mathscr{I}(U, U_2')$. Orthogonalization of the linear homotopy, $(1-t)f' + tg'$ shows that $f' \simeq g'$. □

LEMMA 1.6. *Assume that there is a finite dimensional representation $V \subset U$ such that the projection $\pi : \mathscr{I}(U, U') \longrightarrow \mathscr{I}(V, U')$ is a weak G-equivalence. Then π is a G-homotopy equivalence, and $\mathscr{I}(U, U')$ has the homotopy type of a G-CW complex.*

PROOF. As a union of smooth G-manifolds, $\mathscr{I}(V, U')$ is triangulable as a G-CW complex. Therefore, by the G-Whitehead theorem, there is a G-map $\phi : \mathscr{I}(V, U') \longrightarrow \mathscr{I}(U, U')$ such that $\pi \circ \phi \simeq \mathrm{id}$. The previous lemma gives that $\phi \circ \pi \simeq \mathrm{id}$, completing the proof. □

In practice, $\mathscr{I}(U, U')$ and $\mathscr{I}(V, U')$ have the appropriate behavior on fixed point spaces to be universal $\mathscr{F}$-spaces for some family $\mathscr{F}$ of subgroups of G [38, II.2.11], and this allows one to verify the hypothesis on π. We will shortly use the following example.

LEMMA 1.7. *The space $\mathscr{L}(j) = \mathscr{I}(U^j, U)$, $U \cong \mathbb{R}^\infty$, has the homotopy type of a Σ_j-CW complex.*

PROOF. The previous lemma applies with V replaced by V^j for any non-zero finite dimensional $V \subset U$. □

2. Fine structure of the linear isometries operad

We here prove two deferred technical results about the linear isometries operad that were used in I§8, namely I.8.1 and I.8.4. Actually, we will prove a generalization of I.8.4 and of its consequence I.8.5, and we shall also prove III.1.7. The essential point is to give the analogues of I.5.4 in the cases $i = 0$ and $j = 0$ that it excludes.

PROOF OF I.8.1. We must prove that $\mathscr{L}(2)/\mathscr{L}(1) \times \mathscr{L}(1)$ consists of a single point. Consider points f and g in $\mathscr{L}(2)$. Let V_1 and V_2 be the images of the restrictions of f to the two copies of U in $U \oplus U$, and let W_1 and W_2 be the images of the restrictions of g to these copies of U. Clearly the point f is specified by isomorphisms $U \longrightarrow V_1$ and $U \longrightarrow V_2$. We can find pairwise orthogonal infinite dimensional sub inner product spaces V_1', V_2', W_1', and W_2' of V_1, V_2, W_1, and W_2, respectively. If f' is specified by isomorphisms $U \longrightarrow V_1' \subset V_1$ and $U \longrightarrow V_2' \subset V_2$, then $f' = fk$ for some $k \in \mathscr{L}(1) \times \mathscr{L}(1)$, namely $k = f^{-1}(i_1 \oplus i_2)f'$, where $i_1 : V_1' \longrightarrow V_1$ and $i_2 : V_2' \longrightarrow V_2$ are the inclusions. Via the right action of $\mathscr{L}(1) \times \mathscr{L}(1)$, f is equivalent to such an f', which in turn is equivalent to a point specified by isomorphisms $U \longrightarrow V_1' \oplus W_1'$ and $U \longrightarrow V_2' \oplus W_2'$. By symmetry, the same is true for g, hence f is equivalent to g. $\square$

To explain the analogue of I.5.4 for the case $i = 0$ or $j = 0$, let $\mathscr{L}^0(j)$ denote the subspace of $\mathscr{L}(j)$ that consists of those linear isometries whose images have infinite dimensional orthogonal complements. Note that, for $f \in \mathscr{L}^0(j)$, $\operatorname{im}(f) \oplus \operatorname{im}(f)^{\perp}$ is contained in but not necessarily equal to U. Define

$$\hat{\mathscr{L}}(j) = \mathscr{L}(2) \times_{\mathscr{L}(1)\times\mathscr{L}(1)} \mathscr{L}(0) \times \mathscr{L}(j). \tag{2.1}$$

LEMMA 2.2. *The map*

$$\gamma : \mathscr{L}(2) \times \mathscr{L}(0) \times \mathscr{L}(j) \longrightarrow \mathscr{L}(j)$$

induces a continuous bijection

$$\hat{\gamma} : \hat{\mathscr{L}}(j) \longrightarrow \mathscr{L}^0(j).$$

Both $\hat{\mathscr{L}}(j)$ and $\mathscr{L}^0(j)$ are Σ_j-free and contractible and have the homotopy types of Σ_j-CW complexes, hence γ and $\hat{\gamma}$ are Σ_j-equivariant homotopy equivalences.

Before giving the proof, we explain some consequences. We begin with the proof of III.1.7, and we need the following point-set level result. As in I§1, we let $\mathscr{S}U$ denote the category of spectra indexed on U when U is not clear from context.

LEMMA 2.3. *Let A be a space over $\mathscr{I}(U, U')$ and let $\nu : B \longrightarrow A$ be a surjective map. Then the induced map*

$$\mathscr{S}U'(A \ltimes E, E') \longrightarrow \mathscr{S}U'(B \ltimes E, E')$$

is injective for spectra $E \in \mathscr{U}$ and $E' \in \mathscr{U}'$. Moreover, if $\alpha : E' \longrightarrow E''$ is a spacewise inclusion of spectra, then the following diagram is a pullback of spaces:

$$\begin{array}{ccc} \mathscr{S}U'(A \ltimes E, E') & \xrightarrow{\alpha_*} & \mathscr{S}U'(A \ltimes E, E'') \\ \downarrow{\nu^*} & & \downarrow{\nu^*} \\ \mathscr{S}U'(B \ltimes E, E') & \xrightarrow[\alpha_*]{} & \mathscr{S}U'(B \ltimes E, E''). \end{array}$$

PROOF. This can be verified by use of the point-set level description of maps $A \ltimes E \longrightarrow E'$ given in A.5.4. A more conceptual proof uses the following result to reduce the claims to easily verified space level assertions. □

The point-set level analysis of twisted half-smash products, such as is needed for the previous proof, is facilitated by the point of view developed in the first author's paper [20]. There is an enlarged category of spectra, which we denote by $\mathscr{S}$ in this discussion, that contains all of the $\mathscr{S}U$. An object of $\mathscr{S}$ is just a spectrum in $\mathscr{S}U$ for some U. A map $E \longrightarrow E'$ in $\mathscr{S}$ between spectra $E \in \mathscr{S}U$ and $E' \in \mathscr{S}U'$ is a linear isometry $f : U \longrightarrow U'$ together with a map $g : E \longrightarrow f^*E'$ in $\mathscr{S}U$; we write g for such a map, letting f be understood. The full definition of $\mathscr{S}$ exploits the topology of Grassmannian manifolds to topologize the set $\mathscr{S}(E, E')$ of maps $E \longrightarrow E'$ in such a way that the function

$$\varepsilon : \mathscr{S}(E, E') \longrightarrow \mathscr{I}(U, U')$$

that sends a map g to its underlying linear isometry f is continuous. The twisted half-smash product and twisted function spectra are implicitly built into the topology in view of the following result [20, Thm 0.5]. Let $\mathscr{U}/_{\mathscr{I}(U,U')}$ denote the category of spaces over $\mathscr{I}(U, U')$.

THEOREM 2.4. *There are natural homeomorphisms*

$$\mathscr{S}U'(A \ltimes E, E') \cong \mathscr{U}/_{\mathscr{I}(U,U')}(A, \mathscr{S}(E, E')) \cong \mathscr{S}U(E, F[A, E'))$$

for spaces A over $\mathscr{I}(U, U')$ and spectra $E \in \mathscr{S}U$ and $E' \in \mathscr{S}U'$.

PROOF OF III.1.7. We return to our fixed universe U and write $\mathscr{S} = \mathscr{S}U$. Given a compact R-module L and a sequence of spacewise inclusions of R-modules $M_i \longrightarrow M_{i+1}$, we must prove that the natural map

$$\operatorname{colim} \mathscr{M}_R(L, M_i) \longrightarrow \mathscr{M}_R(L, \operatorname{colim} M_i)$$

is a bijection. Let $L = \mathbb{F}_R K$ for a compact spectrum K. Since colimits of R-modules are computed on the spectrum level and

$$\mathscr{M}_R(\mathbb{F}_R K, M) \cong \mathscr{M}_S(S \wedge_{\mathscr{L}} \mathbb{L}K, M) = \mathscr{S}[\mathbb{L}](S \wedge_{\mathscr{L}} \mathbb{L}K, M),$$

it suffices to show that

$$\operatorname{colim} \mathscr{S}[\mathbb{L}](S \wedge_{\mathscr{L}} \mathbb{L}K, M_i) \longrightarrow \mathscr{S}[\mathbb{L}](S \wedge_{\mathscr{L}} \mathbb{L}K, \operatorname{colim} M_i)$$

is a bijection for a sequence $M_i \longrightarrow M_{i+1}$ of spacewise inclusions of $\mathbb{L}$-spectra. We have $S \wedge_{\mathscr{L}} \mathbb{L}K \cong \hat{\mathscr{L}}(1) \ltimes K$. Fix $f \in \mathscr{L}^0(1)$ such that $\operatorname{im}(f) \oplus \operatorname{im}(f)^\perp = U$ and let g be its preimage in $\hat{\mathscr{L}}(1)$. Clearly any other point $f' \in \mathscr{L}^0(1)$ has the form $e \circ f$ for some $e \in \mathscr{L}(1)$. That is, $\mathscr{L}^0(1)$ consists of a single orbit under the action of $\mathscr{L}(1)$. Obviously this property is inherited by $\hat{\mathscr{L}}(1)$. Thus the map $\nu : \mathscr{L}(1) \longrightarrow \hat{\mathscr{L}}(1)$ defined by $\nu(e) = e \cdot g$ is a surjection. Using I.2.2, we see that ν induces a map of $\mathbb{L}$-spectra

$$\nu \ltimes \mathrm{id} : \mathbb{L}f_*K \longrightarrow S \wedge_{\mathscr{L}} \mathbb{L}K.$$

Consider the following commutative diagram:

$$\begin{array}{ccc}
\operatorname{colim} \mathscr{S}[\mathbb{L}](S \wedge_{\mathscr{L}} \mathbb{L}K, M_i) & \longrightarrow & \mathscr{S}[\mathbb{L}](S \wedge_{\mathscr{L}} \mathbb{L}K, \operatorname{colim} M_i) \\
\downarrow{\scriptstyle \nu^*} & & \downarrow{\scriptstyle \nu^*} \\
\operatorname{colim} \mathscr{S}[\mathbb{L}](\mathbb{L}f_*K, M_i) & \longrightarrow & \mathscr{S}[\mathbb{L}](\mathbb{L}f_*K, \operatorname{colim} M_i) \\
\downarrow{\scriptstyle \cong} & & \downarrow{\scriptstyle \cong} \\
\operatorname{colim} \mathscr{S}(K, f^*M_i) & \longrightarrow & \mathscr{S}(K, f^* \operatorname{colim} M_i).
\end{array}$$

Since ν is a surjection, we can deduce from the preceding lemma and the evident description of $\mathscr{S}[\mathbb{L}](M, N)$ as an equalizer that the top two vertical maps are injections and that, for each i, the ith top square, before passage to colimits on the left, is a pullback. We have $f^*(M)(V) = M(f(V))$, hence $f^* \operatorname{colim} M_i$ is the colimit of the sequence of inclusions $f^*M_i \longrightarrow f^*M_{i+1}$. The spectrum level analogue [38, I.4.8] gives that the bottom horizontal arrow is a bijection. It follows immediately that the top horizontal arrow is a bijection, the cited pullback squares implying the surjectivity. □

The following result generalizes the first statement of I.8.5 from $j = 1$ to $j \geq 1$. It was used in the proof of VII.6.3.

THEOREM 2.5. *Let $j_i \geq 1$ and let Y_i, $1 \leq i \leq n$, be a tame Σ_{j_i}-spectrum indexed on U^{j_i}, such as $Y_i = (X_i)^{j_i}$ for a tame spectrum X indexed on U. Then*

$$\bigwedge_{i=1}^{n} \lambda : \bigwedge\nolimits_{\mathscr{L}} S \wedge_{\mathscr{L}} (\mathscr{L}(j_i) \ltimes Y_i) \longrightarrow \bigwedge\nolimits_{\mathscr{L}} \mathscr{L}(j_i) \ltimes Y_i$$

is a $(\Sigma_{j_1} \times \cdots \times \Sigma_{j_n})$-equivariant homotopy equivalence of spectra. If the Y_i have the homotopy types of CW Σ_{j_i}-spectra, then $\bigwedge_{\mathscr{L}} \mathscr{L}(j_i) \ltimes Y_i$ has the homotopy type of a CW $(\Sigma_{j_1} \times \cdots \times \Sigma_{j_n})$-spectrum, and its orbit spectrum has the homotopy type of a CW spectrum. In particular, if X is a tame spectrum indexed on U, then

$$\lambda : S \wedge_{\mathscr{L}} \mathbb{B}X \longrightarrow \mathbb{B}X \quad \text{and} \quad \lambda : S \wedge_{\mathscr{L}} \mathbb{C}X \longrightarrow \mathbb{C}X$$

are homotopy equivalences of spectra, and similarly for smash products over $\mathscr{L}$ of such maps.

PROOF. By the associativity and commutativity of $\wedge_{\mathscr{L}}$ and the isomorphism $S \wedge_{\mathscr{L}} S \cong S$,

$$\bigwedge\nolimits_{\mathscr{L}} S \wedge_{\mathscr{L}} (\mathscr{L}(j_i) \ltimes Y_i) \cong S \wedge_{\mathscr{L}} (\bigwedge\nolimits_{\mathscr{L}} \mathscr{L}(j_i) \ltimes Y_i)).$$

Using I.5.4 and I.5.6, we find that

$$\bigwedge_{\mathscr{L}} \mathscr{L}(j_i) \ltimes Y_i \cong \mathscr{L}(j_1 + \cdots + j_n) \ltimes (Y_1 \wedge \cdots \wedge Y_n).$$

Let $j = j_1 + \cdots + j_n$ and $Z = Y_1 \wedge \cdots \wedge Y_n$. By definition and inspection (see I.5.1 and the proofs of I.3.2 and I.8.3),

$$S \wedge_{\mathscr{L}} (\mathscr{L}(j) \ltimes Z) = \mathscr{L}(2) \ltimes_{\mathscr{L}(1)\times\mathscr{L}(1)} (\mathscr{L}(0) \ltimes S^0) \wedge (\mathscr{L}(j) \ltimes Z) \cong \hat{\mathscr{L}}(j) \ltimes Z$$

and, under our isomorphisms, the smash product of maps λ coincides with

$$\lambda = \gamma \ltimes \mathrm{id} : \hat{\mathscr{L}}(j) \ltimes Z \longrightarrow \mathscr{L}(j) \ltimes Z.$$

The claimed homotopy equivalence follows from the equivariant form of I.2.5; see A.7.4 and A§9. The statements about CW homotopy type follow from A.7.3, the equivariant form of I.2.6, and standard results on smash products and passage to orbits [38, II.3.8, VI.5.2, and I.5.6]. □

PROOF OF LEMMA 2.2. We obtain a homeomorphism $\mathscr{L}(j) \longrightarrow \mathscr{L}(1)$ of left $\mathscr{L}(1)$-spaces by composing on the right with an isomorphism $U^j \longrightarrow U$. It follows that, except for the equivariance statements, the result will be true in general if it is true when $j = 1$. It is clear that both spaces are Σ_j-free since $\mathscr{L}(j)$ is, and an elaboration of Lemma 1.6 gives the assertion about Σ_j-CW homotopy types. Thus we assume that $j = 1$ in the rest of the proof.

Let $i : \{0\} \longrightarrow U$ and $i_2 = i \oplus \mathrm{id} : U \longrightarrow U \oplus U$ be the obvious isometries. Then, for $f \in \mathscr{L}(1)$ and $g \in \mathscr{L}(2)$,

$$\gamma(g; i, f) = g \circ i_2 \circ f.$$

To see the surjectivity of $\hat{\gamma}$, let $h \in \mathscr{L}^0(1)$, let V be the orthogonal complement of the image of h, and choose an isomorphism $j : U \oplus V \longrightarrow V$. Then h is the composite

$$U \xrightarrow{i_2} U \oplus U \xrightarrow{\mathrm{id}\oplus h} U \oplus h(U) \subset U \oplus V \oplus h(U) \xrightarrow{j \oplus \mathrm{id}} V \oplus h(U) \subset U.$$

Let $g = (j \oplus \mathrm{id}) \circ (\mathrm{id} \oplus h)$. Then $h = g \circ i_2 = \gamma(g; i, \mathrm{id})$.

To see the injectivity, consider (g, f) and (g', f') in $\mathscr{L}(2) \times \mathscr{L}(1)$. Let "$\sim$" be the equivalence relation generated by $(g, f) \sim (g', f')$ if

$$g' = g \circ (j_1 \oplus j_2) \quad \text{and} \quad j_2 \circ f' = f \tag{2.6}$$

for points j_1 and j_2 in $\mathscr{L}(1)$. It suffices to show that

$$(g', f') \sim (g, f) \iff g' \circ i_2 \circ f' = g \circ i_2 \circ f,$$

and the forward implication is clear. The isometry g is given by the orthogonal pair of subspaces $V_1 = (g \circ i_1)(U)$ and $V_2 = (g \circ i_2)(U)$ of U together with isomorphisms $U \longrightarrow V_1$ and $U \longrightarrow V_2$, and we let $h = g \circ i_2 \circ f : U \longrightarrow V_2$. Thus (g, f) determines a triple (V_1, V_2, h) consisting of a pair of orthogonal infinite dimensional subspaces of $U \oplus U$ and a linear isometry $h : U \longrightarrow V_2$. Moreover, every such triple comes from some (g, f), as we see by choosing a linear isometry g such that $V_1 = (g \circ i_1)(U)$ and $V_2 = (g \circ i_2)(U)$ and setting $f = (g \circ i_2)^{-1} \circ h$. Let "$\sim$" be the equivalence relation on such triples generated by $(V_1, V_2, h) \sim$

(V_1', V_2', h') if $V_1' \subset V_1$, $V_2' \subset V_2$, and $h = h'$ as maps $U \longrightarrow U$. If these triples arise from (g, f) and (g', f') and we set

$$j_1 = (g \circ i_1)^{-1} \circ g' \circ i_1 \quad \text{and} \quad j_2 = (g \circ i_2)^{-1} \circ g' \circ i_2,$$

then we find that (2.6) holds and can conclude that $(g', f') \sim (g, f)$. Thus the injectivity will follow if we can show that $(V_1, V_2, h) \sim (V_1', V_2', h')$ for any two triples such that $h = h'$. Choose infinite dimensional subspaces W_1 of V_1 and W_1' of V_1' such that W_1, W_1', and $h(U)$ are mutually orthogonal. Then

$$(V_1, V_2, h) \sim (W_1, h(U), h) \sim (W_1 + W_1', h(U), h) \sim (W_1', h'(U), h') \sim (V_1', V_2', h').$$

This proves the injectivity and thus the bijectivity of $\hat{\gamma}$.

The contractibility of $\mathscr{L}^0(1)$ is clear since it is closed under the homotopies described in Lemma 1.5. To see the contractibility of $\hat{\mathscr{L}}(1)$, write $U = U_1 \oplus U_2$, where U_1 and U_2 are isomorphic to U, and define

$$\mathscr{K}(2) = \{g | g(\{0\} \oplus U) \subset U_2\} \subset \mathscr{L}(2),$$

$$\hat{\mathscr{K}}(1) = \mathscr{K}(2) \times_{\mathscr{L}(1) \times \mathscr{L}(1)} \mathscr{L}(0) \times \mathscr{L}(1),$$

and

$$\mathscr{K}^0(1) = \{f | f(U) \subset U_2\} \subset \mathscr{L}^0(1).$$

Since $\mathscr{K}^0(1) \cong \mathscr{I}(U, U_2)$, it is contractible. Since $\mathscr{L}^0(1)$ is also contractible, the inclusion $\mathscr{K}^0(1) \longrightarrow \mathscr{L}^0(1)$ is obviously a homotopy equivalence. We have the following commutative diagram, in which the vertical arrows are inclusions:

$$\begin{array}{ccc}
\hat{\mathscr{K}}(1) & \xrightarrow{\hat{\beta}} & \mathscr{K}^0(1) \\
\downarrow & & \downarrow \\
\hat{\mathscr{L}}(1) & \xrightarrow[\hat{\gamma}]{} & \mathscr{L}^0(1).
\end{array}$$

Modifying the proof that $\hat{\gamma}$ is a bijection by restricting V_2 to be contained in U_2, we see that $\hat{\beta}$ is also a bijection. Choose linear isometric isomorphisms $k_1 : U \longrightarrow U_1$ and $k_2 : U \longrightarrow U_2$ and define

$$\sigma : \mathscr{K}^0(1) \longrightarrow \mathscr{K}(2) \times \mathscr{L}(0) \times \mathscr{L}(1)$$

by $\sigma(f) = (k_1 \oplus k_2, i, k_2^{-1} \circ f)$. Then σ is a continuous section of

$$\gamma : \mathscr{K}(2) \times \mathscr{L}(0) \times \mathscr{L}(1) \longrightarrow \mathscr{K}^0(1).$$

It follows that $\hat{\beta}$ is a homeomorphism and in particular that $\hat{\mathscr{K}}(1)$ is contractible. We claim that the inclusion $\iota : \mathscr{K}(2) \longrightarrow \mathscr{L}(2)$ is a homotopy equivalence of right $\mathscr{L}(1) \times \mathscr{L}(1)$-spaces. It will follow that the inclusion $\hat{\mathscr{K}}(1) \longrightarrow \hat{\mathscr{L}}(1)$ is a homotopy equivalence, proving the contractibility of $\hat{\mathscr{L}}(1)$.

Define $\rho : \mathscr{L}(2) \longrightarrow \mathscr{K}(2)$ by $\rho(g) = i_2 \circ k_2 \circ g$. To prove our claim, it suffices to find a homotopy $h : i_2 \circ k_2 \simeq \mathrm{id}$ such that $h_t(U_2) \subset U_2$ for all t, for then $g \in \mathscr{K}(2)$ will imply $h_t \circ g \in \mathscr{K}(2)$ and, via right composition with maps g, h

will induce the required homotopies $\mathrm{id} \simeq \iota \circ \rho$ and $\mathrm{id} \simeq \rho \circ \iota$. Trivially, we have the following commutative diagram:

$$\begin{array}{ccc} U_1 \oplus U_2 & \xrightarrow{i_2 \oplus \mathrm{id}} & U_1 \oplus U_1 \oplus U_2 \\ {\scriptstyle k_2}\downarrow & & \downarrow{\scriptstyle \mathrm{id} \oplus k_2} \\ U_2 & \xrightarrow[i_2]{} & U_1 \oplus U_2. \end{array}$$

We can homotope $i_2 \oplus \mathrm{id}$ to $i_1 \oplus \mathrm{id}$ by homotoping i_2 to i_1, after which the right composite becomes $\mathrm{id} \oplus (k_2|_{U_2})$. We can then homotope $k_2|_{U_2}$ to the identity, after which the right composite becomes the identity. It is clear that U_2 is carried into U_2 by these homotopies. □

REMARK 2.7. It seems unlikely to us that $\hat{\gamma}$ is actually a homeomorphism.

3. The unit equivalence for the operadic smash product

We here prove I.6.2, and we restate each of its clauses as a lemma.

LEMMA 3.1. *For $\mathbb{L}$-spectra N, there is a natural weak equivalence of $\mathbb{L}$-spectra*

$$\omega : \mathbb{L}S \wedge_{\mathscr{L}} N \longrightarrow N.$$

PROOF. Let $\mathscr{L}(1)$ act from the right on $\mathscr{L}(2)$ by setting $fe = f \circ (1 \oplus e)$ for $f \in \mathscr{L}(2)$ and $e \in \mathscr{L}(1)$. Regard $\mathscr{L}(2)$ as a space over $\mathscr{L}(1)$ via the map $\sigma_2 : \mathscr{L}(2) \longrightarrow \mathscr{L}(1)$ specified by $\sigma_2(f) = f \circ i_2$, where $i_2 : U \longrightarrow U^2$ is the inclusion of the second summand. Then

$$\mathbb{L}S \wedge_{\mathscr{L}} N \cong \mathscr{L}(2) \ltimes_{\mathscr{L}(1)} N, \tag{3.2}$$

and σ_2 induces the required natural map

$$\omega = \sigma_2 \ltimes \mathrm{id} : \mathscr{L}(2) \ltimes_{\mathscr{L}(1)} N \longrightarrow \mathscr{L}(1) \ltimes_{\mathscr{L}(1)} N \cong N.$$

Observe that, by I.2.5, ω is a homotopy equivalence of spectra when $N = \mathbb{L}S^n$.

We must prove that ω induces an isomorphism on homotopy groups. By adjunction, we may identify $\pi_n(N)$ with $h\mathscr{S}[\mathbb{L}](\mathbb{L}S^n, N)$. First, to prove surjectivity, suppose given a map of $\mathbb{L}$-spectra $\alpha : \mathbb{L}S^n \longrightarrow N$. Write

$$\hat{\alpha} = \mathrm{id} \wedge_{\mathscr{L}} \alpha : \mathbb{L}S \wedge_{\mathscr{L}} \mathbb{L}S^n \longrightarrow \mathbb{L}S \wedge_{\mathscr{L}} N.$$

The following diagram commutes:

$$\begin{array}{ccc} \mathbb{L}S \wedge_{\mathscr{L}} \mathbb{L}S^n & \xrightarrow{\hat{\alpha}} & \mathbb{L}S \wedge_{\mathscr{L}} N \\ {\scriptstyle \omega}\downarrow & & \downarrow{\scriptstyle \omega} \\ \mathbb{L}S^n & \xrightarrow[\alpha]{} & N. \end{array}$$

Since ω on the left is an equivalence, $\alpha \in \mathrm{Im}(\omega_*)$.

To prove injectivity, suppose given an $\mathbb{L}$-map $\beta : \mathbb{L}S^n \to \mathbb{L}S \wedge_{\mathscr{L}} N$ such that $\alpha \equiv \omega \circ \beta \simeq 0$. Define $\hat{\alpha}$ as above. Since $\alpha \simeq 0$ and $\mathbb{L}S \wedge_{\mathscr{L}} (-) : \mathscr{S}[\mathbb{L}] \longrightarrow \mathscr{S}[\mathbb{L}]$ is a homotopy preserving functor, $\hat{\alpha} \simeq 0$. Define

$$\hat{\beta} = \beta \circ \omega : \mathbb{L}S \wedge_{\mathscr{L}} \mathbb{L}S^n \longrightarrow \mathbb{L}S \wedge_{\mathscr{L}} N.$$

Since ω is an equivalence, to prove that $\beta \simeq 0$, it suffices to prove that $\hat{\alpha} \simeq \hat{\beta}$. Here, by the naturality of ω, $\hat{\beta}$ coincides with the composite

$$\mathbb{L}S \wedge_{\mathscr{L}} \mathbb{L}S^n \xrightarrow{\mathrm{id}\wedge\beta} \mathbb{L}S \wedge_{\mathscr{L}} (\mathbb{L}S \wedge_{\mathscr{L}} N) \xrightarrow{\omega} \mathbb{L}S \wedge_{\mathscr{L}} N,$$

while $\hat{\alpha}$ coincides with the composite

$$\mathbb{L}S \wedge_{\mathscr{L}} \mathbb{L}S^n \xrightarrow{\mathrm{id}\wedge\beta} \mathbb{L}S \wedge_{\mathscr{L}} (\mathbb{L}S \wedge_{\mathscr{L}} N) \xrightarrow{\mathrm{id}\wedge\omega} \mathbb{L}S \wedge_{\mathscr{L}} N.$$

Thus it suffices to show that $\omega \simeq \mathrm{id}\wedge\omega$. Let $\mathscr{L}(1)$ act from the right on $\mathscr{L}(3)$ by setting $ge = g \circ (1 \oplus 1 \oplus e)$ for $g \in \mathscr{L}(3)$ and $e \in \mathscr{L}(1)$. Regard $\mathscr{L}(3)$ as a space over $\mathscr{L}(1)$ via the map $\sigma_3 : \mathscr{L}(3) \longrightarrow \mathscr{L}(1)$ specified by $\sigma_3(f) = f \circ i_3$, where $i_3 : U \longrightarrow U^3$ is the inclusion of the third summand. By the proof of the associativity isomorphism I.5.5, we have

$$(3.3) \qquad \mathbb{L}S \wedge_{\mathscr{L}} (\mathbb{L}S \wedge_{\mathscr{L}} N) \cong \mathscr{L}(3) \ltimes_{\mathscr{L}(1)} N.$$

Under the identifications (3.2) and (3.3), the maps ω and $\mathrm{id}\wedge\omega$ in our factorizations of $\hat{\beta}$ and $\hat{\alpha}$ coincide with the maps

$$\sigma_{2,3} \ltimes_{\mathscr{L}(1)} \mathrm{id} : \mathscr{L}(3) \ltimes_{\mathscr{L}(1)} N \longrightarrow \mathscr{L}(2) \ltimes_{\mathscr{L}(1)} N$$

and

$$\sigma_{1,3} \ltimes_{\mathscr{L}(1)} \mathrm{id} : \mathscr{L}(3) \ltimes_{\mathscr{L}(1)} N \longrightarrow \mathscr{L}(2) \ltimes_{\mathscr{L}(1)} N,$$

where

$$\sigma_{2,3} : \mathscr{L}(3) \longrightarrow \mathscr{L}(2) \quad \text{and} \quad \sigma_{1,3} : \mathscr{L}(3) \longrightarrow \mathscr{L}(2)$$

are the maps that restrict $g \in \mathscr{L}(3)$ to the second and third and first and third coordinates, respectively. Thus, it suffices to show that $\sigma_{2,3}$ and $\sigma_{1,3}$ are homotopic as maps of right $\mathscr{L}(1)$-spaces over $\mathscr{L}(1)$. Since the images under $\sigma_{2,3}(g)$ and $\sigma_{1,3}(g)$ of the first copy of U in U^2 are orthogonal and the right action of $\mathscr{L}(1)$ is on the second copy, to which we restrict when mapping to $\mathscr{L}(1)$, we obtain the required homotopy by normalizing the evident linear homotopy

$$h(g,t)(u_1,u_2) = tg(u_1,0,u_2) + (1-t)g(0,u_1,u_2). \quad \square$$

LEMMA 3.4. *The suspension homomorphism $\Sigma : \pi_n(N) \longrightarrow \pi_{n+1}(\Sigma N)$ is an isomorphism for any $\mathbb{L}$-spectrum N and integer n.*

PROOF. We shall construct an explicit inverse isomorphism

$$\Sigma^{-1} : \pi_{n+1}(\Sigma N) \longrightarrow \pi_n(N).$$

We again think of $\pi_n(N)$ as $h\mathscr{S}[\mathbb{L}](\mathbb{L}S^n, N)$. Since the functors Σ and $\mathbb{L}$ commute, we may identify $\mathbb{L}S^{n+1}$ with $\Sigma\mathbb{L}S^n$. Similarly, we have a natural isomorphism

$$\iota : \mathbb{L}S^{-1} \wedge_{\mathscr{L}} \Sigma N \cong \mathbb{L}S \wedge_{\mathscr{L}} N.$$

Since $\mathbb{L}S \wedge_{\mathscr{L}} \mathbb{L}S^n$ and $\mathbb{L}S^n$ are CW $\mathbb{L}$-spectra, the weak equivalence

$$\omega : \mathbb{L}S \wedge_{\mathscr{L}} \mathbb{L}S^n \longrightarrow \mathbb{L}S^n$$

is a homotopy equivalence of $\mathbb{L}$-spectra, and we choose a homotopy inverse ν. Suppose given an $\mathbb{L}$-map $\beta : \Sigma\mathbb{L}S^n \longrightarrow \Sigma N$. We define $\Sigma^{-1}\beta$ to be the composite

$$\mathbb{L}S^n \xrightarrow{\nu} \mathbb{L}S \wedge_{\mathscr{L}} \mathbb{L}S^n \xrightarrow{\iota^{-1}} \mathbb{L}S^{-1} \wedge_{\mathscr{L}} \Sigma\mathbb{L}S^n \xrightarrow{\mathrm{id}\wedge\beta} \mathbb{L}S^{-1} \wedge_{\mathscr{L}} \Sigma N \xrightarrow{\iota} \mathbb{L}S \wedge_{\mathscr{L}} N \xrightarrow{\omega} N.$$

If $\beta = \Sigma\alpha$, then the naturality of ι and ω imply that $\Sigma^{-1}\beta = \alpha \circ \omega \circ \nu \simeq \alpha$. Thus $\Sigma^{-1} \circ \Sigma = \mathrm{id}$. To evaluate $\Sigma \circ \Sigma^{-1}$, consider the following diagram:

$$\begin{array}{ccccc}
\Sigma\mathbb{L}S^n & & \xrightarrow{\quad\Sigma\nu\quad} & & \Sigma(\mathbb{L}S \wedge_{\mathscr{L}} \mathbb{L}S^n) \\
\vdots & & & & \downarrow{\scriptstyle \Sigma\iota^{-1}} \\
\Sigma\mathbb{L}S^n & \xleftarrow{\omega} & \mathbb{L}S \wedge_{\mathscr{L}} \Sigma\mathbb{L}S^n & \xleftarrow{\iota} & \Sigma(\mathbb{L}S^{-1} \wedge_{\mathscr{L}} \Sigma\mathbb{L}S^n) \\
{\scriptstyle \beta}\downarrow & & \downarrow{\scriptstyle \mathrm{id}\wedge\beta} & & \downarrow{\scriptstyle \Sigma(\mathrm{id}\wedge\beta)} \\
\Sigma N & \xleftarrow{\omega} & \mathbb{L}S \wedge_{\mathscr{L}} \Sigma N & \xleftarrow{\iota} & \Sigma(\mathbb{L}S^{-1} \wedge_{\mathscr{L}} \Sigma N) \\
\vdots & & & & \downarrow{\scriptstyle \Sigma\iota} \\
\Sigma N & & \xleftarrow[\quad\Sigma\omega\quad]{} & & \Sigma(\mathbb{L}S \wedge_{\mathscr{L}} N)
\end{array}$$

The upper left dotted arrow is the composite homotopy equivalence dictated by commutativity of the top rectangle. The maps $\Sigma\iota$ and ι appearing in the bottom rectangle differ by an interchange of circle coordinates, hence we obtain a dotted homotopy equivalence making the bottom rectangle homotopy commute by using a map of degree minus one on the circle coordinate. This implies that the composite $\Sigma \circ \Sigma^{-1}$ is an isomorphism, and it follows formally that it must be the identity. □

CHAPTER XII

The monadic bar construction

The monadic bar construction was a central tool in earlier drafts of this paper, but it plays a very minor role in this version. It is nevertheless an important construction. We shall say just enough about it to prove the two deferred results that depend on it and to allow rigorous use of it in later work. The essential point is to prove certain lemmas on cofibrations, one of which played a role in our construction of model structures on the categories of R-algebras and commutative R-algebras.

1. The bar construction and two deferred proofs

Recall the definitions of an action of a monad on a functor and of a monadic bifunctor from II.6.3.

DEFINITION 1.1. For a triple $(F, \mathbb{S}, R)$ consisting of a monad $(\mathbb{S}, \mu, \eta)$ in a category $\mathscr{C}$, an $\mathbb{S}$-algebra (R, ξ), and an $\mathbb{S}$-functor (F, ν) in $\mathscr{C}'$, define a simplicial object $B_*(F, \mathbb{S}, R)$ in $\mathscr{C}'$ by letting the q-simplices $B_q(F, \mathbb{S}, R)$ be $F\mathbb{S}^q R$ (where $\mathbb{S}^q$ denotes $\mathbb{S}$ composed with itself q times); the faces and degeneracies are given by

$$d_i = \begin{cases} \nu\mathbb{S}^{q-1} & \text{if } i = 0 \\ F\mathbb{S}^{i-1}\mu\mathbb{S}^{q-i-1} & \text{if } 1 \le i < q \\ F\mathbb{S}^{q-1}\xi & \text{if } i = q \end{cases}$$

and $s_i = F\mathbb{S}^i\eta\mathbb{S}^{q-i}$. If $\mathbb{S}'$ is a monad in $\mathscr{C}'$ and F is an $(\mathbb{S}', \mathbb{S})$-bifunctor, then $B_*(F, \mathbb{S}, R)$ is a simplicial $\mathbb{S}'$-algebra.

When F takes values in a category with a forgetful functor to $\mathscr{S}$, we write

$$B(F, \mathbb{S}, R) = |B_*(F, \mathbb{S}, R)|.$$

We use a similar notation for pairs when F takes pairs of spectra as values. All of the bar constructions used earlier, such as $B^R(M, A, N)$, can be interpreted as instances of this general construction. In the context of II.6.4, we have the following standard example.

EXAMPLE 1.2. We have a simplicial $\mathbb{S}$-algebra $B_*(\mathbb{S}, \mathbb{S}, R)$ associated to an $\mathbb{S}$-algebra R. Let $\underline{R}$ denote R regarded as a constant simplicial object, $\underline{R}_q = R$ for all q, with each face and degeneracy the identity map. Iterates of μ and ξ give a map $\varepsilon_* : B_*(\mathbb{S}, \mathbb{S}, R) \longrightarrow \underline{R}$ of simplicial $\mathbb{S}$-algebras in $\mathscr{C}$. Similarly, iterates of η give a map $\eta_* : \underline{R} \longrightarrow B_*(\mathbb{S}, \mathbb{S}, R)$ of simplicial objects in $\mathscr{C}$ such that $\varepsilon_* \eta_* = \mathrm{id}$. Moreover, there is a simplicial homotopy $\eta_* \varepsilon_* \simeq \mathrm{id}$ [45, 9.8].

By II.4.1, we have monads $\mathbb{B}$ and $\mathbb{C}$ in $\mathscr{S}$ whose algebras are the A_∞ and E_∞ ring spectra. We shall work in the ground category of spectra, rather than that of $\mathbb{L}$-spectra, for definiteness and because we envision more applications in that setting. Recall that geometric realization carries simplicial A_∞ and E_∞ rings, modules, and algebras to A_∞ and E_∞ rings, modules, and algebras, by X.1.5. We assume that all given spectra are Σ-cofibrant. In the contrary case, we first apply the cylinder construction K to make them so. By the results of X§4, this implies that the spectra of q-simplices in all of our constructions are tame. As we shall explain in the next section, it also implies that our simplicial spectra are proper, so that our homotopical results on geometric realization apply.

DEFINITION 1.3. For an A_∞ ring spectrum R, define an A_∞ ring spectrum UR by

$$UR = B(\mathbb{B}, \mathbb{B}, R).$$

For an E_∞ ring spectrum R, define an E_∞ ring spectrum UR by

$$UR = B(\mathbb{C}, \mathbb{C}, R).$$

The following result is immediate from Example 1.2 and X.1.2.

LEMMA 1.4. *For E_∞ ring spectra R there is a natural map of E_∞ ring spectra $\varepsilon : UR \longrightarrow R$ that is a homotopy equivalence of spectra, and similarly for A_∞ ring spectra.*

We shall prove the following addendum in the next section.

LEMMA 1.5. *The unit $\eta : S \longrightarrow UR$ is a cofibration of $\mathbb{L}$-spectra.*

REMARK 1.6. The A_∞ and E_∞ versions of the lemmas are compatible. If R is an E_∞ ring spectrum, then the natural map

$$B(\mathbb{B}, \mathbb{B}, R) \longrightarrow B(\mathbb{C}, \mathbb{C}, R)$$

of A_∞ ring spectra is a map under S and over R and is therefore a homotopy equivalence of spectra.

We now prove our change of operads result II.4.3. The proof is virtually the same as that given on the space level in [45].

PROOF OF II.4.3. We are given an E_∞ operad $\mathscr{O}$ over $\mathscr{L}$ and an $\mathscr{O}$-spectrum R. Technically, we must assume that the unit element of $\mathscr{O}(1)$ is a nondegenerate basepoint in order to ensure that the arguments of the next section apply to show that the simplicial spectra we use are proper. As in II.4.1, we have a monad $\mathbb{O}$ whose algebras are the $\mathscr{O}$-spectra together with a map $\mathbb{O} \longrightarrow \mathbb{C}$ of monads. We define VR to be the bar construction $B(\mathbb{C}, \mathbb{O}, R)$. By X.4.7, X.4.9, and the equivariant version of I.2.5, $\mathscr{O}X \longrightarrow \mathbb{C}X$ is a homotopy equivalence of spectra for any tame spectrum X. By X.2.4, there result maps of $\mathscr{O}$-spectra

$$R \longleftarrow B(\mathbb{O},\mathbb{O},R) \longrightarrow B(\mathbb{C},\mathbb{O},R) = VR$$

that are homotopy equivalences of spectra. □

We have precisely analogous constructions for modules. By II.6.2, we have a monad $\mathbb{C}[1]$ in the category of pairs of spectra whose algebras are pairs consisting of an E_∞ ring spectrum and a module over it. Its second coordinate is made explicit in II.5.7. We have a similar monad $\mathbb{B}[1]$ in the A_∞ case.

DEFINITION 1.7. For an E_∞ ring spectrum R and an R-module M, define a UR-module UM by

$$(UR;UM) = B(\mathbb{C}[1],\mathbb{C}[1],(R;M))$$

Replacing $\mathbb{C}$ by $\mathbb{B}$, we obtain an analogous functor U on modules over an A_∞ ring spectrum.

LEMMA 1.8. *There is a natural map of UR-modules $\varepsilon : UM \longrightarrow M$ that is a homotopy equivalence of spectra.*

REMARK 1.9. The A_∞ and E_∞ interpretations of the lemma are compatible. If M is a module over an E_∞ ring spectrum R, then the natural map

$$B(\mathbb{B}[1],\mathbb{B}[1],(R;M)) \longrightarrow B(\mathbb{C}[1],\mathbb{C}[1],(R;M))$$

is a map over $(R;M)$ and is thus a pair of homotopy equivalences of spectra.

PROOF OF II.5.2. The argument is much the same as the proof of II.4.3. As in II.5.7, we have a monad $\mathscr{O}[1]$ such that a $\mathscr{O}[1]$-algebra is a pair consisting of an $\mathscr{O}$-spectrum and a module over it, and we have a map of monads $\mathscr{O}[1] \longrightarrow \mathbb{C}[1]$. For an $\mathscr{O}$-spectrum R and an R-module M, we define VM to be the VR-module given by the second coordinate of $B(\mathbb{C}[1],\mathscr{O}[1],(R;M))$. The second coordinate of the weak equivalence

$$(R;M) \longleftarrow B(\mathscr{O}[1],\mathscr{O}[1],(R;M)) \longrightarrow B(\mathbb{C}[1],\mathscr{O}[1],(R;M)) = (VR;VM)$$

is the required weak equivalence between the R-module M and the VR-module VM; it is a homotopy equivalence of spectra. □

2. Cofibrations and the bar construction

We must prove that our simplicial bar constructions are proper and prove Lemma 1.5. Recall the definition of a proper simplicial $\mathbb{L}$-spectrum K_* from X.2.2, and remember that the simplicial filtration of $|K_*|$ is then given by a sequence of cofibrations of $\mathbb{L}$-spectra; it follows that the inclusion $K_0 \longrightarrow |K_*|$ is a cofibration of $\mathbb{L}$-spectra.

PROPOSITION 2.1. *The simplicial bar constructions used to construct the various functors U and V in the previous section are all proper simplicial $\mathbb{L}$-spectra.*

PROOF OF LEMMA 1.5. The unit $S \longrightarrow UR$ is the composite of the inclusion of S as a wedge summand of $\mathbb{C}R$ and the inclusion $\mathbb{C}R \longrightarrow UR$. □

The following two lemmas directly imply Proposition 2.1 in the case of UR. The proofs in the remaining cases are similar. Looking back at $B_*(F,\mathbb{S},R)$, we see that all of its degeneracy operators are maps of the form $F(-)$, so that the inclusion

$$sB_q(F,\mathbb{S},R)\subset B_q(F,\mathbb{S},R)$$

is obtained by applying the functor F to an inclusion that we may write

$$s\mathbb{S}^qR\subset\mathbb{S}^qR. \tag{2.2}$$

As explained after X.2.1, pedantic care with pushouts and coends is needed to be precise about this. The following lemma shows that the functor $\mathbb{C}$ converts cofibrations of spectra to cofibrations of $\mathbb{L}$-spectra. Here we are thinking of $\mathbb{C}$ as playing the role of F in our bar construction. The second part of the lemma is more interesting. It was essential to make sense of the Cofibration Hypothesis of VII§4.

LEMMA 2.3. *The following statements hold.*

(i) *The monads $\mathbb{T}$ and $\mathbb{P}$ in $\mathscr{S}[\mathbb{L}]$ that define A_∞ and E_∞ ring spectra preserve cofibrations of $\mathbb{L}$-spectra. Therefore the monads $\mathbb{B}$ and $\mathbb{C}$ in $\mathscr{S}$ convert cofibrations of spectra to cofibrations of $\mathbb{L}$-spectra.*

(ii) *For any commutative S-algebra R, the monads $\mathbb{T}$ and $\mathbb{P}$ in $\mathscr{M}_R$ that define R-algebras and commutative R-algebras preserve cofibrations of R-modules.*

PROOF. We prove the second statement. The first is similar; its second statement holds because $\mathbb{B}=\mathbb{T}\mathbb{L}$, $\mathbb{C}=\mathbb{P}\mathbb{L}$, and $\mathbb{L}$ carries cofibrations of spectra to cofibrations of $\mathbb{L}$-spectra. Let $f_i : M_i \longrightarrow N_i$ be cofibrations of R-modules, $1\le i\le j$. We claim that $f_1\wedge_R\cdots\wedge_R f_j$ is a cofibration of R-modules. There are retractions of R-modules $r_i : N_i\wedge I_+ \longrightarrow Mf_i$, where $Mf_i = N_i\cup(M_i\wedge I_+)$ is the mapping cylinder of f_i. The diagonal map $\Delta : I\longrightarrow I^j$ is a deformation retraction of spaces, with retraction ρ given by the averaging map $(t_1,\cdots,t_j)\longrightarrow\sum t_i/j$. The following composite is a retraction of R-modules, proving our claim:

$$\begin{array}{c}
(N_1\wedge_R\cdots\wedge_R N_j)\wedge I_+\\
\Big\downarrow{\scriptstyle \mathrm{id}\wedge\Delta_+}\\
(N_1\wedge_R\cdots\wedge_R N_j)\wedge(I^j)_+\cong(N_1\wedge I_+)\wedge_R\cdots\wedge_R(N_j\wedge I_+)\\
\Big\downarrow{\scriptstyle r_1\wedge\cdots\wedge r_j}\\
(Mf_1)\wedge_R\cdots\wedge_R(Mf_j)\cong(N_1\wedge_R\cdots\wedge_R N_j)\cup((M_1\wedge_R\cdots\wedge_R M_j)\wedge(I^j)_+)\\
\Big\downarrow{\scriptstyle \mathrm{id}\cup(\mathrm{id}\wedge\rho_+)}\\
(N_1\wedge_R\cdots\wedge_R N_j)\cup((M_1\wedge_R\cdots\wedge_R M_j)\wedge I_+)=M(f_1\wedge\cdots\wedge f_j).
\end{array}$$

Now let $f_1=\cdots=f_j=f$, say. Then the j-fold $\wedge_R$-power f^j is a cofibration. Moreover, since Δ and ρ are Σ_j-equivariant, f^j is a Σ_j-cofibration. Since passage to orbits carries Σ_j-cofibrations to cofibrations, f^j/Σ_j is also a cofibration of R-modules. By II.7.1, our monads are given by wedges of j-fold smash powers or j-fold symmetric smash powers, and the conclusion follows. □

LEMMA 2.4. *For each $q \geq 1$, the inclusion $s\mathbb{C}^q R \subset \mathbb{C}^q R$ is a cofibration of spectra.*

PROOF. Recall that $\mathbb{C}R = \bigvee_{j\geq 0} \mathscr{L}(j) \ltimes_{\Sigma_j} R^j$. Our standing assumption that R is Σ-cofibrant implies that each R^j is Σ_j-equivariantly Σ-cofibrant, by X.4.7. This allows us to apply A.8.1 and its equivariant version A§9 to show that maps between twisted half-smash products are cofibrations. The unit map $\eta : R \longrightarrow \mathbb{C}R$ is the composite

$$\eta \ltimes \mathrm{id} : R = \{1\} \ltimes R \longrightarrow \mathscr{L}(1) \ltimes R \subset \mathbb{C}R.$$

Here the first map is induced by the inclusion $\{1\} \longrightarrow \mathscr{L}(1)$ and is a cofibration by XI.1.4 and A.8.1, and the second map is the inclusion of a wedge summand. We shall prove that the inclusion $s\mathbb{C}^q R \subset \mathbb{C}^q R$ is a wedge of inclusions of the general form $A' \ltimes_G R^j \longrightarrow A \ltimes_G R^j$, where G is a subgroup of Σ_j and $A' \longrightarrow A$ is a G-cofibration. By A.8.1 and A§9, this will imply the conclusion. Although the combinatorics of the proof are a bit messy, it is easy to see what is going on by writing out the first few cases explicitly. The subspectrum $s\mathbb{C}^q$ is the union of the images of the inclusions $\mathbb{C}^i\eta\mathbb{C}^{q-i-1} : \mathbb{C}^{q-1}R \longrightarrow \mathbb{C}^q R$, $0 \leq i \leq q-1$. We first show by induction on q that $\mathbb{C}^q R$ is a wedge of spectra of the form $A \ltimes_G R^j$. This is obvious if $q = 1$ and we assume it for $q-1$. We first consider $\mathscr{L}(j) \ltimes_{\Sigma_j} Q^j$, where Q is a wedge of spectra Q_v indexed on a totally ordered set $\mathscr{V}$. (See [14, II§2] for more details of this analysis of extended powers of wedges.) Let V run through the set of ordered j-tuples of elements of $\mathscr{V}$; these V can be viewed as canonical elements in the distinct orbits of $\mathscr{V}^j$ under the permutation action of Σ_j. For such a V, let $(v(1), \cdots, v(n))$ be the distinct elements of $\mathscr{V}$ appearing in V, let $v(i)$ appear $j(i)$ times, so that $j = \sum j(i)$, and let $\Sigma_V \subset \Sigma_j$ be the image of $\Sigma_{j(1)} \times \cdots \times \Sigma_{j(n)}$ under the block sum homomorphism. Then

$$\mathscr{L}(j) \ltimes_{\Sigma_j} Q^j \cong \bigvee_V \mathscr{L}(j) \ltimes_{\Sigma_V} (Q_{v(1)}^{j(1)} \wedge \cdots \wedge Q_{v(n)}^{j(n)}).$$

Now suppose that each spectrum Q_v is of the form $X(v) \ltimes_{G(v)} R^{k(v)}$ for some subgroup $G(v)$ of $\Sigma_{k(v)}$ and some $G(v)$-space $X(v)$ over $\mathscr{L}(k(v))$. Then canonical isomorphisms in I.2.2 imply that

$$\mathscr{L}(j) \ltimes_{\Sigma_V} (Q_{v(1)}^{j(1)} \wedge \cdots \wedge Q_{v(n)}^{j(n)}) \cong A(V) \ltimes_{G(V)} R^{k(V)},$$

where $k(V) = \sum j(i)k(v(i))$, $G(V) \subset \Sigma_{k(V)}$ is the image under the canonical homomorphism of the product of wreath products

$$\Sigma_{j(1)} \int G_{v(1)} \times \cdots \times \Sigma_{j(n)} \int G_{v(n)},$$

and

$$A(V) = \mathscr{L}(j) \times X_{v(1)}^{j(1)} \times \cdots \times X_{v(n)}^{j(n)}$$

with its structural map γ to $\mathscr{L}(k(V))$. This implies that $\mathbb{C}^q R$ is a wedge of spectra of the form $A \ltimes_G R^j$, and it implicitly gives a complete inductive description of the relevant spaces A and groups G. The wedge summands are indexed by certain directed trees T. Each vertex of T has a prescribed level i, $1 \leq i \leq q$. There is a unique vertex of level q, there are directed edges from vertices of level i to vertices of level $i-1$, and each vertex of level less than q is the target of exactly one edge:

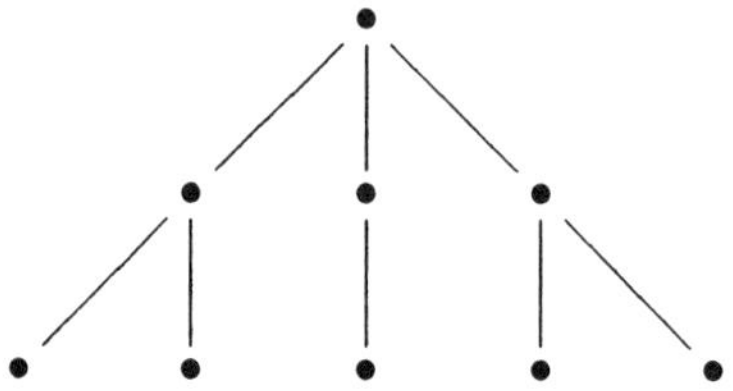

Each vertex is labelled with some $\mathscr{L}(j)$, where, if the vertex has level greater than 1, then the vertex is the source of j edges. (We allow $j = 0$, when the vertex is the source of no edges.) For such a tree T, the space $A(T)$ that is used to construct the corresponding wedge summand is the product of the labelling spaces $\mathscr{L}(j)$, ordered as prescribed by the tree and the inductive specification of the wedge summands given above. The degeneracy subspace lies in those wedge summands whose corresponding trees have all of their labels $\mathscr{L}(1)$ at one or more levels, and it is obtained by replacing $\mathscr{L}(1)$ by the point $\{1\}$ in the labels of vertices at those levels. This proves the lemma. $\square$

CHAPTER XIII

Epilogue: The category of $\mathbb{L}$-spectra under S

In a previous draft of this paper, certain variants of smash products that are defined between $\mathbb{L}$-spectra under S played a central role. In the present version, the only vestige remaining is the definition of free modules over A_∞ ring spectra given in II.5.3.

However, the parallel algebraic theory of [35] still requires such variant tensor products. We imagine that there is a functor, like the singular chain complex functor, from topological A_∞ and E_∞ rings and modules to algebraic ones. Such a construction would require the old definitions. The point is that, in algebra, it seems that one cannot hope to have an analogue of the isomorphism $S \wedge_{\mathscr{L}} S \cong S$. The theory of [35] is based on the algebraic operad $\mathscr{C} = C_*(\mathscr{L})$, where C_* is the singular chain complex functor. Hopkins' lemma, I.5.4, carries over since C_* preserves split coequalizers. However, the relation $\mathscr{L}(2)/(\mathscr{L}(1) \times \mathscr{L}(1)) = \{*\}$ does not carry over, and in fact one cannot have the relation $\mathscr{C}(2) \otimes_{\mathscr{C}(1)\otimes\mathscr{C}(1)} \mathbb{Z} = \mathbb{Z}$ in any E_∞ operad of (connected) chain complexes. Thus the topological theory is intrinsically better behaved algebraically than the parallel algebraic theory.

We explain just enough of the old definitions to give the idea and to explain how the new theory gives homotopical information about the old definitions.

1. The modified smash products $\lhd_{\mathscr{L}}$, $\rhd_{\mathscr{L}}$, and $\star_{\mathscr{L}}$

We return to the prologue and work in the category of $\mathbb{L}$-spectra in this section. We shall leave all proofs as exercises for the reader. They are easy consequences of results in Chapter I. Let $\mathscr{S}[\mathbb{L}]\backslash S$ denote the category of $\mathbb{L}$-spectra under S. We write η generically for the given map $S \longrightarrow M$.

DEFINITION 1.1. Let M be an $\mathbb{L}$-spectrum under S and let N be an $\mathbb{L}$-spectrum. Define the mixed smash product $M \lhd_{\mathscr{L}} N$ to be the pushout displayed in the following diagram of $\mathbb{L}$-spectra:

$$\begin{array}{ccc} S \wedge_{\mathscr{L}} N & \xrightarrow{\eta\wedge\mathrm{id}} & M \wedge_{\mathscr{L}} N \\ \lambda\downarrow & & \downarrow \\ N & \longrightarrow & M \lhd_{\mathscr{L}} N. \end{array} \qquad (1.2)$$

Define $N \rhd_{\mathscr{L}} M$ by symmetry.

If we apply the functor $S \wedge_{\mathscr{L}} (-)$ to the diagram (1.2), we obtain a weakly equivalent pushout diagram whose left arrow is an isomorphism and whose right arrow is therefore also an isomorphism. This implies the basic relation

$$(1.3) \qquad S \wedge_{\mathscr{L}} (M \lhd_{\mathscr{L}} N) \cong (S \wedge_{\mathscr{L}} M) \wedge_S (S \wedge_{\mathscr{L}} N),$$

which allows us to deduce homotopical properties of $\lhd_{\mathscr{L}}$ from homotopical properties of $\wedge_S$. It also implies the following result.

PROPOSITION 1.4. *For any $\mathbb{L}$-spectrum N, the canonical map*

$$N \longrightarrow S \lhd_{\mathscr{L}} N$$

is an isomorphism of $\mathbb{L}$-spectra and the canonical map

$$M \wedge_{\mathscr{L}} N \longrightarrow M \lhd_{\mathscr{L}} N$$

is a weak equivalence of $\mathbb{L}$-spectra.

For any $\mathbb{L}$-spectrum N under S, the canonical map

$$S \wedge_{\mathscr{L}} N \longrightarrow S \rhd_{\mathscr{L}} N$$

is an isomorphism because $\lambda : S \wedge_{\mathscr{L}} S \longrightarrow S$ is an isomorphism. Composing the inverse of this isomorphism with the unit weak equivalence $S \wedge_{\mathscr{L}} N \longrightarrow N$, we obtain the following result.

PROPOSITION 1.5. *For $\mathbb{L}$-spectra N under S, there is a natural weak equivalence of $\mathbb{L}$-spectra $\lambda : S \rhd_{\mathscr{L}} N \longrightarrow N$.*

LEMMA 1.6. *Let M and N be $\mathbb{L}$-spectra. Then*

$$(M \vee S) \lhd_{\mathscr{L}} N \cong (M \wedge_{\mathscr{L}} N) \vee N.$$

The commutativity and associativity of $\wedge_{\mathscr{L}}$ imply the following commutativity and associativity isomorphisms relating $\wedge_{\mathscr{L}}$ and $\lhd_{\mathscr{L}}$; these isomorphisms imply various others. The monad on $\mathscr{S}[\mathbb{L}]$ whose algebras are the $\mathbb{L}$-spectra under S sends M to $M \vee S$, hence the $\mathbb{L}$-spectra $M \vee S$ are the free $\mathbb{L}$-spectra under S. Results like the following one can be proven by first checking them on the $M \vee S$ and then deducing them in general.

LEMMA 1.7. *Let M and M' be $\mathbb{L}$-spectra under S and let N and N' be $\mathbb{L}$-spectra. Then there are natural isomorphisms*

$$M \lhd_{\mathscr{L}} N \cong N \rhd_{\mathscr{L}} M,$$

$$M \lhd_{\mathscr{L}} (N \wedge_{\mathscr{L}} N') \cong (M \lhd_{\mathscr{L}} N) \wedge_{\mathscr{L}} N',$$

and

$$M \lhd_{\mathscr{L}} (N \rhd_{\mathscr{L}} M') \cong (M \lhd_{\mathscr{L}} N) \rhd_{\mathscr{L}} M'.$$

With a view towards generalization to arbitrary ground E_∞ ring spectra, for which $R \wedge_{\mathscr{L},R} R$ will not be isomorphic to R, we give the following definition in a form that does not rely on the isomorphism $S \wedge_{\mathscr{L}} S \cong S$.

DEFINITION 1.8. Let M and N be $\mathbb{L}$-spectra under S. The coproduct of M and N in $\mathscr{S}[\mathbb{L}]\backslash S$ is the pushout $M \cup_S N$. There is an analogous pushout

$$(M \wedge_{\mathscr{L}} S) \cup_{S \wedge_{\mathscr{L}} S} (S \wedge_{\mathscr{L}} N),$$

and the unit maps λ determine a natural map of $\mathbb{L}$-spectra

$$\lambda : (M \wedge_{\mathscr{L}} S) \cup_{S \wedge_{\mathscr{L}} S} (S \wedge_{\mathscr{L}} N) \longrightarrow M \cup_S N.$$

The restrictions to $S \wedge_{\mathscr{L}} S$ of the maps

$$\mathrm{id} \wedge_{\mathscr{L}} \eta : M \wedge_{\mathscr{L}} S \longrightarrow M \wedge_{\mathscr{L}} N \quad \text{and} \quad \eta \wedge_{\mathscr{L}} \mathrm{id} : S \wedge_{\mathscr{L}} N \longrightarrow M \wedge_{\mathscr{L}} N$$

coincide, hence these maps determine a map

$$\theta : (M \wedge_{\mathscr{L}} S) \cup_{S \wedge_{\mathscr{L}} S} (S \wedge_{\mathscr{L}} N) \longrightarrow M \wedge_{\mathscr{L}} N.$$

Define the unital operadic smash product $M \star_{\mathscr{L}} N$ to be the pushout displayed in the following diagram of $\mathbb{L}$-spectra:

$$\begin{array}{ccc} (M \wedge_{\mathscr{L}} S) \cup_{S \wedge_{\mathscr{L}} S} (S \wedge_{\mathscr{L}} N) & \xrightarrow{\lambda} & M \cup_S N \\ {\scriptstyle\theta}\downarrow & & \downarrow \\ M \wedge_{\mathscr{L}} N & \longrightarrow & M \star_{\mathscr{L}} N. \end{array}$$

Then $M \star_{\mathscr{L}} N$ is an $\mathbb{L}$-spectrum under S with unit the composite of the unit $S \longrightarrow M \cup_S N$ and the displayed canonical map $M \cup_S N \to M \star_{\mathscr{L}} N$.

The essential, obvious, point is that S is a strict unit for the product $\star_{\mathscr{L}}$.

REMARK 1.9. Because $S \wedge_{\mathscr{L}} S \cong S$, $\star_{\mathscr{L}}$ can be defined less conceptually but more succinctly as the pushout in the diagram

$$\begin{array}{ccc} (M \wedge_{\mathscr{L}} S) \vee (S \wedge_{\mathscr{L}} N) & \xrightarrow{\lambda} & M \vee N \\ {\scriptstyle\theta}\downarrow & & \downarrow \\ M \wedge_{\mathscr{L}} N & \longrightarrow & M \star_{\mathscr{L}} N. \end{array}$$

An immediate comparison of pushout diagrams gives a natural map

$$M \rhd_{\mathscr{L}} N \longrightarrow M \star_{\mathscr{L}} N,$$

and a diagram chase shows that the product $\star_{\mathscr{L}}$ can be constructed in terms of the product $\rhd_{\mathscr{L}}$.

LEMMA 1.10. *If M and N are $\mathbb{L}$-spectra under S, then the following diagram is a pushout:*

$$(1.11) \qquad \begin{array}{ccc} S \rhd_{\mathscr{L}} N & \xrightarrow{\eta \rhd \mathrm{id}} & M \rhd_{\mathscr{L}} N \\ {\scriptstyle\lambda}\downarrow & & \downarrow \\ N & \longrightarrow & M \star_{\mathscr{L}} N. \end{array}$$

If M is an $\mathbb{L}$-spectrum and N is an $\mathbb{L}$-spectrum under S, then

$$(M \vee S) \star_{\mathscr{L}} N \cong (M \rhd_{\mathscr{L}} N) \vee N.$$

Applying the functor $S \wedge_{\mathscr{L}} (-)$ to the diagram (1.11) and using (1.3), we find that

$$S \wedge_{\mathscr{L}} (M \star_{\mathscr{L}} N) \cong (S \wedge_{\mathscr{L}} M) \wedge_S (S \wedge_{\mathscr{L}} N). \tag{1.12}$$

Again, homotopical properties of $\star_{\mathscr{L}}$ can be deduced from homotopical properties of $\wedge_S$, and we have the following result.

PROPOSITION 1.13. *The canonical map*

$$M \rhd_{\mathscr{L}} N \longrightarrow M \star_{\mathscr{L}} N$$

is a weak equivalence of $\mathbb{L}$-spectra.

LEMMA 1.14. *Let M and N be $\mathbb{L}$-spectra. Then*

$$(M \vee S) \star_{\mathscr{L}} (N \vee S) \cong (M \wedge_{\mathscr{L}} N) \vee M \vee N \vee S.$$

LEMMA 1.15. *The following associativity relation holds, where M and M' are $\mathbb{L}$-spectra under S and N is an $\mathbb{L}$-spectrum:*

$$(M \star_{\mathscr{L}} M') \lhd_{\mathscr{L}} N \cong M \lhd_{\mathscr{L}} (M' \lhd_{\mathscr{L}} N).$$

THEOREM 1.16. *The category $\mathscr{S}[\mathbb{L}]\backslash S$ is symmetric monoidal under $\star_{\mathscr{L}}$. The categories of monoids and commutative monoids in $\mathscr{S}[\mathbb{L}]\backslash S$ are isomorphic to the categories of A_∞ ring spectra and E_∞ ring spectra.*

We have a mixed function $\mathbb{L}$-spectrum (but not a unital operadic one).

DEFINITION 1.17. Let M be an $\mathbb{L}$-spectrum under S and N be an $\mathbb{L}$-spectrum. Define $F^{\rhd}_{\mathscr{L}}(M, N)$ to be the $\mathbb{L}$-spectrum displayed in the following pullback diagram:

$$\begin{array}{ccc} F^{\rhd}_{\mathscr{L}}(M,N) & \longrightarrow & F_{\mathscr{L}}(M,N) \\ \Big\downarrow & & \Big\downarrow{\scriptstyle \eta^*} \\ N & \longrightarrow & F_{\mathscr{L}}(S,N); \end{array}$$

here the bottom arrow is adjoint to $\lambda\tau : N \wedge_{\mathscr{L}} S \cong S \wedge_{\mathscr{L}} N \longrightarrow N$.

PROPOSITION 1.18. *Let M be an $\mathbb{L}$-spectrum under S and L and N be $\mathbb{L}$-spectra. Then*

$$\mathscr{S}[\mathbb{L}](M \lhd_{\mathscr{L}} L, N) \cong \mathscr{S}[\mathbb{L}](L \rhd_{\mathscr{L}} M, N) \cong \mathscr{S}[\mathbb{L}](L, F^{\rhd}_{\mathscr{L}}(M, N)).$$

2. The modified smash products $\lhd_R$, $\rhd_R$, and $\star_R$

We assume that R is an A_∞ ring spectrum in this section, and we understand modules in the sense of II.3.3. Thus the unit weak equivalences are not required to be isomorphisms. The cell and CW theory of such R-modules is developed in the same fashion as for modules over S-algebras. The appropriate definition of a smash product over R in this context reads as follows.

DEFINITION 2.1. Let R be an A_∞ ring spectrum and let M be a right and N be a left R-module. Define $M \wedge_{\mathscr{L},R} N$ to be the coequalizer displayed in the following diagram of $\mathscr{L}$-spectra:

$$(M \rhd_{\mathscr{L}} R) \wedge_{\mathscr{L}} N \cong M \wedge_{\mathscr{L}} (R \lhd_{\mathscr{L}} N) \underset{\mathrm{id}\wedge\nu}{\overset{\mu\wedge\mathrm{id}}{\rightrightarrows}} M \wedge_{\mathscr{L}} N \longrightarrow M \wedge_{\mathscr{L},R} N,$$

where μ and ν are the given actions of R on M and N; the canonical isomorphism of the terms on the left is implied by Lemma 1.7.

When $R = S$, $M \rhd_{\mathscr{L}} S \cong M$, $S \lhd_{\mathscr{L}} N \cong N$, and we are coequalizing the same isomorphism. Therefore our new $M \wedge_{\mathscr{L},S} N$ coincides with our old $M \wedge_{\mathscr{L}} N$. We have used the notation $\wedge_{\mathscr{L},R}$ to emphasize the conceptual point that we are here generalizing $\wedge_{\mathscr{L}}$ rather than $\wedge_S$.

REMARK 2.2. We have given the definition in the form most convenient for proofs, because the displayed coequalizer is split. However it is equivalent to define $M \wedge_{\mathscr{L},R} N$ more intuitively as the coequalizer displayed in the diagram

$$M \wedge_{\mathscr{L}} R \wedge_{\mathscr{L}} N \underset{\mathrm{id}\wedge\nu}{\overset{\mu\wedge\mathrm{id}}{\rightrightarrows}} M \wedge_{\mathscr{L}} N \longrightarrow M \wedge_{\mathscr{L},R} N.$$

We can define modified smash products $\lhd_R$, $\rhd_R$, and $\star_R$ when one or both of the variables comes with a given map of R-modules $\eta : R \longrightarrow M$, copying Definitions 1.1 and 1.8 with S replaced by R.

DEFINITION 2.3. Let M be a right R-module under R and let N be a left R-module. Define the mixed smash product $M \lhd_R N$ to be the pushout displayed in the following diagram of $\mathbb{L}$-spectra:

$$\begin{array}{ccc} R \wedge_{\mathscr{L},R} N & \xrightarrow{\eta\wedge\mathrm{id}} & M \wedge_{\mathscr{L},R} N \\ {\scriptstyle\lambda}\downarrow & & \downarrow \\ N & \longrightarrow & M \lhd_R N. \end{array}$$

Define $\rhd_R$ by symmetry. Observe that the displayed pushout is a diagram of R-modules when R is an E_∞ ring spectrum.

DEFINITION 2.4. Let M be a right and N a left R-module under R. Define the unital smash product $M \star_R N$ to be the pushout displayed in the following diagram of $\mathbb{L}$-spectra:

$$\begin{array}{ccc} (M \wedge_{\mathscr{L},R} R) \cup_{R\wedge_{\mathscr{L},R}R} (R \wedge_{\mathscr{L},R} N) & \xrightarrow{\lambda} & M \cup_R N \\ \downarrow{\scriptstyle\theta} & & \downarrow \\ M \wedge_{\mathscr{L},R} N & \longrightarrow & M \star_R N. \end{array}$$

Here, as in Definition 1.8, λ is induced by the unit maps λ of M and N and θ is induced by the structure maps η of M and N. Observe that the displayed pushout is a diagram of R-modules when R is an E_∞ ring spectrum.

All of the results of the previous section apply verbatim in the context of R-modules, except for those that depend on the isomorphism $S \wedge_{\mathscr{L}} S \cong S$.

LEMMA 2.5. *Let M be a right and N be a left R-module. Then*

$$(M \vee R) \lhd_R N \cong (M \wedge_{\mathscr{L},R} N) \vee N$$

and

$$(M \vee R) \star_R (N \vee R) \cong (M \wedge_{\mathscr{L},R} N) \vee M \vee N \vee R.$$

If N is an R-module under R, then

$$(M \vee R) \star_R N \cong (M \rhd_R N) \vee N.$$

If M and N are R-modules under R, then the following diagram is a pushout, where the unit map $\lambda : R \rhd_R N \longrightarrow N$ is constructed by a comparison of pushout diagrams:

$$\begin{array}{ccc} R \rhd_R N & \xrightarrow{\eta \rhd \mathrm{id}} & M \rhd_R N \\ \downarrow{\scriptstyle\lambda} & & \downarrow \\ N & \longrightarrow & M \star_R N. \end{array}$$

Using this, we can deduce alternative expressions for these products in terms of coequalizer diagrams like that which defines $\wedge_{\mathscr{L},R}$.

LEMMA 2.6. *For a right R-module M under R and a left R-module N, $M \lhd_R N$ can be identified with the coequalizer displayed in the diagram*

$$(M \star_{\mathscr{L}} R) \lhd_{\mathscr{L}} N \cong M \lhd_{\mathscr{L}} (R \lhd_{\mathscr{L}} N) \underset{\mathrm{id}\lhd\nu}{\overset{\mu\lhd\mathrm{id}}{\rightrightarrows}} M \lhd_{\mathscr{L}} N \longrightarrow M \lhd_R N.$$

For a right R-module M under R and a left R-module N under R, $M \star_R N$ can be identified with the coequalizer displayed in the diagram

$$(M \star_{\mathscr{L}} R) \star_{\mathscr{L}} N \cong M \star_{\mathscr{L}} (R \star_{\mathscr{L}} N) \underset{\mathrm{id}\star\nu}{\overset{\mu\star\mathrm{id}}{\rightrightarrows}} M \star_{\mathscr{L}} N \longrightarrow M \star_R N.$$

PROPOSITION 2.7. *The following statements hold.*

(i) *For any R-module N, the canonical map of R-modules*

$$N \longrightarrow R \lhd_R N$$

is an isomorphism.

(ii) *For any R-module M under R, the canonical map of R-modules*

$$M \wedge_{\mathscr{L},R} N \longrightarrow M \lhd_R N$$

is a weak equivalence.

(iii) *For any R-module N under R, the canonical map of R-modules*

$$\lambda : R \rhd_R N \longrightarrow N$$

is a weak equivalence.

(iv) *For any R-modules M and N under R, the canonical map of R-modules*

$$M \rhd_R N \longrightarrow M \star_R N$$

is a weak equivalence.

We have various commutativity and associativity isomorphisms that involve several A_∞ rings, as in III.3.4. Note that an (R, R')-bimodule is the same thing as an $(R \star_{\mathscr{L}} R'^{op})$-module, and an $(R \star_{\mathscr{L}} R'^{op})$-module under $(R \star_{\mathscr{L}} R'^{op})$ is both a right R-module under R and a left R'-module under R'.

PROPOSITION 2.8. *Let M be an (R, R')-bimodule, N be an (R', R'')-bimodule, and P be an (R'', R''')-bimodule.*

(i) *If M is an R'-module under R', then*

$$M \lhd_{R'} N \cong N \rhd_{R'^{op}} M$$

as $R \star_{\mathscr{L}} R''^{op}$-modules and

$$M \lhd_{R'} (N \wedge_{R''} P) \cong (M \lhd_{R'} N) \wedge_{R''} P$$

as (R, R''')-bimodules.

(ii) *If M is an R'-module under R' and P is an R''-module under R'', then*

$$M \lhd_{R'} (N \rhd_{R''} P) \cong (M \lhd_{R'} N) \rhd_{R''} P$$

as (R, R''')-bimodules.

(iii) *If M is an R'-module under R' and N is an $(R' \star_{\mathscr{L}} R''^{op})$-module under $(R' \star_{\mathscr{L}} R''^{op})$, then*

$$(M \star_{R'} N) \lhd_{R''} P \cong M \lhd_{R'} (N \lhd_{R''} P)$$

as (R, R''')-bimodules.

(iv) *If M is an R'-module under R', N is an $(R' \star_{\mathscr{L}} R''^{op})$-module under $(R' \star_{\mathscr{L}} R''^{op})$, and P is an R''-module under R'', then*

$$(M \star_{R'} N) \star_{R''} P \cong M \star_{R'} (N \star_{R''} P)$$

as (R, R''')-bimodules.

PROOF. This is a formal exercise in the commutation of coequalizers with coequalizers, starting with the analogous isomorphisms for our various smash products over S. One writes down three-by-three diagrams of coequalizers and uses that the coequalizer of coequalizers is a coequalizer vertically and horizontally to deduce the stated isomorphisms. The top left-hand corner of the diagram needed for the isomorphism of (iii), for example, is

$$(((M \star_{\mathscr{L}} R') \star_{\mathscr{L}} N) \star_{\mathscr{L}} R'') \lhd_{\mathscr{L}} P \cong M \lhd_{\mathscr{L}} (R' \lhd_{\mathscr{L}} (N \lhd_{\mathscr{L}} (R'' \lhd_{\mathscr{L}} P))). \quad \square$$

Similarly, all other results of III§§3-4 carry over directly to the present context, and we reach the following conclusion in the commutative case.

THEOREM 2.9. *The category of R-modules under R is symmetric monoidal under $\star_R$.*

We can define R-algebras and commutative R-algebras to be monoids and commutative monoids in this symmetric monoidal category. Of course, if we apply the functor $S \wedge_{\mathscr{L}} (-)$ to such algebras we obtain weakly equivalent algebras over the S-algebra $S \wedge_{\mathscr{L}} R$.

Again, we have a mixed function R-spectrum giving an adjunction like Proposition 1.18.

DEFINITION 2.10. Let M be an R-module under R and let N be an R-spectrum. Define $F_R^{\rhd}(M, N)$ to be the $\mathbb{L}$-spectrum displayed in the following pullback diagram:

$$\begin{array}{ccc} F_R^{\rhd}(M,N) & \longrightarrow & F_R(M,N) \\ \downarrow & & \downarrow \eta^* \\ N & \longrightarrow & F_R(R,N); \end{array}$$

here the bottom arrow is adjoint to $\lambda\tau : N \wedge_R R \cong R \wedge_R N \longrightarrow N$. If R is an E_∞ ring spectrum, then $F_R^{\rhd}(M, N)$ is an R-module.

REMARK 2.11. Geometric realization behaves as expected. If L_* and L'_* are simplicial R-modules under R and K_* is a simplicial R-module, then

$$|L_*| \lhd_{\mathscr{L}} |K_*| \cong |L_* \lhd_{\mathscr{L}} K_*|$$

and

$$|L_*| \star_{\mathscr{L}} |L'_*| \cong |L_* \star_{\mathscr{L}} L'_*|.$$

Similarly our work on enriched model categories carries over to the present framework. There is an analogue of VII.2.8 for the category of R-modules under R. Here we must enrich over the category of unbased spaces. Recall that a colimit is said to be connected if it is indexed on a diagram whose domain category is connected.

PROPOSITION 2.12. *The category of R-modules under R is topologically cocomplete and complete. The cotensors $F(X_+, E)$ and all other indexed limits are created in $\mathscr{S}$; ordinary connected colimits are also created in $\mathscr{S}$. For an R-module N, the functor $M \lhd_R N$ on R-modules M under R preserves connected* *colimits. For an R-module N under R, the functor $M \star_R N$ on R-modules M under R preserves connected colimits.*

PROOF. The colimit, $\operatorname{colim}_{relR} D$ of a diagram D of R-modules under R is computed from its colimit as a diagram of R-modules via the pushout

$$\begin{array}{ccc} \operatorname{colim} \underline{R} & \longrightarrow & R \\ \downarrow & & \downarrow \\ \operatorname{colim} D & \longrightarrow & \operatorname{colim}_{relR} D; \end{array}$$

the right vertical arrow gives the unit. Here $\underline{R}$ is the constant diagram at R. The top horizontal arrow is an isomorphism if the domain category of D is connected, and the relative and ordinary colimits colimits then agree. Since this holds for coequalizers, VII.2.6 and 2.8 imply the first statement. The remaining statements follow since the products $\lhd$ and $\star$ are defined in terms of $(-) \wedge_{\mathscr{L},R} N$ and pushouts, which preserve colimits of R-modules. $\square$

THEOREM 2.13. *For any E_∞ ring spectrum R the categories of R-algebras and of commutative R-algebras are topologically cocomplete and complete. Their cotensors and all other indexed limits are created in $\mathscr{S}$.*

Similarly, all categories in sight admit model structures.

THEOREM 2.14. *The categories of modules over an A_∞ ring spectrum R and of algebras and commutative algebras over an E_∞ ring spectrum R are topological model categories. In all cases, the weak equivalences and q-fibrations are the maps which are weak equivalences or Serre fibrations of underlying spectra.*

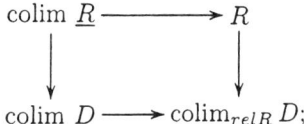

APPENDIX A

Twisted half-smash products and function spectra

by Michael Cole

1. Introduction

Let U, U' be universes, $\mathscr{I}(U,U')$ the space of linear isometries $U \longrightarrow U'$, A a space, $\alpha : A \longrightarrow \mathscr{I}(U,U')$ a map, and let $E \in \mathscr{S}U$, $E' \in \mathscr{S}U'$ be spectra. The twisted half-smash product $\alpha \ltimes E \in \mathscr{S}U'$ and the twisted function spectrum $F[\alpha, E') \in \mathscr{S}U$ are fundamental constructions which underly the foundations of stable homotopy theory that were introduced in [38] and developed further in this book. These functors specialize to the change of universe functors that are necessary to define internal smash products and function spectra, and they provide a simple way to prove that the different internal smash products so obtained are canonically and coherently equivalent upon passage to the stable category.

In connection with the research presented in this book, the theory of twisted half-smash products has undergone major clarification and sharpening of the results originally given in [38]. In particular, the improved "untwisting theorem", given as Theorem 5.5 below, has led to corresponding improvements of the theorems concerning the homotopy invariance properties of $\alpha \ltimes E$ and $F[\alpha, E')$ in the variable α — the crucial point being that for tame spectra E, the homotopical properties of $\alpha \ltimes E$ depend only on the homotopical properties of the space A rather than on the particular map α. In particular, it is vital to the theory that for tame spectra E the functor $\alpha \ltimes E$ is well-behaved homotopically before passage to the stable category.

We present new definitions of twisted half-smash products and function spectra which have considerable advantages, both conceptually and expositionally, compared to the definitions in [38]. In particular, our definitions do not require choosing arbitrary cofinal sequences of representations and therefore our treatment avoids the technical complications concerning the behavior of colimits found in the approach of [38]. We shall recast much of the material of [38, VI§§1-3] in a simpler, more user-friendly, form.

The treatment of twisted half-smash products and function spectra in [38] is written in the equivariant context, with a compact Lie group G acting on all spaces and spectra in sight. We present our definitions and theorems nonequivariantly here and note that virtually everything that we say carries over mutatis mutandis to the equivariant case. We will point out the small exceptions to this at the end.

We give preliminary definitions and constructions in Sections 2-4, reaching the definition of twisted half-smash products and function spectra in Section 5. That section also gives a simple proof of the untwisting theorem, which describes an isomorphism $\alpha \ltimes E \cong E \wedge A_+$ when the spectrum E is a finite desuspension of a space. Our proof relies on an idea of Neil Strickland. We give some formal properties of our functors in Section 6 and consider their homotopical properties in Section 7. In Section 8 we prove a cofibration theorem that plays an important technical role in the theory of S-modules. We shall derive it from the untwisting theorem. We conclude in Section 9 with a brief discussion of the equivariant versions of our constructions; more details will appear in [16].

2. The category $\mathscr{S}(U';U)$

To define our constructions we introduce a category $\mathscr{S}(U';U)$. An object $\mathscr{E} \in \mathscr{S}(U';U)$ is a family of spectra $\mathscr{E}_V \in \mathscr{S}U'$, one for each indexing space $V \subset U$. We require isomorphisms $\rho_{V,W} : \Sigma^{W-V}\mathscr{E}_W \longrightarrow \mathscr{E}_V$, $V \subset W$, that satisfy the evident transitivity relation. Here we are mixing universes since we are suspending the spectrum $\mathscr{E}_W$ which is indexed on U' by the indexing space $W - V$ which is in U, but in any case $\Sigma^{W-V}\mathscr{E}_W$ just means the usual smash product of the spectrum $\mathscr{E}_W$ with the space S^{W-V}. We think of $\mathscr{E}$ as the spectrum $\mathscr{E}_0 \in \mathscr{S}U'$, an object stable with respect to suspension by indexing spaces in U', but also equipped with a choice of compatible desuspensions by indexing spaces in U.

EXAMPLE 2.1. If $U' = U$ and X is a space, we get an associated object $\mathscr{E}(X) \in \mathscr{S}(U;U)$ by setting $\mathscr{E}(X)_V = \Sigma^\infty_V X$ and considering the canonical natural isomorphisms $\Sigma^{W-V}\Sigma^\infty_W X \cong \Sigma^\infty_V X$ as structure maps.

EXAMPLE 2.2. Generalizing Example 2.1, let $f : U \longrightarrow U'$ be a linear isometry and X a space. The specification $\mathscr{E}_f(X)_V = \Sigma^\infty_{fV} X$ defines an object $\mathscr{E}_f(X) \in \mathscr{S}(U';U)$ with structure maps given by $\Sigma^{W-V}\Sigma^\infty_{fW}X \cong \Sigma^{fW-fV}\Sigma^\infty_{fW}X \cong \Sigma^\infty_{fV}X$.

We will prove shortly that a map $\alpha : A \longrightarrow \mathscr{I}(U,U')$ gives rise to an object $\mathscr{M}\alpha \in \mathscr{S}(U';U)$ for which $\mathscr{M}\alpha_0 \cong \Sigma^\infty A_+$. The construction is natural in spaces over $\mathscr{I}(U,U')$. If α is the constant map at $f \in \mathscr{I}(U,U')$ then $\mathscr{M}\alpha$ is the object discussed in Example 2.2 with $X = A_+$.

We will define a smash product

$$\wedge : \mathscr{S}(U';U) \times \mathscr{S}U \longrightarrow \mathscr{S}U'$$

and a function spectrum

$$F(-,-) : \mathscr{S}(U';U)^{\mathrm{op}} \times \mathscr{S}U' \longrightarrow \mathscr{S}U$$

satisfying an adjunction isomorphism

$$\mathscr{S}U'(\mathscr{E} \wedge E, E') \cong \mathscr{S}U(E, F(\mathscr{E}, E')), \tag{2.3}$$

where $E \in \mathscr{S}U$, $E' \in \mathscr{S}U'$, $\mathscr{E} \in \mathscr{S}(U'; U)$. For a map $\alpha : A \longrightarrow \mathscr{I}(U, U')$ and spectra $E \in \mathscr{S}U$, $E' \in \mathscr{S}U'$ we will define

$$\alpha \ltimes E = \mathscr{M}\alpha \wedge E$$

and

$$F[\alpha, E') = F(\mathscr{M}\alpha, E')$$

and show that this agrees with the definitions of [38].

3. Smash products and function spectra

We first record the following obvious fact.

PROPOSITION 3.1. *For an object $\mathscr{E} \in \mathscr{S}(U'; U)$ and a space $X \in \mathscr{T}$ there is a smash product $\mathscr{E} \wedge X \in \mathscr{S}(U'; U)$ defined by $(\mathscr{E} \wedge X)_V = \mathscr{E}_V \wedge X$ and the evident structure maps. There is an adjunction isomorphism*

$$\mathscr{S}(U'; U)(\mathscr{E} \wedge X, \mathscr{D}) \cong \mathscr{T}(X, \mathscr{S}(U'; U)(\mathscr{E}, \mathscr{D}))$$

where the morphism set $\mathscr{S}(U'; U)(\mathscr{E}, \mathscr{D})$ is given the evident topology and basepoint.

Although not obvious from the definitions, it can be shown that there is a "function object" $F(X, \mathscr{D}) \in \mathscr{S}(U'; U)$ such that the adjunction extends in the expected way.

We now define the object $F(\mathscr{E}, E')$.

DEFINITION 3.2. For objects $\mathscr{E} \in \mathscr{S}(U'; U)$ and $E' \in \mathscr{S}U'$ the spectrum $F(\mathscr{E}, E') \in \mathscr{S}U$ is given by $F(\mathscr{E}, E')(V) = \mathscr{S}U'(\mathscr{E}_V, E')$. Abbreviating $\mathscr{S}' = \mathscr{S}U'$, the structure maps are given by the sequences of isomorphisms

$$\mathscr{S}'(\mathscr{E}_V, E') \cong \mathscr{S}'(\Sigma^{W-V}\mathscr{E}_W, E') \cong \mathscr{S}'(\mathscr{E}_W, \Omega^{W-V}E') \cong \Omega^{W-V}\mathscr{S}'(\mathscr{E}_W, E').$$

For $\mathscr{E} \in \mathscr{S}(U'; U)$ and $E \in \mathscr{S}U$, our definition of $\mathscr{E} \wedge E \in \mathscr{S}U'$ is dictated by the desired adjunction (2.3). Recall that for a spectrum $E \in \mathscr{S}U$ and prespectrum $D \in \mathscr{P}U$, the morphism set $\mathscr{P}U(D, E)$ may be described by

$$\mathscr{P}U(D, E) = \lim_{V \subset U} \mathscr{T}(DV, EV), \tag{3.3}$$

where the limit is taken over the maps

$$\mathscr{T}(DW, EW) \longrightarrow \mathscr{T}(\Sigma^{W-V}DV, EW) \cong \mathscr{T}(DV, \Omega^{W-V}EW) \cong \mathscr{T}(DV, EV).$$

It follows that

$$\begin{aligned} \mathscr{S}U(E, F(\mathscr{E}, E')) &= \lim \mathscr{T}(EV, F(\mathscr{E}, E')V) \\ &= \lim \mathscr{T}(EV, \mathscr{S}U'(\mathscr{E}_V, E')) \\ &= \lim \mathscr{S}U'(\mathscr{E}_V \wedge EV, E') \\ &= \mathscr{S}U'(\operatorname{colim} \mathscr{E}_V \wedge EV, E'), \end{aligned}$$

where the colimit is taken over the maps

$$(3.4) \qquad \mathscr{E}_V \wedge EV \cong \Sigma^{W-V}\mathscr{E}_W \wedge EV \cong \mathscr{E}_W \wedge \Sigma^{W-V}EV \longrightarrow \mathscr{E}_W \wedge EW.$$

Hence our definition of $\mathscr{E} \wedge E$:

DEFINITION 3.5. If $\mathscr{E} \in \mathscr{S}(U';U)$ and $E \in \mathscr{S}U$, then $\mathscr{E} \wedge E \in \mathscr{S}U'$ is the spectrum $\operatorname{colim}_{V\subset U} \mathscr{E}_V \wedge EV$ where the colimit is taken over the maps (3.4).

We have contrived our definitions to make the following true.

PROPOSITION 3.6. *There is an adjunction isomorphism*

$$\mathscr{S}U'(\mathscr{E} \wedge E, E') \cong \mathscr{S}U(E, F(\mathscr{E}, E')).$$

The following result is easy.

PROPOSITION 3.7. *If $\mathscr{E} \in \mathscr{S}(U';U)$, $E \in \mathscr{S}U$, and $X \in \mathscr{T}$ there are natural isomorphisms*

$$(\mathscr{E} \wedge E) \wedge X \cong \mathscr{E} \wedge (E \wedge X) \cong (\mathscr{E} \wedge X) \wedge E.$$

In practice if a spectrum E is the spectrification LD of a prespectrum D, it is often useful to describe $\mathscr{E} \wedge E$ as a colimit involving the spaces DV rather than the spaces EV. A simple adjunction argument together with (3.3) proves the following result.

PROPOSITION 3.8. *If $D \in \mathscr{P}U$ is a prespectrum and $\mathscr{E} \in \mathscr{S}(U';U)$ then*

$$\mathscr{E} \wedge LD \cong \operatorname{colim}_{V\subset U} \mathscr{E}_V \wedge DV$$

where the colimit is taken over the maps (3.4) (replacing E by D).

In particular, if E is the desuspension spectrum $\Sigma^\infty_V X$ of a space X, this has the following consequence.

PROPOSITION 3.9. *For a fixed indexing space $V \subset U$ there is an isomorphism*

$$\mathscr{E} \wedge \Sigma^\infty_V X \cong \mathscr{E}_V \wedge X$$

that is natural in $\mathscr{E}$ and X.

PROOF. By Proposition 3.8 we see that

$$\mathscr{E} \wedge \Sigma^\infty_V X \cong \operatorname{colim}_{W\subset U} \mathscr{E}_W \wedge \Sigma^{W-V} X.$$

But clearly, for $V \subset W$ the structure map

$$\mathscr{E}_V \wedge X \cong \Sigma^{W-V}\mathscr{E}_W \wedge X \cong \mathscr{E}_W \wedge \Sigma^{W-V} X$$

is an isomorphism. Hence the colimit stabilizes at V and the claim follows. $\square$

4. The object $\mathscr{M}\alpha \in \mathscr{S}(U';U)$

Let $\alpha : A \longrightarrow \mathscr{I}(U, U')$ be a map. For any $V \subset U$ and $V' \subset U'$, let $A_{V,V'}$ denote the pullback

$$\begin{array}{ccccc} A_{V,V'} & & \longrightarrow & & \mathscr{I}(V,V') \\ \downarrow & & & & \downarrow \\ A & \xrightarrow[\alpha]{} & \mathscr{I}(U,U') & \longrightarrow & \mathscr{I}(V,U') \end{array}$$

Thus $A_{V,V'}$ is the set of $a \in A$ such that $\alpha(a)V \subset V'$. Note that $A_{V,V'}$ will often be empty (for example, if V' is too small). Note that $A_{0,V'} = A$ for any V'. If $V \subset W$ then $A_{W,V'} \subset A_{V,V'}$. If $V' \subset W'$ then $A_{V,V'} \subset A_{V,W'}$ and, for fixed V, $A = \bigcup_{V' \subset U'} A_{V,V'}$. In particular, if A is compact, or more generally has compact image, then, for fixed V, we see that $A = A_{V,V'}$ for large V'.

Now for any V, V' let $\eta(\alpha)_{V,V'}$ denote the vector bundle over $A_{V,V'}$ with total space

$$E(\eta(\alpha)_{V,V'}) = \{(a, v) \in A_{V,V'} \times V' \mid v' \perp \alpha(a)V\}.$$

Let $T\alpha_{V,V'}$ be the Thom space of $\eta(\alpha)_{V,V'}$. In the case that $A_{V,V'} = \emptyset$ then $\eta(\alpha)_{V,V'}$ is the empty bundle $\emptyset \longrightarrow \emptyset$ and, by convention, the Thom space is a single point.

OBSERVATION 4.1. For fixed V, $\{T\alpha_{V,V'}\}$ is a prespectrum indexed over U', which we will denote by $\mathscr{T}\alpha_V$. The structure maps $\Sigma^{W'-V'} T\alpha_{V,V'} \longrightarrow T\alpha_{V,W'}$ are induced by the evident vector bundle morphisms

$$\eta(\alpha)_{V,V'} \oplus (W' - V') \cong \eta(\alpha)_{V,W'}|_{A_{V,V'}} \longrightarrow \eta(\alpha)_{V,W'}.$$

OBSERVATION 4.2. For fixed V' we have maps $\Sigma^{W-V} T\alpha_{W,V'} \longrightarrow T\alpha_{V,V'}$ that are induced by the evident vector bundle morphisms

$$\eta(\alpha)_{W,V'} \oplus (W - V) \cong \eta(\alpha)_{V,V'}|_{A_{W,V'}} \longrightarrow \eta(\alpha)_{V,V'}.$$

Thus for each $V \subset U$ we have a prespectrum $\mathscr{T}\alpha_V \in \mathscr{P}U'$ and we have maps of prespectra $\Sigma^{W-V} \mathscr{T}\alpha_W \longrightarrow \mathscr{T}\alpha_V$ that satisfy the evident transitivity condition. Let $\mathscr{M}\alpha_V \in \mathscr{S}U'$ be the spectrification of $\mathscr{T}\alpha_V$. Then we have maps of spectra $\Sigma^{W-V} \mathscr{M}\alpha_W \longrightarrow \mathscr{M}\alpha_V$ that satisfy the transitivity condition.

PROPOSITION 4.3. *The maps $\Sigma^{W-V} \mathscr{M}\alpha_W \longrightarrow \mathscr{M}\alpha_V$ are isomorphisms of spectra. Hence $\mathscr{M}\alpha$ is an object in the category $\mathscr{S}(U';U)$.*

PROOF. For a compact $K \subset A$ let $\alpha|_K$ be the composite $K \longrightarrow A \xrightarrow{\alpha} \mathscr{I}(U, U')$. Since, as always, we are working in the category of compactly generated spaces, A is topologized as the union of its compact subspaces and it follows that in the category of spaces over $\mathscr{I}(U, U')$, $\alpha = \operatorname{colim}_K \alpha|_K$ where the colimit is taken over all compact $K \subset A$. All constructions we have made are natural in spaces over $\mathscr{I}(U, U')$, and they commute with filtered colimits. In particular, $\mathscr{M}\alpha_V = \operatorname{colim}_K \mathscr{M}(\alpha|_K)_V$ and, of course, Σ^{W-V} commutes with colimits. Hence it suffices to show that the map $\Sigma^{W-V} \mathscr{M}\alpha_W \longrightarrow \mathscr{M}\alpha_V$ is an isomorphism when A is compact.

Assume then that A is compact. Fix $V \subset W \subset U$ and choose V' large enough that $A_{V,V'} = A_{W,V'} = A$. Then for $V' \subset W' \subset U'$ the structure maps $\Sigma^{W'-V'} T\alpha_{V,V'} \longrightarrow T\alpha_{V,W'}$ are isomorphisms. Hence $\mathscr{T}\alpha_V$ agrees cofinally with the desuspension prespectrum $\{\Sigma^{W'-V'} T\alpha_{V,V'}\}$ and thus $\mathscr{M}\alpha_V \cong \Sigma^\infty_{V'} T\alpha_{V,V'}$. Similarly $\mathscr{M}\alpha_W \cong \Sigma^\infty_{V'} T\alpha_{W,V'}$ and we have the sequence of isomorphisms

$$\Sigma^{W-V}\mathscr{M}\alpha_W \cong \Sigma^{W-V}\Sigma^\infty_{V'} T\alpha_{W,V'} \cong \Sigma^\infty_{V'}\Sigma^{W-V} T\alpha_{W,V'} \cong \Sigma^\infty_{V'} T\alpha_{V,V'} \cong \mathscr{M}\alpha_V.$$

□

LEMMA 4.4. *There is an isomorphism* $\mathscr{M}\alpha_0 \cong \Sigma^\infty A_+$ *that is natural in* α.

PROOF. It is immediate from the definitions that $\eta(\alpha)_{0,V'}$ is the trivial bundle $A \times V'$ over A and hence $T\alpha_{0,V'} \cong \Sigma^{V'} A_+$. □

Together with the fact that $\Sigma^V \mathscr{M}\alpha_V \cong \mathscr{M}\alpha_0$, this leads one to suspect that $\mathscr{M}\alpha_V$ is isomorphic to a shift desuspension of the space A_+. That turns out to be the case and, in our version of twisted half-smash products, this is the explanation for the untwisting theorem which we will discuss in the next section.

LEMMA 4.5 (UNTWISTING). *Let* $V \subset U$ *and* $V' \subset U'$ *be indexing spaces such that* $V \cong V'$ *and let* $\alpha : A \longrightarrow \mathscr{I}(U, U')$ *be a map. There is an isomorphism* $\mathscr{M}\alpha_V \cong \Sigma^\infty_{V'} A_+$ *that is natural in* α.

PROOF. Let $\mathscr{O}(U')$ be the orthogonal group of linear isometric isomorphisms $U' \longrightarrow U'$ that are the identity off a finite subspace. Since the restriction map $r' : \mathscr{O}(U') \longrightarrow \mathscr{I}(V', U')$ is a bundle over the contractible paracompact space $\mathscr{I}(V', U')$, it admits a section $s' : \mathscr{I}(V', U') \longrightarrow \mathscr{O}(U')$. Fixing an isomorphism $f : V \longrightarrow V'$, we obtain a homeomorphism $f^* : \mathscr{I}(V', U') \longrightarrow \mathscr{I}(V, U')$. Hence we have a composite

$$s = s' \circ (f^*)^{-1} : \mathscr{I}(V, U') \longrightarrow \mathscr{O}(U')$$

with the property that $s(g)(fv) = gv$ for any $g \in \mathscr{I}(V, U')$ and $v \in V$.

Now $\alpha : A \longrightarrow \mathscr{I}(U, U')$ passes to a map $\alpha : A \longrightarrow \mathscr{I}(V, U')$ which we also write as α. For $W' \subset U'$, let $A_{[V,W']}$ denote $\emptyset$ unless $V' \subset W'$, in which case $A_{[V,W']}$ denotes the pullback

$$\begin{array}{ccc} A_{[V,W']} & \longrightarrow & \mathscr{O}(W') \\ \downarrow & & \downarrow \\ A \xrightarrow{\ \alpha\ } \mathscr{I}(V, U') & \xrightarrow{\ s\ } & \mathscr{O}(U'). \end{array}$$

Thus when $V' \subset W'$, $A_{[V,W']}$ is the set of all $a \in A$ such that $s(a) \in \mathscr{O}(W')$. Note that $A_{[V,W']} \subset A_{V,W'}$ and $A = \bigcup_{W' \subset U'} A_{[V,W']}$. Let $\zeta(\alpha)_{[V,W']}$ denote the trivial bundle $A_{[V,W']} \times (W' - V')$. Taking Thom spaces, we obtain a prespectrum $\mathscr{T}\alpha_{[V]} \in \mathscr{S}U'$ with

$$\mathscr{T}\alpha_{[V]}V' = T(\zeta(\alpha)_{[V,W']}) \cong \Sigma^{W'-V'}(A_{[V,W']})_+$$

and the obvious structure maps. Let $\mathscr{M}\alpha_{[V]}$ be the spectrification of $\mathscr{T}\alpha_{[V]}$. Note that $\mathscr{M}\alpha_{[V]}$ is functorial in α and commutes with filtered colimits of spaces over $\mathscr{I}(U, U')$. We will show that 1) $\mathscr{M}\alpha_{[V]} \cong \Sigma^{\infty}_{V'}A_+$ and 2) $\mathscr{M}\alpha_{[V]} \cong \mathscr{M}\alpha_V$, naturally in α.

For 1) we observe that $\mathscr{M}\alpha_{[V]} = \operatorname{colim}_K \mathscr{M}(\alpha|_K)_{[V]}$ where the colimit is taken over all compact $K \subset A$. Hence it suffices to establish 1) when A is compact. But for compact A, $A_{[V,W']} = A$ for W' large enough, so we see that $\mathscr{T}\alpha_{[V]}$ agrees cofinally with the desuspension prespectrum $\{\Sigma^{W'-V'}A_+\}$. The conclusion follows.

For 2) we construct bundle morphisms $\zeta(\alpha)_{[V,W']} \longrightarrow \eta(\alpha)_{V,W'}$. Recall that

$$E(\eta(\alpha)_{V,W'}) = \{(a, w') \in A_{V,W'} \times W' \mid w' \perp \alpha(a)V\}.$$

We define a map

$$\theta_{W'} : A_{[V,W']} \times (W' - V') \longrightarrow E(\eta(\alpha)_{V,W'})$$

by $\theta_{W'}(a, w') = (a, s(\alpha(a))(w'))$. The properties of s ensure that these maps are well-defined and induce a map of prespectra $\theta : \mathscr{T}\alpha_{[V]} \longrightarrow \mathscr{T}\alpha_V$ on passage to Thom spaces. To show that the resulting map of spectra $\theta : \mathscr{M}\alpha_{[V]} \longrightarrow \mathscr{M}\alpha_V$ is an isomorphism, the colimit argument again lets us reduce the problem to the case when A is compact. But for compact A, with W' large enough so that $A_{[V,W']} = A$, the bundle map $\theta_{W'}$ is an isomorphism. Therefore the map of prespectra $\theta : \mathscr{T}\alpha_{[V]} \longrightarrow \mathscr{T}\alpha_V$ is cofinally a spacewise isomorphism, and the conclusion follows. □

Observe that the isomorphism $\mathscr{M}\alpha_V \cong \Sigma^{\infty}_{V'}A_+$ depends on the choice of f and of the section s'. The idea of exploiting such a section, although proposed in the context of a very different proof of the untwisting theorem, is due to Neil Strickland.

5. Twisted half-smash products and function spectra

DEFINITION 5.1. For $\alpha : A \longrightarrow \mathscr{I}(U, U')$, $E \in \mathscr{S}U$, and $E' \in \mathscr{S}U'$, the twisted half-smash product $\alpha \ltimes E \in \mathscr{S}U'$ is the spectrum $\mathscr{M}\alpha \wedge E$ and the twisted function spectrum $F[\alpha, E') \in \mathscr{S}U$ is the spectrum $F(\mathscr{M}\alpha, E')$.

PROPOSITION 5.2. *The above definitions agree with the definitions of* [38].

PROOF. Consider first the case of A compact. Then for a given $V \subset U$ one can find $V' \subset U'$ large enough that $\mathscr{M}\alpha_V \cong \Sigma^{\infty}_{V'}T\alpha_{V,V'}$. Thus

$$F[\alpha, E')V = \mathscr{S}U'(\mathscr{M}\alpha_V, E') \cong \mathscr{S}U'(\Sigma^{\infty}_{V'}T\alpha_{V,V'}, E') \cong F(T\alpha_{V,V'}, E'V'),$$

which is the definition of [38]. Therefore, by uniqueness of adjoints, the definitions of $\alpha \ltimes E$ agree for compact A. For general A one simply observes that our constructions and the definitions of [38] behave "properly" with respect to colimits. Specifically,

$$\alpha \ltimes E = \operatorname{colim}_K \alpha|_K \ltimes E \quad \text{and} \quad F[\alpha, E') = \lim_K F[\alpha|_K, E'),$$

where K runs through the compact subspaces of A. □

The difference between our approach and the approach of [38] is this: The definitions in [38] of $\alpha \ltimes E$ and $f[\alpha, E')$ are simple and concrete when A is compact, but they depend on arbitrary choices of cofinal sequences of indexing spaces in U and U'. Hence naturality in α is difficult to prove. To get around this, the authors of [38] develop an elaborate theory of "connections" that allows them to show that the definitions of $\alpha \ltimes E$ and $F[\alpha, E')$ are independent of the choices involved and, therefore, that the constructions are natural in α. Only after proving all this is it possible to define $\alpha \ltimes E$ for non-compact A as the relevant spectrum-level colimit and $F[\alpha, E')$ as the limit. The advantage of our treatment is that the objects $\alpha \ltimes E$ and $F[\alpha, E')$ are defined once and for all for arbitrary A and naturality in α is immediate from the definitions.

We show in the following pair of results that twisted half-smash products and function spectra generalize change of universe functors and are in a sense built up out of them.

PROPOSITION 5.3. *If $f : U \longrightarrow U'$ is a linear isometry regarded as a map $* \longrightarrow \mathscr{I}(U, U')$ then $f \ltimes E = f_*E$ and $F[f, E') = f^*E'$, where $f_* : \mathscr{S}U \longrightarrow \mathscr{S}U'$ and $f^* : \mathscr{S}U' \longrightarrow \mathscr{S}U$ are the standard change of universe functors. In particular, if $U = U'$ and* id *denotes the identity map then* id $\ltimes E \cong E$.

PROOF. For any $V \subset U$ we have $T\alpha_{V,fV} \cong \{f\}_+ \cong S^0$ and thus $\mathscr{M}f_V = \Sigma^\infty_{fV} S^0$. Therefore

$$F[f, E')V = \mathscr{S}U'(\Sigma^\infty_{fV} S^0, E') \cong F(S^0, E'(fV)) \cong E'(fV) = (f^*E')V.$$

The structure maps work properly and thus $F[f, E') = f^*E'$. The fact that $f \ltimes E = f_*E$ follows by the uniqueness of adjoints, or by an easy inspection. □

One should think of $\alpha \ltimes E$ intuitively as the union over $a \in A$ of the spectra $\alpha(a)_*E$, suitably topologized. Similarly, $F[\alpha, E')$ should be thought of as a suitably structured object that arises from the collection of spectra $\{\alpha(a)^*E'\}$. More precisely, we have the following result [38, VI.2.7], which admits an easy proof in our setup.

PROPOSITION 5.4. *Let $\alpha : \mathscr{I}(U, U')$ be a map and let $E \in \mathscr{S}U$ and $E' \in \mathscr{S}U'$ be spectra. A map $\xi : \alpha \ltimes E \longrightarrow E'$ determines and is determined by maps $\xi(\alpha) : E \longrightarrow \alpha(a)^*E'$ for points $a \in A$ such that the functions*

$$\zeta_{V,V'} : T\alpha_{V,V'} \wedge EV \longrightarrow E'V'$$

specified by $\zeta_{V,V'}((a, v') \wedge y) = \sigma(\xi(a)(y) \wedge v')$ for $a \in A_{V,V'}$, $v' \in V' - \alpha(a)V$, and $y \in EV$ are continuous, where σ denotes the structure map

$$\sigma_{\alpha(a)V,V'} : \Sigma^{V'-\alpha(a)V} E'(\alpha(a)V) \longrightarrow E'V'.$$

PROOF. Since for any $V \subset U$, $\mathscr{M}\alpha_V$ is the spectrification of $\mathscr{T}\alpha_V = \{T\alpha_{V,V'}\}$ we can write $\mathscr{M}\alpha_V$ as the spectrum level colimit

$$\mathscr{M}\alpha_V = \mathrm{colim}_{V' \subset U'} \Sigma^\infty_{V'} T\alpha_{V,V'}.$$

Therefore

$$\begin{aligned}
\mathscr{S}U'(\alpha \ltimes E, E') &= \mathscr{S}U'(\operatorname{colim}_{V\subset U} \mathscr{M}\alpha_V \wedge EV, E') \\
&= \lim_{V\subset U} \mathscr{S}U'(\mathscr{M}\alpha_V \wedge EV, E') \\
&= \lim_{V\subset U} \mathscr{S}U'(\operatorname{colim}_{V'\subset U'} \Sigma^\infty_{V'} T\alpha_{V,V'} \wedge EV, E') \\
&= \lim_{V\subset U} \lim_{V'\subset U'} \mathscr{S}U'(\Sigma^\infty_{V'} T\alpha_{V,V'} \wedge EV, E') \\
&= \lim_{V\subset U} \lim_{V'\subset U'} \mathscr{T}(T\alpha_{V,V'} \wedge EV, E'V').
\end{aligned}$$

This gives the connection between maps of spectra $\alpha \ltimes E \longrightarrow E'$ and families of maps $\zeta_{V,V'} : T\alpha_{V,V'} \wedge EV \longrightarrow E'V'$.

By restriction to points $a \in A$, a given map $\xi : \alpha \ltimes E \longrightarrow E'$ induces maps $\alpha(a)_* E \longrightarrow E'$ which are adjoint to maps $\xi(a) : E \longrightarrow \alpha(a)^* E'$. By projection from the double limit, ξ induces the map $\zeta_{V,V'} : T\alpha_{V,V'} \wedge EV \longrightarrow E'V'$, which may be checked to have the claimed description.

Conversely, given maps $\xi(a) : E \longrightarrow \alpha(a)^* E'$ satisfying the hypotheses, the maps $\zeta_{V,V'} : T\alpha_{V,V'} \wedge EV \longrightarrow E'V'$ specified by $\zeta_{V,V'}((a, v') \wedge y) = \sigma(\xi(a)(y) \wedge v')$ are easily checked to be compatible with the maps over which the double limit is taken, and therefore they specify a map of spectra $\alpha \ltimes E \longrightarrow E'$. □

The following untwisting theorem is an important sharpening of the result originally given in [38] and is vital to the theory of S-modules. The original proof of Elmendorf, Kriz, May, and Mandell, although not long or difficult, is rather technical in that it relies heavily on Elmendorf's category of spectra [20].

THEOREM 5.5 (UNTWISTING). *Let $V \subset U$ and $V' \subset U'$ be indexing spaces such that $V \cong V'$, let $\alpha : A \longrightarrow \mathscr{I}(U, U')$ be a map, and let X be a based space. There are isomorphisms*

$$\alpha \ltimes \Sigma^\infty_V X \cong A_+ \wedge \Sigma^\infty_{V'} X$$

and

$$\Omega^\infty_V F[\alpha, E') \cong F(A_+, \Omega^\infty_{V'} E')$$

that are natural in α and X.

PROOF. By Proposition 3.9 there is an isomorphism $\mathscr{E} \wedge \Sigma^\infty_V X \cong \mathscr{E}_V \wedge X$ that is natural in spaces $X \in \mathscr{T}$ and objects $\mathscr{E} \in \mathscr{S}(U'; U)$. Thus

$$\alpha \ltimes \Sigma^\infty_V X = \mathscr{M}\alpha \wedge \Sigma^\infty_V X \cong \mathscr{M}\alpha_V \wedge X.$$

By the untwisting lemma, $\mathscr{M}\alpha_V \cong \Sigma^\infty_{V'} A_+$, so it follows that

$$\alpha \ltimes \Sigma^\infty_V X \cong (\Sigma^\infty_{V'} A_+) \wedge X \cong A_+ \wedge \Sigma^\infty_{V'} X.$$

The isomorphism $\Omega^\infty_V F[\alpha, E') \cong F(A_+, \Omega^\infty_{V'} E')$ follows by uniqueness of adjoints. □

6. Formal properties of twisted half-smash products

We here give three basic formal properties of twisted half-smash products and function spectra. They have been used over and over in the text of the book. The first is an easy direct consequence of Proposition 3.7.

PROPOSITION 6.1. *For a map $\alpha : A \longrightarrow \mathscr{I}(U,U')$, spectra $E \in \mathscr{S}U$ and $E' \in \mathscr{S}U'$ and a based space X, there are natural isomorphisms*

$$(\alpha \ltimes E) \wedge X \cong \alpha \ltimes (E \wedge X)$$

and

$$F[\alpha, F(X, E')) \cong F(X, F[\alpha, E')).$$

Moreover, if Y is an unbased space and we denote by $\alpha \times Y$ the map

$$A \times Y \xrightarrow{\pi} A \xrightarrow{\alpha} \mathscr{I}(U,U'),$$

there are natural isomorphisms

$$(\alpha \times Y) \ltimes E \cong (\alpha \ltimes E) \wedge Y_+ \cong \alpha \ltimes (E \wedge Y_+).$$

The following two results relate twisted half-smash products to the naturality properties of spaces of linear isometries.

PROPOSITION 6.2. *Let $\alpha : A \longrightarrow \mathscr{I}(U,U')$ and $\beta : B \longrightarrow \mathscr{I}(U',U'')$ be maps and let $\beta \times_c \alpha$ denote the composite*

$$B \times A \xrightarrow{\beta \times \alpha} \mathscr{I}(U',U'') \times \mathscr{I}(U,U') \xrightarrow{c} \mathscr{I}(U,U''),$$

where c denotes composition. Then there are isomorphisms

$$(\beta \times_c \alpha) \ltimes E \cong \beta \ltimes (\alpha \ltimes E)$$

and

$$F[\beta \times_c \alpha, E'') \cong F[\alpha, F[\beta, E''))$$

that are natural in $E \in \mathscr{S}U$ and $E'' \in \mathscr{S}U''$.

PROPOSITION 6.3. *Let $\alpha : A \longrightarrow \mathscr{I}(U_1,U_1')$ and $\beta : B \longrightarrow \mathscr{I}(U_2,U_2')$ be maps and let $\alpha \times_\oplus \beta$ denote the composite*

$$A \times B \xrightarrow{\alpha \times \beta} \mathscr{I}(U_1,U_1') \times \mathscr{I}(U_2,U_2') \xrightarrow{\oplus} \mathscr{I}(U_1 \oplus U_2, U_1' \oplus U_2').$$

There is a natural isomorphism

$$(\alpha \times_\oplus \beta) \ltimes (E_1 \wedge E_2) \cong (\alpha \ltimes E_1) \wedge (\beta \ltimes E_2),$$

where $\wedge$ denotes the external smash product of spectra.

PROOF OF PROPOSITION 6.2. For $\mathscr{D} \in \mathscr{S}(U'';U')$ and $\mathscr{E} \in \mathscr{S}(U';U)$, we define the composition smash product $\mathscr{D} \wedge_c \mathscr{E} \in \mathscr{S}(U'';U)$ by $(\mathscr{D} \wedge_c \mathscr{E})_V = \mathscr{D} \wedge \mathscr{E}_V$. It is elementary to check that

$$(\mathscr{D} \wedge_c \mathscr{E}) \wedge E \cong \mathscr{D} \wedge (\mathscr{E} \wedge E)$$

for $E \in \mathscr{S}U$. Thus it suffices to show that $\mathscr{M}(\beta \times_c \alpha) \cong \mathscr{M}\beta \wedge_c \mathscr{M}\alpha$.

We begin by constructing bundle morphisms

$$\eta(\beta)_{V',V''} \times \eta(\alpha)_{V,V'} \longrightarrow \eta(\beta \times_c \alpha)_{V,V''}.$$

By definition,

$$E(\eta(\beta)_{V',V''}) = \{(b, v'') \in B_{V',V''} \times V'' \mid v'' \perp \beta(b)V'\},$$
$$E(\eta(\alpha)_{V,V'}) = \{(a, v') \in A_{V,V'} \times V' \mid v' \perp \alpha(a)V\},$$

and

$$E(\eta(\beta \times_c \alpha)_{V,V''}) = \{(b, a, v'') \in (B \times A)_{V,V''} \times V'' \mid v'' \perp \beta(b)\alpha(a)V\}.$$

It is easily seen that $B_{V',V''} \times A_{V,V'} \subset (B \times A)_{V,V''}$. Consider the map

$$E(\eta(\beta)_{V',V''}) \times E(\eta(\alpha)_{V,V'}) \longrightarrow E(\eta(\beta \times_c \alpha)_{V,V''})$$

that takes $(b, v'') \times (a, v')$ to $(b, a, v'' + \beta(b)v')$. Passing to Thom spaces, we obtain maps

$$T\beta_{V',V''} \wedge T\alpha_{V,V'} \longrightarrow T(\beta \times_c \alpha)_{V,V''}$$

which define a map of prespectra

$$\mathscr{T}\beta_{V'} \wedge T\alpha_{V,V'} \longrightarrow \mathscr{T}(\beta \times_c \alpha)_V$$

and therefore a map of spectra

$$\mathscr{M}\beta_{V'} \wedge T\alpha_{V,V'} \longrightarrow \mathscr{M}(\beta \times_c \alpha)_V.$$

One checks that, for $V' \subset W'$, the diagram

$$\begin{array}{ccc}
\mathscr{M}\beta_{V'} \wedge T\alpha_{V,V'} & \longrightarrow & \mathscr{M}\beta_{W'} \wedge T\alpha_{V,W'} \\
& \searrow \quad \swarrow & \\
& \mathscr{M}(\beta \times_c \alpha)_V &
\end{array}$$

commutes, where the top row is the composite isomorphism

$$\mathscr{M}\beta_{V'} \wedge T\alpha_{V,V'} \cong \Sigma^{W'-V'}\mathscr{M}\beta_{W'} \wedge T\alpha_{V,V'}$$
$$\cong \mathscr{M}\beta_{W'} \wedge \Sigma^{W'-V'}T\alpha_{V,V'} \cong \mathscr{M}\beta_{W'} \wedge T\alpha_{V,W'}.$$

Thus we obtain a map of spectra

$$(\mathscr{M}\beta \wedge_c \mathscr{M}\alpha)_V = \mathscr{M}\beta \wedge \mathscr{M}\alpha_V \cong \operatorname{colim}_{V' \subset U'} \mathscr{M}\beta_{V'} \wedge T\alpha_{V,V'} \longrightarrow \mathscr{M}(\beta \times_c \alpha)_V.$$

One checks that these maps are compatible with the structure maps of our objects and so define a morphism $\theta : \mathscr{M}\beta \wedge_c \mathscr{M}\alpha \longrightarrow \mathscr{M}(\beta \times_c \alpha)$ in the category $\mathscr{S}(U'';U)$.

To show that θ is an isomorphism, we invoke the familiar colimit argument. We have

$$\mathscr{M}\beta \wedge_c \mathscr{M}\alpha = \operatorname{colim}_{K,L} \mathscr{M}\beta|_L \wedge_c \mathscr{M}\alpha|_K$$

and

$$\mathscr{M}(\beta \times_c \alpha) = \operatorname{colim}_{K,L} \mathscr{M}(\beta \times_c \alpha)|_{L\times K},$$

where the colimits are taken over all compact $K \subset A$ and all compact $L \subset B$. Thus it suffices to show that θ is an isomorphism when A and B are compact. In that case, for a fixed V, we may choose V' large enough that $A_{V,V'} = A$ and we may then choose V'' large enough that $B_{V',V''} = B$. It follows that $(B \times A)_{V,V''} = B \times A$ and the map $T\beta_{V',V''} \wedge T\alpha_{V,V'} \longrightarrow T(\beta \times_c \alpha)_{V,V''}$ is a homeomorphism. Therefore

$$\begin{aligned}(\mathscr{M}\beta \wedge_c \mathscr{M}\alpha)_V &= \mathscr{M}\beta \wedge \mathscr{M}\alpha_V \cong \mathscr{M}\beta \wedge \Sigma^\infty_{V'} T\alpha_{V,V'}\\ &\cong \mathscr{M}\beta_{V'} \wedge T\alpha_{V,V'} \cong (\Sigma^\infty_{V''} T\beta_{V',V''}) \wedge T\alpha_{V,V'}\\ &\cong \Sigma^\infty_{V''} T\beta_{V',V''} \wedge T\alpha_{V,V'} \cong \Sigma^\infty_{V''} T(\beta \times_c \alpha)_{V,V''}\\ &\cong \mathscr{M}(\beta \times_c \alpha)_V. \quad \square\end{aligned}$$

PROOF OF PROPOSITION 6.3. For $\mathscr{D} \in \mathscr{S}(U_1'; U_1)$ and $\mathscr{E} \in \mathscr{S}(U_2'; U_2)$ we define the external direct sum smash product $\mathscr{D} \wedge_\oplus \mathscr{E} \in \mathscr{S}(U_1' \oplus U_2'; U_1 \oplus U_2)$ by

$$(\mathscr{D} \wedge_\oplus \mathscr{E})_{V_1 \oplus V_2} = \mathscr{D}_{V_1} \wedge \mathscr{E}_{V_2}.$$

Since the set $\{V_1 \oplus V_2\}$ is cofinal in $U_1 \oplus U_2$ this specifies an object in the category $\mathscr{S}(U_1' \oplus U_2'; U_1 \oplus U_2)$. It is easily checked that, for spectra $E_i \in \mathscr{S}U_i$,

$$(\mathscr{D} \wedge_\oplus \mathscr{E}) \wedge (E_1 \wedge E_2) \cong (\mathscr{D} \wedge E_1) \wedge (\mathscr{E} \wedge E_2).$$

Thus it suffices to show that $\mathscr{M}\alpha \wedge_\oplus \mathscr{M}\beta \cong \mathscr{M}(\alpha \times_\oplus \beta)$.

By definition,

$$E(\eta(\alpha)_{V_1,V_1'}) = \{(a, v_1') \in A_{V_1,V_1'} \times V_1' \mid v_1' \perp \alpha(a)V_1\},$$

$$E(\eta(\beta)_{V_2,V_2'}) = \{(b, v_2') \in B_{V_2,V_2'} \times V_2' \mid v_2' \perp \beta(b)V_2\},$$

and

$$E(\eta(\alpha \times_\oplus \beta)_{V_1\oplus V_2, V_1'\oplus V_2'}) =$$
$$\{(a, b, v_1', v_2') \in (A \times B)_{V_1\oplus V_2, V_1'\oplus V_2'} \times (V_1' \oplus V_2') \mid v_1' + v_2' \perp \alpha(a)V_1 \oplus \beta(b)V_2\}.$$

It follows immediately that there is a bundle isomorphism

$$\eta(\alpha)_{V_1,V_1'} \times \eta(\beta)_{V_2,V_2'} \cong \eta(\alpha \times_\oplus \beta)_{V_1\oplus V_2, V_1'\oplus V_2'}.$$

Thus we have isomorphisms of Thom spaces

$$T\alpha_{V_1,V_1'} \wedge T\beta_{V_2,V_2'} \cong T(\alpha \times_\oplus \beta)_{V_1\oplus V_2, V_1'\oplus V_2'}$$

and therefore an isomorphism of prespectra

$$\mathscr{T}\alpha_{V_1} \wedge \mathscr{T}\beta_{V_2} \cong \mathscr{T}(\alpha \times_\oplus \beta)_{V_1\oplus V_2}.$$

It follows that

$$(\mathcal{M}\alpha \wedge_{\oplus} \mathcal{M}\beta)_{V_1 \oplus V_2} \cong \mathcal{M}\alpha_{V_1} \wedge \mathcal{M}\beta_{V_2} \cong \mathcal{M}(\alpha \times_{\oplus} \beta)_{V_1 \oplus V_2}.$$

□

7. Homotopical properties of $\alpha \ltimes E$ and $F[\alpha, E')$

We now turn our attention to the homotopy preservation properties of our constructions. It is evident from the definitions that $\alpha \ltimes E$ and $F[\alpha, E')$ preserve homotopies in E and E' and homotopies over $\mathcal{I}(U, U')$ in α. However, it is vital to the theory that many homotopical properties of $\alpha \ltimes E$ and $F[\alpha, E')$ depend only on the homotopical properties of the space A.

PROPOSITION 7.1. *If A has the homotopy type of a CW complex, then the functor $\alpha \ltimes E$ preserves CW homotopy types in E and the functor $F[\alpha, E')$ preserves weak equivalences in E'.*

PROOF. It is a standard categorical observation that the two assertions are equivalent since $\alpha \ltimes -$ and $F[\alpha, -)$ are an adjoint pair of functors. Let $E_1' \longrightarrow E_2'$ be a weak equivalence in $\mathcal{S}U'$. Then each $E_1'V' \longrightarrow E_2'V'$ is a weak equivalence of spaces. Since A has CW homotopy type, $F(A_+, E_1'V') \longrightarrow F(A_+, E_2'V')$ is a weak equivalence for any $V' \subset U'$. It follows from the untwisting theorem that $F[\alpha, E_1') \longrightarrow F[\alpha, E_2')$ is a spacewise weak equivalence and therefore a weak equivalence of spectra. □

COROLLARY 7.2. *If $\alpha : A \longrightarrow \mathcal{I}(U, U')$ is a map and A has CW homotopy type then the functors $\alpha \ltimes -$ and $F[\alpha, -)$ pass to an adjoint pair of functors on the stable categories.*

As always, a functor such as $\alpha \ltimes E$ that does not preserve weak equivalences in E is defined on the stable category by first replacing E with a CW approximation. We can strengthen Proposition 6.1 by obtaining a CW structure on $\alpha \ltimes E$ from CW structures on A and E.

PROPOSITION 7.3. *Let $\alpha : A \longrightarrow \mathcal{I}(U, U')$ be a map, where A is a CW complex with skeletal filtration $\{A^n\}$. Let E be a CW spectrum with skeletal filtration $\{E^n\}$ and sequential filtration $\{E_n\}$. Then $\alpha \ltimes E$ is a CW spectrum with skeletal filtration*

$$(\alpha \ltimes E)^n = \bigcup_{p+q=n} (\alpha|_{A^p}) \ltimes E^q, \ n \in \mathbb{Z},$$

and sequential filtration

$$(\alpha \ltimes E)_n = \bigcup_{p+q=n} (\alpha|_{A^p}) \ltimes E_q, \ n \geq 0.$$

PROOF. We induct up the sequential filtration. Since $E_0 = *$, $(\alpha \ltimes E)_0 = *$. Assume that $(\alpha \ltimes E)_{n-1}$ is a CW spectrum. Let $\varepsilon : D^p \longrightarrow A$ be a p-cell with restriction σ to S^{p-1}. Let $CS^{r-1} \longrightarrow E_q$ be an r-cell of E_q with attaching map $S^{r-1} \longrightarrow E_{q-1}$, where $p+q=n$ and $r \in \mathbb{Z}$. Then $(\alpha \ltimes E)_n$ is obtained from $(\alpha \ltimes E)_{n-1}$ by attaching "twisted cells" of the form $(\alpha \circ \varepsilon) \ltimes CS^{r-1}$ along attaching maps

$$(\alpha \circ \varepsilon) \ltimes S^{r-1} \cup (\alpha \circ \sigma) \ltimes CS^{r-1} \longrightarrow (\alpha \ltimes E)_{n-1}.$$

It follows from the untwisting theorem and inspection on the space level that the pair $((\alpha\circ\varepsilon)\ltimes CS^{r-1}, (\alpha\circ\varepsilon)\ltimes S^{r-1})$ is isomorphic to the pair (CS^{p+r-1}, S^{p+r-1}). Therefore $(\alpha \ltimes E)_n$ is constructed from $(\alpha \ltimes E)_{n-1}$ by attaching genuine cells along cellular maps and is thus a CW spectrum. $\square$

Recall that a prespectrum D is said to be Σ-cofibrant if the structure maps $\Sigma^{W-V}DV \longrightarrow DW$ are cofibrations. A spectrum is Σ-cofibrant if it is isomorphic to LD for some Σ-cofibrant prespectrum D. A spectrum is tame if it is homotopy equivalent to a Σ-cofibrant spectrum. Roughly speaking, tame spectra are to spacewise homotopy equivalences of spectra as spectra with CW homotopy type are to weak equivalences of spectra. Thus, if D is tame and $f : E_1 \longrightarrow E_2$ is a map of spectra such that each $fV : E_1V \longrightarrow E_2V$ is a homotopy equivalence, then

$$f_* : h\mathscr{S}U(D, E_1) \longrightarrow h\mathscr{S}U(D, E_2)$$

is a bijection. It follows formally that a spacewise homotopy equivalence of tame spectra is a genuine homotopy equivalence. Pursuing the analogy further, the cylinder construction KE may be thought of as a "Σ-cofibrant approximation" to the spectrum E, and there is a map $KE \longrightarrow E$ that is a spacewise homotopy equivalence. Categorically, it is generally true that if $L : \mathscr{S}U \longrightarrow \mathscr{S}U'$ and $R : \mathscr{S}U' \longrightarrow \mathscr{S}U$ are a left-right adjoint pair of functors, then L preserves tameness if and only if R preserves spacewise homotopy equivalences.

THEOREM 7.4. *Let $\phi : A \longrightarrow B$ be a homotopy equivalence, let $\beta : B \longrightarrow \mathscr{I}(U, U')$ be a map, and let $\alpha : A \longrightarrow \mathscr{I}(U, U')$ be the composite $\beta \circ \phi$. If $E \in \mathscr{S}U$ is tame, then the map $\phi \ltimes E : \alpha \ltimes E \longrightarrow \beta \ltimes E$ is a homotopy equivalence. For any $E' \in \mathscr{S}U'$, the map $F[\phi, E') : F[\beta, E') \longrightarrow F[\alpha, E')$ is a spacewise homotopy equivalence.*

PROOF. It follows from the untwisting theorem that for any spectrum $E' \in \mathscr{S}U'$, the maps $F[\phi, E')(V)$ are homotopy equivalences. Thus for any tame spectrum E and any spectrum E' the map

$$F[\phi, E')_* : h\mathscr{S}U(E, F[\beta, E')) \longrightarrow h\mathscr{S}U(E, F[\alpha, E'))$$

is a bijection. By adjunction, this says that the map

$$(\phi \ltimes E)^* : h\mathscr{S}U'(\beta \ltimes E, E') \longrightarrow h\mathscr{S}U'(\alpha \ltimes E, E')$$

is a bijection. It is now formal that $\phi \ltimes E$ is a homotopy equivalence. $\square$

8. The cofibration theorem

COROLLARY 7.5. *Let $\alpha_1, \alpha_2 : A \longrightarrow \mathscr{I}(U,U')$ be maps. If $E \in \mathscr{S}U$ is tame, then $\alpha_1 \ltimes E$ and $\alpha_2 \ltimes E$ are homotopy equivalent. If $E' \in \mathscr{S}U'$ is any spectrum, then $F[\alpha_1, E)$ and $F[\alpha_2, E)$ are weakly equivalent.*

PROOF. We note that since $\mathscr{I}(U,U')$ is contractible, α_1 and α_2 are homotopic. Let $H : A \times I \longrightarrow \mathscr{I}(U,U')$ be a homotopy and apply Theorem 6.4 with $B = A \times I$ and $\phi = i_t : A \longrightarrow A \times I$ for $t = 0, 1$. □

Since CW spectra are tame, this implies the following result.

COROLLARY 7.6. *If A has CW homotopy type, the functors $\alpha \ltimes -$ obtained by varying the map α are canonically and coherently equivalent upon passage to the stable category. Similarly the functors $F[\alpha, -)$ are canonically and coherently equivalent as functors on the stable category.*

8. The cofibration theorem

In this section we prove the following analog for cofibrations of Theorem 7.4. The original proof of Elmendorf, Kriz, May, and Mandell relied heavily on the properties of Elmendorf's category of spectra [20], using bundle theoretic arguments about the morphism sets in that category. We shall show that it actually follows formally from the untwisting theorem and some elementary homotopy theory.

THEOREM 8.1. *Let $\phi : A \longrightarrow B$ be a cofibration, let $\beta : B \longrightarrow \mathscr{I}(U,U')$ be a map, and let $\alpha : A \longrightarrow \mathscr{I}(U,U')$ be the composite $\beta \circ \phi$. If $E \in \mathscr{S}U$ is Σ-cofibrant or is a CW spectrum, then $\phi \ltimes E : \alpha \ltimes E \longrightarrow \beta \ltimes E$ is a cofibration.*

PROOF. Consider a test diagram

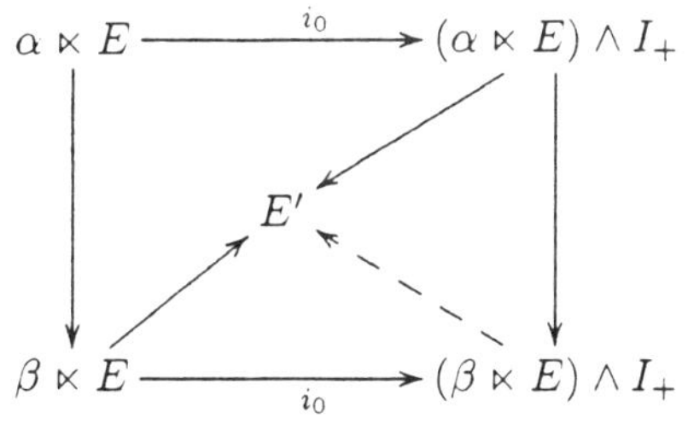

for which we must prove that the dotted arrow exists making the diagram commute. By adjunction, we may consider instead the test diagram

$$\begin{array}{ccc} E & \xrightarrow{f} & F[\beta, E') \\ i_0 \downarrow & \nearrow h & \downarrow F[\phi, E') \\ E \wedge I_+ & \xrightarrow{g} & F[\alpha, E') \end{array}$$

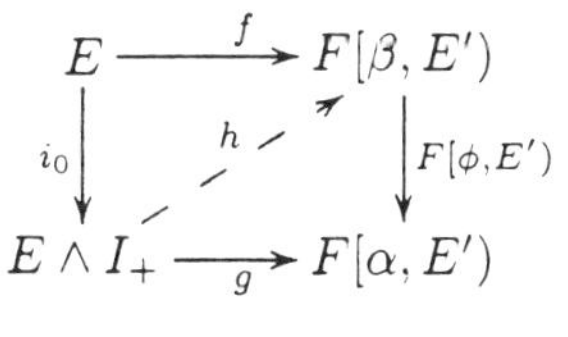

and try to prove that h exists. We are not claiming that $F[\phi, E') : F[\beta, E') \longrightarrow F[\alpha, E')$ is a fibration of spectra, but we will show that it does have the covering homotopy property (CHP) with respect to Σ-cofibrant spectra and CW spectra.

It is easily seen that $F[\phi, E')$ is a spacewise fibration. For if $V \subset U$ and $V' \subset U'$ are such that $V \cong V'$, then $F[\phi, E')$ is equivalent to a map

$$F(\phi_+, E'V') : F(B_+, E'V') \longrightarrow F(A_+, E'V'),$$

which is clearly a fibration. It follows now that $F[\phi, E')$ has the CHP with respect to spectra of the form $\Sigma^\infty_V X$ for any $V \subset U$. In particular, $F[\phi, E')$ has the CHP with respect to sphere spectra and hence $F[\phi, E')$ may be thought of as a spectrum level Serre fibration. The standard arguments now show that $F[\phi, E')$ has the CHP with respect to CW spectra.

The argument for Σ-cofibrant spectra is a bit more delicate. Let $E = LD$ where D is a Σ-cofibrant prespectrum. We want to use the cofibration condition on the structure maps $\Sigma^{W-V}DV \longrightarrow DW$ to construct the lift h for the diagram of prespectra

$$\begin{array}{ccc} D & \xrightarrow{f} & F[\beta, E') \\ {\scriptstyle i_0}\downarrow & \overset{h}{\nearrow} & \downarrow{\scriptstyle F[\phi, E')} \\ D \wedge I_+ & \xrightarrow[g]{} & F[\alpha, E'). \end{array}$$

We do this by choosing a cofinal sequence $V_0 \subset V_1 \subset \dots$ of representations in U and then arguing that after having constructed the map $hV_i : DV_i \wedge I_+ \longrightarrow F[\beta, E')(V_i)$ we can construct the map hV_{i+1} in a compatible way so as to obtain a map of prespectra.

The problem reduces to the following: Define

$$k : DV_{i+1} \wedge \{0\}_+ \cup \Sigma^{V_{i+1}-V_i} DV_i \wedge I_+ \longrightarrow F[\beta, E')(V_{i+1})$$

to be the pushout of the maps

$$DV_{i+1} \xrightarrow{fV_{i+1}} F[\beta, E')(V_{i+1})$$

and

$$\Sigma^{V_{i+1}-V_i} DV_i \wedge I_+ \xrightarrow{\Sigma^{V_{i+1}-V_i} hV_i} \Sigma^{V_{i+1}-V_i} F[\beta, E')(V_i) \longrightarrow F[\beta, E')(V_{i+1}).$$

Then $hV_{i+1} : DV_{i+1} \wedge I_+ \longrightarrow F[\beta, E')(V_{i+1})$ must be a solution to the lifting problem

$$\begin{array}{ccc} DV_{i+1} \wedge \{0\}_+ \cup \Sigma^{V_{i+1}-V_i} DV_i \wedge I_+ & \xrightarrow{k} & F[\beta, E')(V_{i+1}) \\ \downarrow & \overset{hV_{i+1}}{\nearrow} & \downarrow{\scriptstyle F[\phi, E')(V_{i+1})} \\ DV_{i+1} \wedge I_+ & \xrightarrow[gV_{i+1}]{} & F[\alpha, E')(V_{i+1}). \end{array}$$

By the untwisting theorem, if $V' \subset U'$ is such that $V_{i+1} \cong V'$ then we can replace $F[\phi, E')(V_{i+1})$ with the map

$$F(\phi_+, E'V') : F(B_+, E'V') \longrightarrow F(A_+, E'V').$$

The existence of the lift hV_{i+1} follows from our next lemma. $\square$

LEMMA 8.2. *If $X \longrightarrow Y$ is a cofibration of based spaces, $A \longrightarrow B$ is a cofibration of unbased spaces, and Z is a based space, then any diagram of the form*

$$\begin{array}{ccc} Y \wedge \{0\}_+ \cup X \wedge I_+ & \longrightarrow & F(B_+, Z) \\ \downarrow & \nearrow & \downarrow \\ Y \wedge I_+ & \longrightarrow & F(A_+, Z) \end{array}$$

can be completed by the dotted arrow.

PROOF. By playing with adjunctions we can replace the above diagram with an equivalent diagram

$$\begin{array}{ccc} B_+ \wedge \{0\}_+ \cup A_+ \wedge I_+ & \longrightarrow & F(Y, Z) \\ \downarrow & \nearrow & \downarrow \\ B_+ \wedge I_+ & \longrightarrow & F(X, Z). \end{array}$$

But this diagram of based spaces is equivalent to the diagram of unbased spaces

$$\begin{array}{ccc} B \times \{0\} \cup A \times I & \longrightarrow & F(Y, Z) \\ \downarrow & \nearrow & \downarrow \\ B \times I & \longrightarrow & F(X, Z). \end{array}$$

The based fibration $F(Y, Z) \longrightarrow F(X, Z)$ is also an unbased fibration and the solution now follows from the fact that $(B \times I, B \times \{0\} \cup A \times I)$ is a DR pair. □

9. Equivariant twisted half-smash products

As we have stated previously, all of our definitions and results on twisted half-smash products and function spectra generalize to the equivariant context with little change. In this section we discuss briefly the few exceptions to this assertion. More details will appear in [16].

The basic source material on categories of equivariant spectra is found in [38] and we briefly summarize that setup. For a compact Lie group G, a G-universe U is defined to be a countably infinite dimensional real inner product space on which G acts smoothly through linear isometries. We require that U contain a trivial representation and we require that whenever $V \subset U$ is a finite dimensional representation then U must contain an isomorphic copy of $V^{\oplus\infty}$. At the one extreme, a G universe is called complete if it contains copies of all the irreducible G-representations; at the other extreme, a G-universe is called G-trivial if it contains only the trivial representations.

By a G-spectrum indexed on a G-universe U we mean a spectrum E indexed on the finite dimensional subrepresentations of U such that each component space EV is a G-space and the structure maps $\Sigma^{W-V}EV \longrightarrow EW$ are G-maps. The resulting category is denoted $G\mathscr{S}U$. Thus for non-isomorphic G-universes U, U' we have associated non-equivalent categories of G-spectra that pass to non-equivalent equivariant stable categories.

For G-universes U, U', we let G act on $\mathscr{I}(U, U')$ by conjugation. For a G-map $\alpha : A \longrightarrow \mathscr{I}(U, U')$ and G-spectra $E \in G\mathscr{S}U$, $E' \in G\mathscr{S}U'$ there is a twisted half-smash product $\alpha \ltimes E \in G\mathscr{S}U'$ and a twisted function spectrum $F[\alpha, E') \in G\mathscr{S}U$ whose definitions and properties are wholly analogous to the nonequivariant constructions. Sections 2 to 6 generalize in a straightforward way that requires little comment. The category $G\mathscr{S}(U'; U)$ is defined in the evident way. In the equivariant version of Definition 3.2, for an object $\mathscr{E} \in G\mathscr{S}(U'; U)$ and G-spectrum $E' \in G\mathscr{S}U'$ the G-space $F(\mathscr{E}, E')(V) = \mathscr{S}U'(\mathscr{E}_V, E')$ must be understood to mean the space of all nonequivariant maps of spectra $\mathscr{E}_V \longrightarrow E'$ with G acting by conjugation. The equivariant versions of the untwisting results, Lemma 4.5 and Theorem 5.5, are true, except that the existence of a $V' \subset U'$ isomorphic to the given G-indexing space $V \subset U$ is a non-trivial hypothesis that may not be satisfied.

Fix a copy of $\mathbb{R}^\infty$ in U. It is a crucial fact [38, I.4.6] that for a map $f : E_1 \longrightarrow E_2$ of G-spectra indexed on U to be a spacewise weak equivalence of G-spaces, it suffices that $f\mathbb{R}^n : E_1\mathbb{R}^n \longrightarrow E_1\mathbb{R}^n$ be a weak equivalence for all n. Since all G-universes contain the trivial representations, the untwisting theorem applies to the trivial representations, and we see that the equivariant versions of Proposition 7.1 and Corollary 7.2 hold. The equivariant version of Proposition 7.3 only holds when G is finite. In analogy with [38, I.4.6], it can be shown that in order for a map of G-spectra to be a spacewise homotopy equivalence, it is sufficient to consider only the trivial representations. Hence the equivariant version of Theorem 7.4 holds. While $\mathscr{I}(U, U')$ is not a G-contractible G-space, it does have the property that if A is a G-CW complex then any two maps $\alpha_1, \alpha_2 : A \longrightarrow \mathscr{I}(U, U')$ are G-homotopic. Hence the equivariant versions of Corollaries 7.5 and 7.6 hold.

The equivariant version of Theorem 8.1 holds in full generality for the case that E is a G-CW spectrum. The reason is that the sphere spectra relevant to a G-CW structure are of the form $S^n \wedge (G/H)_+$ where n is an integer. Hence the problem of lifting cells over the map $F[\phi, E')$ only requires consideration of the maps $F[\phi, E')(\mathbb{R}^n)$ for which the untwisting theorem applies. For general universes U, U', if E is tame, but not G-CW, I do not know whether the map $\phi \ltimes E : \alpha \ltimes E \longrightarrow \beta \ltimes E$ must always be a cofibration. Our proof of Theorem 8.1, like the original proof given by Elmendorf, Kriz, May, and Mandell, works equivariantly only if we add the additional assumption that the universe U' is "as big or bigger" than U in the sense that there exists a G-linear isometry $U \longrightarrow U'$.

Bibliography

[1] J. F. Adams. Lectures on generalised cohomology. Springer Lecture Notes Vol 99, 1969, 1–138.

[2] J. F. Adams. Stable homotopy and generalized homology. The University of Chicago Press. 1974. (Reprinted 1995.)

[3] J. F. Adams. A variant of E. H. Brown's representability theorem. Topology, 10(1971), 185–198.

[4] M. Barr and C. Wells. Toposes, triples, and theories. Springer-Verlag. 1985.

[5] D. Blanc. New model categories from old. Preprint. 1994.

[6] J. M. Boardman. Stable homotopy theory. Thesis, Warwick 1964; mimeographed notes from Warwick and Johns Hopkins Universities, 1965-1970.

[7] J. M. Boardman. Conditionally convergent spectral sequences. Preprint. Johns Hopkins University, 1981.

[8] M. Bökstedt. Topological Hochschild homology. Preprint, 1990.

[9] M. Bökstedt. The topological Hochschild homology of $\mathbb{Z}$ and $\mathbb{Z}/p$. Preprint, 1990.

[10] M. Bökstedt, W. C. Hsiang, and I. Madsen. The cyclotomic trace and algebraic K-theory of spaces. Invent. Math. 111(1993), 465-539.

[11] A. K. Bousfield. The localization of spaces with respect to homology. Topology 14(1975), 133–150.

[12] A. K. Bousfield. The localization of spectra with respect to homology. Topology 18(1979), 257–281.

[13] E. H. Brown, Jr. Abstract homotopy theory. Trans. Amer. Math. Soc. 119 (1965), 79–85.

[14] R. R. Bruner, J. P. May, J. E. McClure, and M. Steinberger. H_∞ ring spectra and their applications. Springer Lecture Notes Vol 1176. 1986.

[15] H. Cartan and S. Eilenberg. Homological algebra. Princeton University Press. 1956.

[16] M. Cole. Twisted half-smash products and function spectra. Chapter XXII of "Equivariant homotopy and cohomology theory", by J. P. May, et al. NSF-CBMS Regional Conference Monograph. To appear in 1996.

[17] W. G. Dwyer, M. J. Hopkins, and D. M. Kan. The homotopy theory of cyclic sets. Trans. Amer. Math. Soc. 291(1985), 281-289.

[18] W. G. Dwyer and J. Spalinski. Homotopy theories and model categories. A handbook of algebraic topology, edited by I. M. James. Elsevier Science. 1995.

[19] S. Eilenberg and J. C. Moore. Homology and fibrations, I. Comm. Math. Helv. 40(1966), 199–236.

[20] A. D. Elmendorf. The Grassmannian geometry of spectra. J. Pure and Applied Algebra, 54(1988), 37–94.

[21] A. D. Elmendorf. Geometric realization of simplicial spectra. Manuscripta Mathematica. To appear.

[22] A. D. Elmendorf, J. P. C. Greenlees, I. Kriz, and J. P. May. Commutative algebra in stable

homotopy theory and a completion theorem. Mathematical Research Letters 1(1994), 225-239.

[23] A. D. Elmendorf, I. Kriz, M. A. Mandell, and J. P. May. Modern foundations for stable homotopy theory. A handbook of algebraic topology, edited by I. M. James. Elsevier Science. 1995.

[24] D. Grayson. Localization for Flat Modules in Algebraic K-Theory. J. Alg. 61(1979), 463-496.

[25] J. P. C. Greenlees and J. P. May. Generalized Tate cohomology. Memoirs Amer. Math. Soc. 1995.

[26] J. P. C. Greenlees and J. P. May. Completions in algebra and topology. A handbook of algebraic topology, edited by I. M. James. Elsevier Science. 1995

[27] J. P. C. Greenlees and J. P. May. Equivariant stable homotopy theory. A handbook of algebraic topology, edited by I. M. James. Elsevier Science. 1995.

[28] J. P. C. Greenlees and J. P. May. A completion theorem for MU-module spectra. Preprint, 1995.

[29] V. K. A. M. Gugenheim and J. P. May. On the theory and applications of differential torsion products. Memoirs Amer. Math. Soc. 142, 1974.

[30] L. Hesselholt and I. Madsen. On the K-theory of finite algebras over Witt vectors of perfect fields. Aarhus Universitet Preprint Series 1995, No. 2.

[31] H. Hochschild. On the cohomology groups of an associative algebra. Annals of Math. 46(1945), 58-67.

[32] M. J. Hopkins. Notes on E_∞ ring spectra. Typed notes. 1993.

[33] G. M. Kelly. Basic concepts of enriched category theory. London Math. Soc. Lecture Note Series Vol. 64. Cambridge University Press. 1982.

[34] I. Kriz. Towers of E_∞ ring spectra with an application to BP. Preprint, 1993.

[35] I. Kriz. and J. P. May. Operads, algebras, modules, and motives. Astérisque Vol. 233. 1995.

[36] L. G. Lewis. When is the natural map $X \longrightarrow \Omega\Sigma X$ a cofibration? Trans. Amer. Math. Soc. 273(1982), 147-155.

[37] L. G. Lewis, Jr. Is there a convenient category of spectra? J. Pure and Applied Algebra 73(1991). 233-246.

[38] L. G. Lewis, Jr., J. P. May, and M. Steinberger (with contributions by J. E. McClure). Equivariant stable homotopy theory. Springer Lecture Notes in Mathematics Volume 1213. 1986.

[39] J. Lillig. A union theorem for cofibrations. Arch. Math. 24(1973), 410-415.

[40] T.-Y. Lin. Adams type spectral sequences and stable homotopy modules. Indiana Math. J. 25(1976). 135-158.

[41] F. E. J. Linton. Coequalizers in categories of algebras. Springer Lecture Notes in Mathematics 80, 1969, 75-90.

[42] J. L. Loday. Opérations sur l'homologie cyclique des algébras commutatives. Invent. Math. 96(1989), 205-230.

[43] S. Mac Lane. Categories for the Working Mathematician. Springer-Verlag. 1971.

[44] J. P. May. Simplicial objects in algebraic topology, 1967. Reprinted by the University of Chicago Press, 1982 and 1992.

[45] J. P. May. The Geometry of Iterated Loop Spaces. Springer Lecture Notes in Mathematics Volume 271. 1972.

[46] J. P. May. E_∞ spaces, group completions, and permutative categories. London Math. Soc. Lecture Notes No.11, 1974, 61-93.

[47] J. P. May. Classifying spaces and fibrations. Memoirs Amer. Math. Soc. 155. 1975.

[48] J. P. May (with contributions by F. Quinn, N. Ray, and J. Tornehave). E_∞ ring spaces and E_∞ ring spectra. Springer Lecture Notes in Mathematics Volume 577. 1977.

[49] J. P. May. A_∞ Ring Spaces and Algebraic K-Theory. Springer Lecture Notes in Mathematics Volume 658, 1978, 240-315.

[50] J. P. May. Multiplicative infinite loop space theory. J. Pure and Applied Algebra, 26(1982). 1-69.

[51] J. P. May. Derived categories in algebra and topology. In the Proceedings of the Eleventh

International Conference on Topology, Rendiconti dell Istituto Matematico dell Università di Trieste 25(1993), 363-377.

[52] J. P. May, et al. Equivariant homotopy and cohomology theory. NSF-CBMS Regional Conference Monograph. To appear in 1996.

[53] R. McCarthy. On Fundamental Theorems of Algebraic K-Theory. Topology 32(1993), 325–328.

[54] J. E. McClure, R. Schwänzl, and R. Vogt. $THH(R) \cong R \otimes S^1$ for E_∞ ring spectra. J. Pure and Applied Algebra. To appear.

[55] J. E. McClure and R. E. Staffeldt. On the topological Hochschild homology of bu, I. Amer. J. Math. 115(1993), 1-45.

[56] A. Neeman. On a theorem of Brown and Adams. Preprint, 1995.

[57] D. G. Quillen. Homotopical algebra. Springer Lecture Notes in Mathematics Volume 43. 1967.

[58] D. G. Quillen. Higher Algebraic K-Theory I. Springer Lecture Notes in Mathematics Volume 341, 1973, 85–147.

[59] A. Robinson. Derived tensor products in stable homotopy theory. Topology 22(1983), 1-18.

[60] A. Robinson. Spectra of derived module homomorphisms. Math. Proc. Camb. Phil. Soc. 101(1987), 249-257.

[61] A. Robinson. The extraordinary derived category. Math. Z. 196(1987), 231-238.

[62] A. Robinson. Obstruction theory and the strict associativity of Morava K-theories. London Math. Soc. Lecture Note Series, vol. 139, 1989, 111-126.

[63] A. Robinson. Composition products in RHom, and ring spectra of derived endomorphisms. Springer Lecture Notes Vol 1370, 1989, 374-386.

[64] R. Schwänzl and R. M. Vogt. The categories of A_∞ and E_∞ monoids and ring spaces as closed simplicial and topological model categories. Arch. Math. 56(1991), 405-411.

[65] R. Schwänzl and R. M. Vogt. Basic Constructions in the K-Theory of Homotopy Ring Spaces. Trans. Amer. Mat. Soc 341(1994), 549-584.

[66] G. Segal. Categories and Cohomology Theories. Topology 13(1974), 293–312.

[67] R. E. Staffeldt. On Fundamental Theorems of Algebraic K-Theory. K-Theory 2(1989), 511–532.

[68] R. W. Thomason and T. Trobaugh. Higher Algebraic K-Theory of Schemes. The Grothendieck Festschrift, v. III, 1990, 247-435.

[69] J. L. Verdier. Catégories dérivées. Springer Lecture Notes Vol 569, 1977, 262-311.

[70] F. Waldhausen. Algebraic K-Theory of Topological Spaces II. Springer Lecture Notes in Mathematics Volume 763, 1979, 356–394.

[71] F. Waldhausen. Algebraic K-Theory of Spaces. Springer Lecture Notes in Mathematics Volume 1126. 1985, 318–419.

[72] J. Wolbert. Towards an algebraic classification of module spectra. J. Pure and Applied Algebra. To appear.

[73] N. Yoneda. On the homology theory of modules. J. Fac. Sci. Univ. Tokyo Sec. I, 7(1954), 193-227.

Index

Selected Titles in This Series

(Continued from the front of this publication)

11 **J. Cronin,** Fixed points and topological degree in nonlinear analysis, 1964
10 **R. Ayoub,** An introduction to the analytic theory of numbers, 1963
9 **Arthur Sard,** Linear approximation, 1963
8 **J. Lehner,** Discontinuous groups and automorphic functions, 1964
7.2 **A. H. Clifford and G. B. Preston,** The algebraic theory of semigroups, Volume II, 1961
7.1 **A. H. Clifford and G. B. Preston,** The algebraic theory of semigroups, Volume I, 1961
6 **C. C. Chevalley,** Introduction to the theory of algebraic functions of one variable, 1951
5 **S. Bergman,** The kernel function and conformal mapping, 1950
4 **O. F. G. Schilling,** The theory of valuations, 1950
3 **M. Marden,** Geometry of polynomials, 1949
2 **N. Jacobson,** The theory of rings, 1943
1 **J. A. Shohat and J. D. Tamarkin,** The problem of moments, 1943